GROUND WATER QUALITY AND AGRICULTURAL PRACTICES

Deborah M. Fairchild

CRC Press
Taylor & Francis Group
Boca Raton London New York

CRC Press is an imprint of the
Taylor & Francis Group, an **informa** business

First published 1987 by Lewis Publishers

Published 2020 by CRC Press
Taylor & Francis Group
6000 Broken Sound Parkway NW, Suite 300
Boca Raton, FL 33487-2742

First issued in paperback 2020

ISBN 13: 978-0-367-58029-2 (pbk)
ISBN 13: 978-0-87371-036-7 (hbk)

Visit the Taylor & Francis Web site at
http://www.taylorandfrancis.com

and the CRC Press Web site at
http://www.crcpress.com

Library of Congress Cataloging-in-Publication Data

Ground water quality and agricultural practices.

    Includes bibliographies and index.
      1. Water, Underground—United States—Quality.
2. Irrigation water—United States. I. Fairchild,
D. M. (Deborah M.)
TD223.G738   1987   363.7′394   86-27333
ISBN 0-87371-036-3

To my daughter, Jean Ruth Fairchild

DEBORAH M. FAIRCHILD is a senior environmental scientist and has been with the Environmental and Ground Water Institute for the past five years. She received a BS in biology and chemistry from the State University of New York at Oswego in 1975. She worked as a chemist for the U.S. Department of Agriculture while obtaining an MS in soil microbiology from West Virginia University in 1979. Since coming to the University of Oklahoma, she has worked on many ground water availability and quality studies focused on petroleum production and agriculture.

In addition, she has worked on several environmental impact statements and has served as conference organizer, information and data management specialist, and quality assurance officer for laboratory studies. Her professional activities have resulted in over 30 publications and presentations. In 1985, Ms. Fairchild received the Distinguished Service Award from the University of Oklahoma.

PREFACE

Ground water has become a more important resource over the past decade due to increases in ground water usage and the realization that once contaminated it is difficult, expensive, and sometimes impossible to clean up. The most prevalent sources of ground water contamination are (1) waste disposal, (2) storage, transportation, and handling of commercial materials, (3) mining operations, and (4) nonpoint sources such as highway deicing and agricultural activities. Most of these pollution sources have been addressed extensively through legislation, permitting programs, research, and a plethora of conferences and short courses.

However, the relationships between agricultural practices and ground water quality have been gaining increased attention and have not been addressed as extensively as other pollution sources. Irrigation return-flow, use of pesticides, fertilizers, and manure, changes in vegetative cover through conservation tillage, and application of waste effluents have all been known to cause changes in ground water quality.

The Environmental and Ground Water Institute (EGWI) recognized the need for information exchange among professionals dealing with this emerging problem. A national conference on Ground Water Quality and Agricultural Practices was held May 1 – 2, 1986, at the University of Oklahoma in Norman. The conference was conducted in cooperation with the Oklahoma State University Center for Water Research, the U.S. Soil Conservation Service, and the Oklahoma Department of Agriculture. A call for papers brought in several dozen abstracts which were screened for appropriateness. The conference addressed ground water usage, agricultural chemical usage, ground water pollution sources and evaluation, and protection and management. Attendees were soil scientists, hydrologists, geologists, engineers, agronomists, and scientists from government agencies, industry, universities, consulting firms, and professional and trade organizations. This book contains twenty-seven chapters resulting from the presentations made at the two-day conference.

Special thanks are extended to all the authors and co-authors, without whom the conference and this book would not have been possible. The editor also appreciates the efforts of Mrs. Wilma Clark, Mrs. Mittie Durham, and Mrs. Leslie Rard for their tireless efforts in conference publicity and registration. The assistance of Dr. Larry Canter, Sun Company Professor of Ground Water Hydrology and Director, EGWI, in manuscript editing and proofing is acknowledged. Finally, and most important, the editor is indebted to Mrs. Mittie Durham for her typing skills, perfectionist attitude, and sincere commitment to the preparation of this manuscript.

The editor also gratefully acknowledges the support and encouragement of the College of Engineering at the University of Oklahoma relative to writing endeavors.

Deborah M. Fairchild
Senior Environmental Scientist
Environmental and Ground Water
    Institute
University of Oklahoma
Norman, Oklahoma

CONTENTS

LIST OF CONTRIBUTORS

Samuel F. Atkinson, Assistant Professor of Environmental Sciences, Institute of Applied Sciences, North Texas State University, Denton, Texas.

Daniel D. Badger, Professor, Department of Agricultural Economics, Oklahoma State University, Stillwater, Oklahoma.

Brian Baker, Post-Doctoral Fellow, Department of Agriculture and Research Economics, University of California, Berkeley, California.

W.A. Berg, Soil Scientist, U.S. Department of Agriculture, Agricultural Research Service, Woodward, Oklahoma.

Brian Borofka, Project Manager, IT Corporation, Monroeville, Pennsylvania.

C.R. Cail, Soil Scientist, U.S. Department of Agriculture, Soil Conservation Service, Woodward, Oklahoma.

Larry Canter, Sun Company Professor of Ground Water Hydrology, and Director, Environmental and Ground Water Institute, University of Oklahoma, Norman, Oklahoma.

Thomas Victor Cech, Executive Director, Central Colorado Water Conservancy District, Greeley, Colorado.

Fu-Hsain Chang, Associate Professor of Environmental Microbiology, Center for Environmental Studies, Bemidji State University, Bemidji, Minnesota.

C.S. Cousins-Leatherman, Environmental Specialist, Planning Bureau, Iowa Department of Natural Resources, Des Moines, Iowa.

Debra Curry, Hydrologist, Office of Ground Water Coordination, U.S. Environmental Protection Agency Region II, New York, New York.

Thanh H. Dao, Supervisory Soil Scientist, U.S. Department of Agriculture, Agricultural Research Service, El Reno, Oklahoma.

Deborah Fairchild, Senior Environmental Scientist, Environmental and Ground Water Institute, University of Oklahoma, Norman, Oklahoma.

Akos Fekete, Acting Supervisor of Planning and Program Development, New Jersey Department of Environmental Protection, Residuals Management Section, Trenton, New Jersey.

B.K. (Jim) Gopal, Senior Water Resources Engineer, Oklahoma Water Resources Board, Oklahoma City, Oklahoma.

George R. Hallberg, Chief, Iowa Geological Survey, Iowa City, Iowa.

J. Ross Harris, Jr., Environmental Quality Specialist, Extension Service, University of Delaware, Dover, Delaware.

A. G. Hornsby, Associate Professor, Department of Soil Science, University of Florida, Gainesville, Florida.

Bernard E. Hoyer, Associate State Geologist, Iowa Geological Survey, Iowa City, Iowa.

Marc Hult, Hydrogeologist, U.S. Geological Survey, Water Resources Division, St. Paul, Minnesota.

D.M. Hungerford, District Conservationist, U.S. Department of Agriculture, Soil Conservation Service, Buffalo, Oklahoma.

Gordon V. Johnson, Professor of Soil Science, Oklahoma State University, Stillwater, Oklahoma.

James Kaap, Extension Soil, Water, and Waste Management Specialist, Iowa State University, Ames, Iowa.

Richard Kelley, Environmental Specialist, Planning Bureau, Iowa Department of Natural Resources, Des Moines, Iowa.

Jerry Kotas, Director, National Survey of Pesticides, U.S. Environmental Protection Agency, Office of Drinking Water, Washington, D.C.

Robert D. Libra, Hydrogeologist, Iowa Geological Survey, Iowa City, Iowa.

Joe Makuch, Water Quality Specialist, Chesapeake Bay Project, Department of Agricultural Engineering, Pennsylvania State University, University Park, Pennsylvania.

Harry Mapp, Professor, Department of Agricultural Economics, Oklahoma State University, Stillwater, Oklahoma.

Steve G. McLin, Assistant Professor, Department of Civil Engineering and Environmental Science, University of Oklahoma, Norman, Oklahoma.

J.W. Naney, Geologist, U.S. Department of Agriculture, Agricultural Research Service, Durant, Oklahoma.

Nancy N. Noben, Graduate Assistant, Center for Environmental Studies, Bemidji State University, Bemidji, Minnesota.

N.L. Nofziger, Associate Professor, Department of Agronomy, Oklahoma State University, Stillwater, Oklahoma.

Steven Padgitt, Extension Sociologist, Iowa State University, Ames, Iowa.

Helen Petit-Chase, Section Chief, Residuals Management Section, Department of Environmental Protection, Trenton, New Jersey.

G.A. Sample, Soil Scientist, U.S. Department of Agriculture, Soil Conservation Service, Buffalo, Oklahoma.

S.J. Smith, Soil Scientist, U.S. Department of Agriculture, Agricultural Research Service, Durant, Oklahoma.

Bobby Stewart, Acting Director, Southern Plains Area of the Agricultural Research Service, U.S. Department of Agriculture, Bushland, Texas.

Robert K. Thomas, Jr., Senior Water Resources Geologist, Oklahoma Water Resources Board, Oklahoma City, Oklahoma.

Donald E. Thomason, Jr., former Graduate Research Assistant, Department of Agricultural Economics, Oklahoma State University, Stillwater, Oklahoma.

Robert L. Westerman, Professor of Soil Fertility and Plant Nutrition, Department of Agronomy, Oklahoma State University, Stillwater, Oklahoma.

Lawrence Wetter, Planning Engineer, U.S. Department of Agriculture, U.S. Soil Conservation Service, Water Resources Planning, Salina, Kansas.

Mitchell Woodward, Regional Manure Management Agent, Pennsylvania Cooperative Extension Service, Lancaster County, Pennsylvania.

LIST OF TABLES

LIST OF FIGURES

CHAPTER 1

THE U.S.D.A. AGRICULTURAL RESEARCH SERVICE
COMMITMENT TO GROUND WATER RESEARCH

by

Bobby A. Stewart

It is a complicated task--keeping up with the research carried out by a national organization like the Agricultural Research Service (ARS). Therefore, I am going to take a few minutes to make some general comments about our planning strategy in this present climate of increasing fiscal constraint. Then, I will discuss the ARS commitment to ground water research.

As a thumbnail sketch of ARS, we have about 8,200 employees--2,700 of them scientists--working on about 3,000 different projects at a given time. This research to serve the American farmer and rancher is being carried out at 130 different locations in the United States and eight foreign countries.

Our current budget of $478 million has remained relatively constant, when adjusted for inflation, over the past decade. So, our six-year implementation planning strategy has been developed around the concept of constant dollar funding assumptions. By this, I mean that intended increases in funding are equaled by intended decreases in funding. This constant dollar assumption forces some very difficult decisions about which programs expand and which programs receive less support or are drastically cut. This approach is helping us accommodate the constraints imposed by Gramm-Rudman-Hollings. We expect to experience further belt tightening down the road.

The projects in the ARS planning strategy reflect our emphasis on mission-oriented, fundamental, long-range, high-risk research. They also reflect our responsiveness to action agencies, user groups, and the U.S. Congress. The ARS Program Plan includes six objectives that form the strategy: research to improve soil and water conservation; plant production and protection; animal production and protection; commodity conversion and delivery; human nutrition; and integration of systems, using computer technology.

RESEARCH ON SOIL AND WATER CONSERVATION

There is increasing evidence that the nation's ground water resources are becoming affected by man's activities, including those associated with agriculture. Advances in analytical chemistry often allow detection of chemical concentrations at levels so low that information is lacking on toxicological significance. However, monitoring results indicate that agricultural chemicals are present in ground water.

Even if concerted efforts are made, the extent of the problem may not be adequately assessed for several years. Since about one-half the total United

States population and 95 percent of rural households depend on ground water for drinking supply, research programs are needed immediately for three objectives:

(1)   to define the extent of the problem;

(2)   to develop procedures for estimating the potential loading of chemicals to ground water systems; and

(3)   to develop alternative management systems to alleviate or minimize the occurrence of agricultural chemicals in ground water.

Movement through the vadose and ground water zones can be very slow, anaerobic conditions exist, and biological and chemical processes that degrade contaminants may be much less effective in ground water than in surface soils. Therefore, strategies that prevent chemical entry into ground water will be much more effective and less costly than ground water clean-up strategies.

Agriculture has a tremendous stake in preventing ground water contamination. Chemicals are indispensable in modern agriculture, but their improper use may lead to serious water quality problems that could impair ground water use for crop and animal production and domestic water supply.

When all agricultural chemicals and quantities used are considered, the creation of water quality problems is probably limited to a relatively few compounds or combinations of conditions. The potential for pesticide leaching appears related to exceeding limits on solubility, volatility, hydrolysis, and photolysis of the pesticide as well as the ground water recharge rate. However, the problems identified arouse public concern for the entire spectrum of chemical use, and this public concern has profound implications for American agricultural practices. This, then, is the crux of the dilemma. It is impossible to meet the rapidly growing world needs for agricultural products without the use of chemicals.

Agricultural chemical usage comes with the responsibility for developing farm systems that have minimum impact on ground water quality. Since agriculture is a major ground water user for irrigation, livestock and drinking water, agriculture has a vested interest in protecting its supply.

Information, technology, and analytical and predictive methods are not sufficient to provide needed strategies for ground water resource protection. Solving the problem requires the development of methods to identify chemical and salt sources, transport routes and control measures. We need to determine the efficacy of these measures in controlling chemical and salt loss to ground water caused by agricultural activities.

Although there has been considerable research on the fate of chemicals in soils, streams, and impounded waters, serious gaps exist in data bases, basic concepts, and the understanding of processes that define ground water quality. Proper implementation of regulatory decisions and the development of alternate management systems require that these gaps be filled.

Models based on sound concepts and validated field data will be needed to predict potential chemical loading to ground water systems under different management and climatic scenarios. These models must also answer questions relative to efficacy and crop production requirements. A proposed management system cannot be considered as a viable alternative unless basic production goals are met.

Only a small fraction of the enormous variety of chemicals used in agriculture is responsible for documented ground water quality problems. The problems identified are serious, and they demand our attention--but they could have been much worse. That they are not more aggravated is due in large part to recent progress made in our knowledge of their environmental reactions and appropriate application rates and techniques.

The registration process now administered by the U.S. Environmental Protection Agency prevents most pesticide impacts before they occur. Persistence, mobility, and other environmental effects are all limited--or at least defined--by current registration procedures. The pesticide registration process has, in general, accomplished its goal of keeping chemicals with potentially serious environmental impacts out of the marketplace.

The ARS has been engaged in research on the linkage between agriculture and environmental quality for more than a quarter of a century. Contributions to environmental improvement have come both from research on natural resource management and from advances in crop production and protection technologies.

Current ARS research continues to focus on improved soil and water management and on more effective and efficient use of agricultural chemicals. A substantial component of the ARS program contributes indirectly to the development of technologies for improving ground water quality. Ongoing research will provide a sound scientific basis for evaluating the effect of changes in cropping, culture, tillage, and water management on ground water quality. It will also provide better guidelines for improving the formulation and application of agricultural chemicals.

The current ARS commitment to soil and water conservation research represents about 13 percent of the total agency budget--or a little over 60 million dollars. The ARS planning strategy calls for an increase to 14 percent of the budget for this research over the next 6 years. The present ARS commitment to ground water quality research is about 12 million dollars, or about 20 percent of the funds reserved for soil and water research. The ARS planning strategy calls for increased emphasis on ground water quality research in the years ahead.

RESEARCH ON NUTRIENTS, PESTICIDES, SALINITY, AND MODELING

Nutrient research at a number of ARS locations relates indirectly to ground water quality. For example, at some locations scientists are discovering ways to more efficiently use plant nutrients. This research increases the producer's profit and improves ground water quality by minimizing excess nutrient content in the root zone. Results of current irrigation and drainage research will better define the effect of water management techniques on nutrient movement in the soil. At some locations, scientists are conducting research on nitrogen transformations in the root zone, near-stream zone, wetlands and impoundments. Such nitrogen transformations are crucial to the movement and impact of nutrients in ground water systems.

Research is also underway to determine the effect of nutrient transport and reactions in ground water systems on the nutrient content of streams and down-gradient aquifers. ARS is beginning to explore the geochemistry of deep ground water as affected by land use. Also, models are being developed to describe nutrient movement in saturated and unsaturated flow systems and to better define mass balances.

ARS has a basic research program on the detection, movement, sorption, binding, plant uptake, microbial metabolism, and volatility of pesticides in soils and related systems. Unique pesticide problems exist at certain specific geographic locations. At several locations, ARS is conducting limited research directly related to pesticide movement in the vadose zone and underlying aquifers. This research provides soil-water partition coefficients useful for predicting pesticide leaching into the root zone and eventually into ground water.

Herbicide persistence, leaching, and possible ground water contamination are being investigated under conservation tillage conditions. The possible transmission of pesticides via ground water into the Chesapeake Bay is being investigated. Public water supplies and reservoirs in central Maryland are being monitored for the presence of herbicides.

Control of salts in the root zones of irrigated lands is necessary if the productivity of these lands is to be maintained. Leaching is the only practical method of removing excess salt. The leachate or deep percolation water necessary to remove these salts moves and returns to the river as subsurface return flow, or it moves down to the underlying ground water. The ARS program on ground water salinity is directed toward improved understanding and modeling of water and chemical transport. The objective is to more accurately assess and control the contribution of irrigated agriculture to the total salt loading of aquifers.

The principal processes under investigation include the kinetics of salt precipitation and dissolution in soils, ion exchange, carbonate kinetics, and the dispersion and diffusion of salts in the soil profile. Techniques are being developed to use isotope ratios to trace the origin of leachates from natural or man-made sources. In addition, the potential for reducing the "leaching fraction" of applied irrigation water is being evaluated.

ARS is involved in analytically simulating most of the physical and chemical processes affecting chemical movement in the environment. Many of these models are related to ground water quality, even though not specifically developed for that purpose. Major models such as EPIC (Erosion Productivity Impact Calculator), SWRRB (Simulation for Water Resources in Rural Basins), SPUR (Simulation of Production and Utilization of Rangelands), CREAMS (Chemicals, Runoff, and Erosion from Agricultural Management Systems), and SWAM (Small Watershed Model) all have components that relate to ground water quality. All of them simulate the movement of water in the root zone; and the SWAM model includes a complete ground water quality submodel.

Most ARS scientists include models of the system they are studying as part of their research programs. The models related to ground water quality include pesticide models that simulate adsorption-desorption, volatilization, microbial decomposition, and related processes. Transport in both the vadose and ground water zones has been simulated. The movement and concentration of salts in both natural and irrigated lands have been modeled.

The transport processes of these subsystems are dependent upon accurate representation of water movement since it is the transporting medium. The movement of water is well described in most mediums and subsystems; infiltration, unsaturated and saturated flow, in two and three dimensional format and in stochastic as well as deterministic settings. Many of the models have been developed principally as descriptions of the physical processes, but others are being developed with the planner or manager in mind.

SUMMARY

In summing up, I would remind you that the complexity of the area of ground water quality research means that there are a large number of topics that must be resolved. Among agriculturally related ground water quality problems, I believe three stand out as most important. They are:

(1)   occurrence of pesticides in ground water;

(2)   occurrence of nutrients in ground water; and

(3)   irrigation driven salt-water migration.

Major efforts are needed to ensure long-term good ground water quality. The ARS will play a major role in future ground water quality research and is well qualified to conduct multidisciplinary research, to identify ground water quality issues and develop economical viable management systems to alleviate present concerns and prevent the development of future problems.

CHAPTER 2

NATIONAL SURVEY OF PESTICIDES IN DRINKING WATER WELLS

by

Jerry Kotas

The U.S. Environmental Protection Agency (EPA) is planning to conduct a nationwide survey of pesticides in drinking water wells over the next two years. This project summary explains the reasons for conducting the survey, how the survey will be designed and conducted, and the current status of the survey planning effort as of July 1, 1986.

WHY IS A SURVEY NEEDED?

Pesticides present in drinking water can pose dangers to human health if ingested. Current indications are that some pesticide contamination of drinking water wells does exist. However, no one knows exactly which pesticides are present in wells, how high their concentrations are, and which wells are contaminated. To date, some analyses of pesticides in ground water have been undertaken, but they were limited to a small number of pesticides and specific geographic areas. For example, the State of California completed a ground water monitoring program in 1983. Four pesticides, EDB, DBCP, Simazine, and Carbofuran, were monitored in the San Joaquin Valley and Riverside County.

Since 1975 urban water systems have been required to monitor for six pesticides: endrin, lindane, methoxychlor, toxaphene, 2,4-D, and 2,4,5-T. Recent evidence indicates a larger problem of pesticides in ground water; for example, EPA found that 17 pesticides were detectable in the ground water of 23 states as a result of normal land application (Office of Ground Water Protection, 1986).

The National Survey of Pesticides in Drinking Water Wells is a major component of the agency's overall effort to understand and characterize the problem of agricultural chemicals in ground water. The survey will provide a nationwide assessment of pesticide contamination in drinking water wells, estimates of potential populations at risk, and an understanding of how pesticide use and hydrogeology relate to contamination.

With adequate survey information on the concentrations of different pesticides in wells around the country, EPA's regulatory programs can be better designed to target pesticides of concern and to develop further regulatory initiatives. The Federal Insecticide, Fungicide and Rodenticide Act (FIFRA) gives the agency authority to regulate the marketing and use of pesticides. If certain pesticides are shown to pose potential hazards by their ability to leach into ground water, they could be subject to a range of further regulatory actions, including labeling direction changes, use restrictions, or suspension or cancellation of their registration. Information from the survey will also be used by the agency to implement requirements of the Safe Drinking Water Act

(SDWA). New maximum contaminant levels and monitoring requirements can be proposed for pesticides shown to pose a hazard in public drinking water.

GOALS OF THE SURVEY

The National Pesticide Survey is being designed to meet two major objectives set by EPA:

(1) to obtain sufficient information to characterize pesticide contamination in the drinking water wells of the nation; and

(2) to determine how pesticide concentrations in drinking water wells correlate with patterns of pesticide usage and with ground water vulnerability.

Ground water vulnerability is a composite description of geologic and hydrogeologic characteristics that indicate ground water pollution potential. Factors that affect ground water occurrence and availability will also influence the pollution potential of an aquifer. These characteristics can be combined to estimate the pollution of ground water resources.

The focus of the survey is on the quality of drinking water in wells rather than in ground water, surface water, or drinking water at the tap. The survey is not designed to estimate the risk to human health resulting from pesticides in drinking water. Estimating pesticide exposure from contaminated drinking water would require a different survey and research design. The study will, however, provide substantial data to develop inferences about populations potentially at risk from exposure to pesticides in drinking water, and it will yield a wealth of information on the pesticides present in private and community drinking water wells.

HOW THE SURVEY WILL BE CONDUCTED

EPA is nearing completion of the planning stages of the project. Some additional work remains to be done in refining the research design and in coordinating activities with the many participants involved. Key components of the survey are described here.

The survey is expected to be conducted over a period of two years, from September, 1986, to September, 1988. The survey itself will be conducted in two steps, beginning (in late 1986) with a pilot survey of a limited number of drinking water wells and followed by the full survey about six to nine months later.

The survey design requires development of four major components: (1) a sample of wells that is representative of the types of drinking water wells in the nation; (2) analytical methods to measure the types and amounts of possible pesticide contamination; (3) health advisories that establish the levels at which pesticide concentrations may pose a health problem; and (4) a questionnaire to collect key information to analyze additional factors affecting pesticide contamination. These four components are all at various stages of development and will be tested in the pilot survey.

Statistical Design

The statistical design of the survey is complex, primarily because there is no comprehensive tabulation of private (rural domestic) drinking water wells in the United States and only limited information for community well systems. The process of identifying and selecting representative wells for sampling is organized into three stages, as follows:

Stage 1: EPA will classify all U.S. counties using specified measures of ground water vulnerability (obtained from the DRASTIC model developed by the agency) and measures of pesticide usage. From this classification scheme, EPA will select 100-150 representative counties for the sample frame.

Stage 2: At this stage, the counties will be separated further into Census enumeration districts; these districts, in turn, will be stratified by crop patterns and ground water vulnerability to ensure a representative selection. Within these districts, using Census data and other sources, EPA will identify and select household clusters that use private wells. EPA will also select 500 community well systems for pesticide sampling.

Stage 3: In the final step of the selection process, EPA will identify private wells and characterize their use and structure on the basis of interviews with householders. About 500 private wells will be selected for pesticide sampling.

EPA expects to complete Stage 1 of the survey design, selection of counties, during July, 1986. Work is continuing on the development of the final selection requirements for stages 2 and 3.

Analytical Methods

The water samples to be taken from the wells will be analyzed for the presence of over 70 priority and nonpriority pesticide analytes. Pesticides to be analyzed in the study were selected on the basis of expected leaching potential, occurrence, production volume, and other considerations. Final selection of the analytes will be made when the analytical methods are fully developed and the Quality Assurance Plan completed.

The agency is developing six multiresidue methods to detect and quantify the occurrence of pesticides. These methods should be completed by September, 1986. They are intended to detect several analytes and should enable EPA, states, and industry to efficiently analyze for pesticides expected to leach into ground water. Two regional laboratories will be Quality Assurance and Referee Laboratories.

EPA is performing laboratory validation of these methods. This effort will validate detection limits, determine the precision and accuracy of different methods, and analyze sample preservation requirements. In addition, work on the Quality Assurance Plan for the survey is nearing completion. EPA staff are preparing specifications for sampling procedures and sample custody, data analysis, quality control, and performance audits.

Health Advisories

Development of Health Advisories for the analytes is underway as well. Pesticide Survey Guidance levels for 60 priority pesticides will be developed from information collected on physiochemical properties, uses, chemical fate, health effects, treatment, and existing criteria and guidelines. External review drafts of the health advisories will be available in September.

Questionnaire

The questionnaire is being designed in several stages. The agency will first identify the specific information to be collected, determine potential sources of this information, and evaluate alternative sources and measurement strategies. The major categories of information to be collected are: locations of wells, well uses and construction characteristics, pesticide use in relation to the well, available water samples, hydrogeologic characteristics, demographic characteristics, and economic and crop characteristics.

Pilot Survey

The pilot survey is intended to provide field testing of the major components in the project, some of which have been developed specifically for this project, as well as practical experience in conducting the full survey. Pilot studies are now considered virtually essential in any large-scale or benchmark effort.

The pilot survey will involve sampling on the order of 50-100 drinking water wells from a total of three to five states. The states selected will also be ones included in the full survey. For example, counties in central California exhibiting moderate pesticide use and a high degree of ground water vulnerability may be selected for the pilot project.

Project Participants

The Office of Pesticide Programs and the Office of Drinking Water Headquarters staff are jointly sponsoring the survey. Successful completion of the project will require the support and cooperation of regional programs and laboratories and the help of many other organizations. In particular, the cooperation of state environmental agencies and the participation of community well system operators and private well owners will be essential to implementing the study.

Communications

EPA is currently developing a communications document that will outline what types of information will be collected in the survey and when and how this information will be released. EPA expects to issue further updates on this project as work progresses. To be included on the mailing list for the project update, or for further information on the survey, please contact:

Jerry Kotas, Director
National Survey of Pesticides in Drinking Water Wells
U.S. Environmental Protection Agency
401 M Street, S.W. (WH-550)
Washington, D.C. 20460
Telephone (202) 382-5367

SELECTED REFERENCE

Office of Ground Water Protection, Pesticides in Groundwater: Background Document, U. S. Environmental Protection Agency, May, 1986, Washington, D.C.

CHAPTER 3

WATER CONSERVATION FOR MORE CROP PRODUCTION
IN THE GREAT PLAINS

by

Lawrence H. Wetter

Precipitation is highly variable and limited over much of the Great Plains. Evapotranspiration potential exceeds available moisture most of the time. Despite the water deficiency, excellent dryland crop production is possible with good soil and water management to increase infiltration and minimize needless evaporation.

Western Kansas farmers have more than tripled their average per acre dryland crop yields over the last fifty years while applying approximately half of the needed conservation treatment. The added crop production now obtained with water saved by conservation farming exceeds the added production gained by irrigating with pumped ground water.

Water conservation on both dryland and irrigated cropland can help achieve significant remaining crop yield potential while easing the demand on declining ground water for crop production. This will give agriculture a promising future in the subhumid and semiarid Great Plains.

PRECIPITATION AND RUNOFF

Kansas is typical of the Great Plains states in that both rainfall and runoff are highly variable. Figure 3.1 shows average annual precipitation for Kansas. In the western half it varies from less than 18 inches to 26 inches and averages 21 inches. Precipitation amounts vary greatly from year to year. Figure 3.2 shows the average precipitation for the western half of Kansas for each year from 1937 through 1981.

Figure 3.3 shows the chances of getting various precipitation amounts in any year at a typical western Kansas location. It shows a 20 percent chance that precipitation will be less than 16 inches or more than 26 inches. Another variation is of monthly distribution within a year as shown by Figure 3.4. Seventy-five percent comes during the six-month period (April through September) which roughly corresponds to the growing period.

Individual storm size also varies widely. Seventy percent of all precipitation comes in amounts of one inch or less per day. Of the 21 inches that normally fall on western Kansas each year, an average of only six-tenths of an inch runs off. Figure 3.5 shows mean annual runoff for Kansas. Figure 3.6 is a graph of annual surface runoff versus annual precipitation for different runoff curve numbers (CN). It is a simple way of estimating changes in runoff through installation of conservation measures. Curve number reductions are translated directly into reductions in surface runoff. Table 3.1 is a standard table of Soil

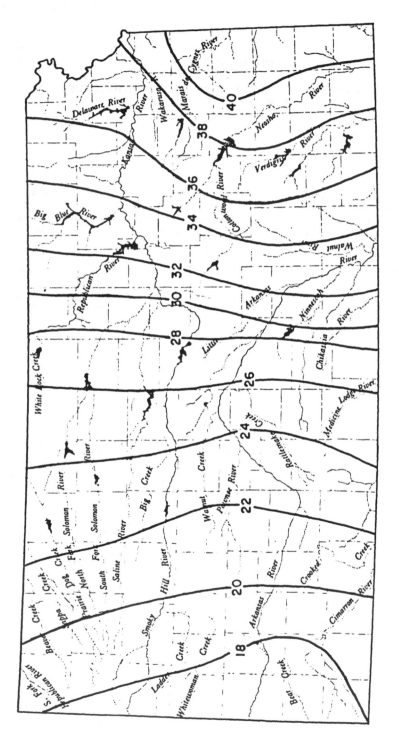

Figure 3.1:   Mean Annual Precipitation in Inches--1941-1970

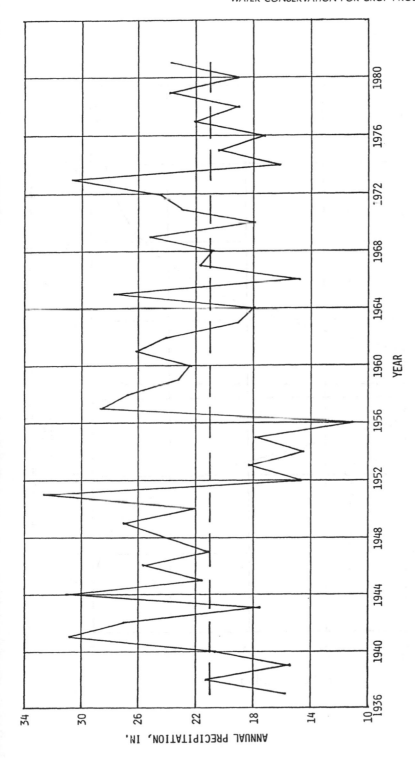

Figure 3.2:   Annual Precipitation Average for 46 Counties of Western Kansas

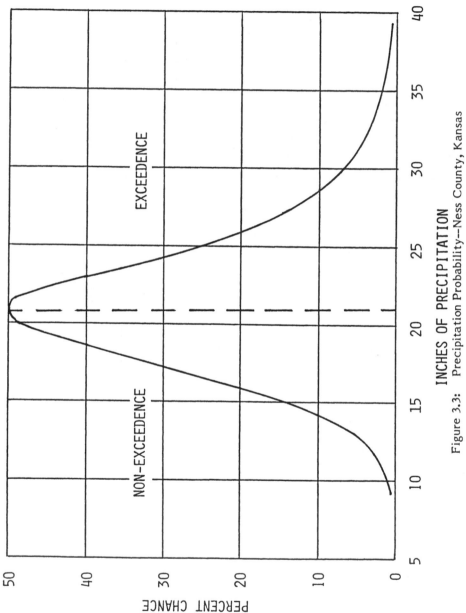

INCHES OF PRECIPITATION

Figure 3.3: Precipitation Probability—Ness County, Kansas

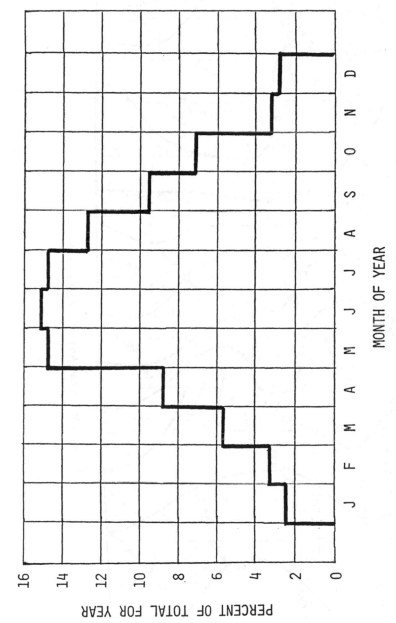

Figure 3.4:   Average Monthly Distribution of Precipitation in Western Kansas

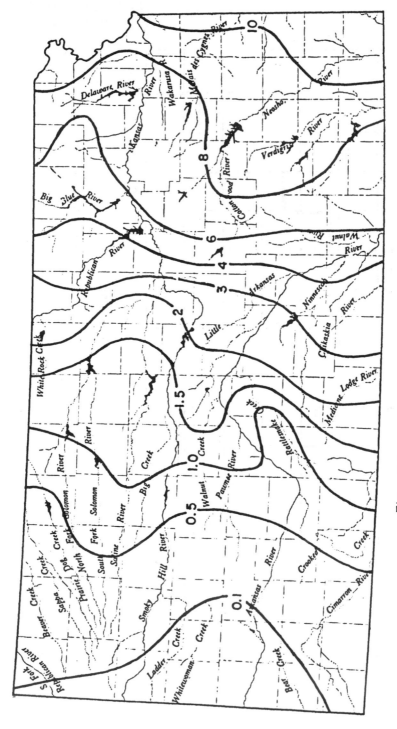

Figure 3.5:   Mean Annual Runoff in Inches

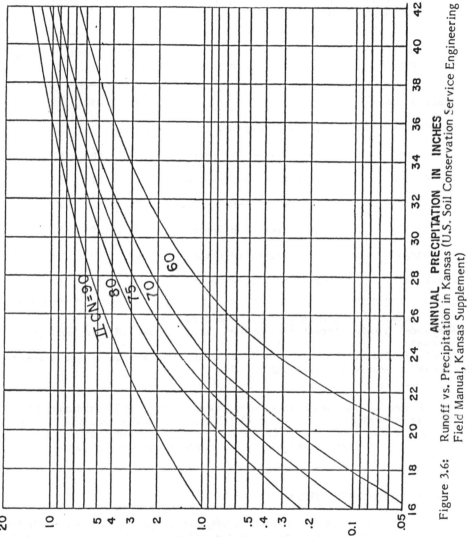

Figure 3.6: Runoff vs. Precipitation in Kansas (U.S. Soil Conservation Service Engineering Field Manual, Kansas Supplement)

Conservation Service (SCS) curve numbers which can be used for various conservation practices.

Table 3.1:   Runoff Curve Numbers for Use in Kansas (Antecedent Moisture Condition II) (Soil Conservation Service Field Manual, Kansas Supplement)

Land Use or Cover[1]	Condition or Practice[1]	Hydrologic Soil Group A	B	C	D
Miscellaneous[2]		72	82	87	89
Cultivated	Gradient Terraces	62	71	78	81
Cultivated	Storage Type Terraces[3]	40	60	67	70
Cultivated	No Residue, No Terrace	72	81	88	91
Cultivated	With Residue, No Terrace	66	77	84	88
Grassland	Poor Vegetative Cover	68	79	86	89
Grassland	Fair Vegetative Cover	49	69	79	84
Grassland	Good Vegetative Cover[4]	39	61	74	80

[1] Use estimated long-term land use and condition
[2] Includes roads, farmsteads, urban, etc. (about 3% for most rural areas)
[3] Includes flat pothole areas and other areas with significant storage
[4] Includes meadow and woodland

Figure 3.7 modified from work by Rawls, Onstad, and Richardson (1980) at the Agricultural Research Service (ARS) and SCS relates reduction in curve numbers to crop residue cover on the soil. Recent work in western Kansas by Steichen and LaForce (1982) of Kansas State University substantiates this curve.

CONSERVATION PRACTICES

In eastern Kansas the terraces, ponds, planned grazing systems, and fields with conservation tillage hold back water which otherwise would have runoff during or immediately following rains. The water then drains slowly through or over the soil to the streams, thus maintaining base flow between rains and increasing available surface water supplies. U.S. Geological Survey streamflow records and observations by SCS verify these flow-duration increases in eastern Kansas streams in recent years.

In western Kansas good rangeland management, conservation farming

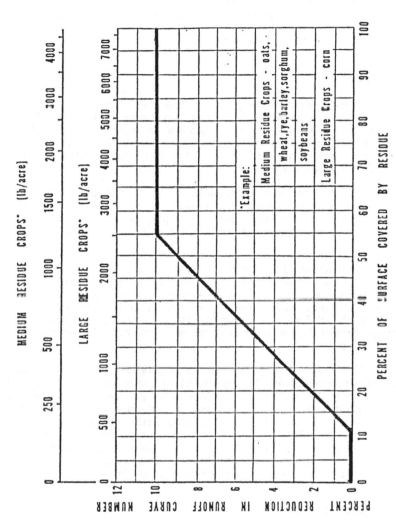

Figure 3.7: Crop Residue Effect on U.S. Soil Conservation Service Runoff Curve Number (Rawls, Onstad, and Richardson, 1980)

methods such as residue management, minimum tillage, no-till, storage-type level terraces, and flat channel terraces hold water where it falls.   Crop production increases but runoff decreases and, for all practical purposes, stops entirely in average years in many areas.   In western Kansas average annual runnoff is only one-half inch or less while precipitation averages about 20 inches.   So, little is gained if half or even all of the runoff is retained.   Runoff and deep percolation constitute about three percent of total precipitation.   The remaining 97 percent leaves by evaporation and transpiration (evapotranspiration), as shown by Figure 3.8.

EVAPOTRANSPIRATION

Evapotranspiration includes evaporation from earth, water, plant, and other surfaces; and transpiration by living organisms, primarily plants. Reduction of evaporation holds the most promise in moisture conservation. Figure 3.9 shows that normal annual evapotranspiration (consumptive use) for the State of Kansas varies from 17 inches to 31 inches.   Figure 3.10 shows the relationship again for western Kansas.

Much research is being done on evaporation and transpiration losses from fields with various cropping systems.   Researchers are studying how soil moisture varies by soil type under different cropping systems.   Mulches and standing stubble affect wind speed and soil temperatures, which, in turn, govern evaporation rates.

The maximum amount of evapotranspiration that can occur in an area is governed by its climate.   Temperature and wind are dominant factors.   This maximum value, called potential evapotranspiration, varies only slightly across Kansas.   Figure 3.11 shows values from 27 inches per year in northwest Kansas to 33 inches in southeast Kansas according to Thornthwaite (1952).   The actual amount of evapotranspiration which can occur on an average, long-term basis will not exceed this potential.

Figure 3.12 shows mean annual precipitation minus potential evapotranspiration across Kansas.   This is also an indication of minimum annual runoff potential.   Only the eastern third of Kansas has positive values.   There is a surplus of precipitation in the east and a deficit in the west.   This is dictated by climate and what we do will change it very little.   It is important to understand this because only in deficit areas can conservation practices stop streamflow in the long term.   Where there is surplus precipitation there will be runoff.   Streams will continue to flow; and while conservation practices may decrease total runoff, they will enhance streamflow in the long term.

REDUCING EVAPOTRANSPIRATION

Researchers Greb (1979), Smika (1983), and others at the Agricultural Research Service's Central Great Plains Research Station near Akron, Colorado, have reported extensively on research with dryland wheat and other crops.   Greb defined fallow efficiency as the percentage of precipitation received during the fallow period which is retained in the soil.   Table 3.2 and Figure 3.13 show how fallow efficiency varies with tillage systems and mulch rates for medium residue crops like wheat, oats, sorghum, and soybeans.

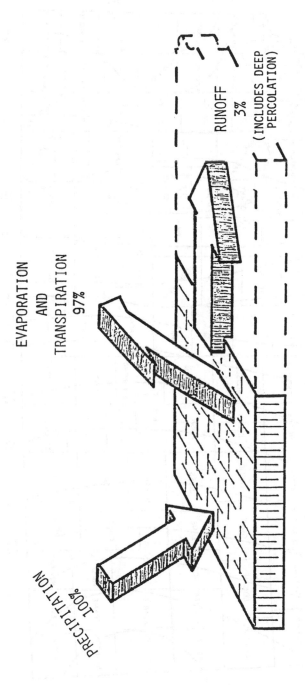

Figure 3.8:   Simplified Western Kansas Hydrology

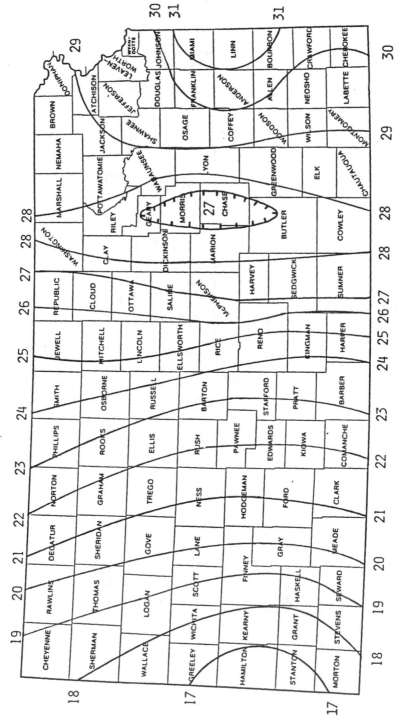

Figure 3.9:   Normal Annual Evapotranspiration, Inches (Precipitation Minus Runoff)

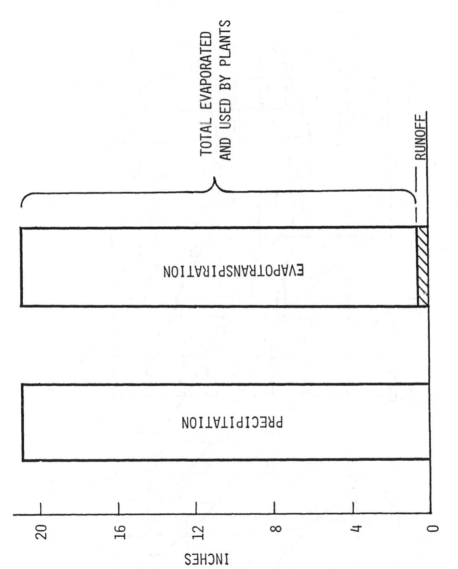

Figure 3.10:   Western Kansas Water Use

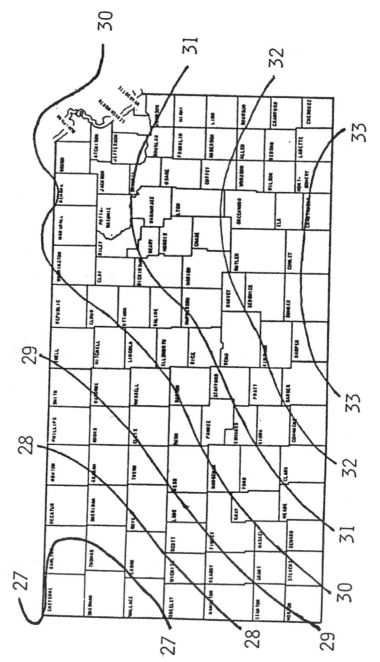

Figure 3.11: Potential Average Annual Evapotranspiration in Inches

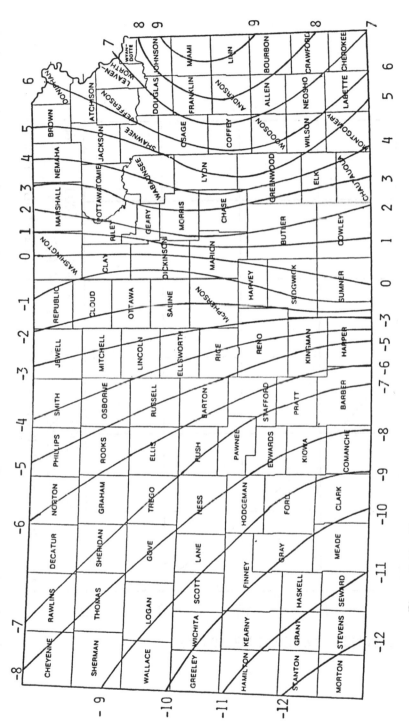

Figure 3.12: Mean Annual Precipitation Minus Potential Evapotranspiration in Inches

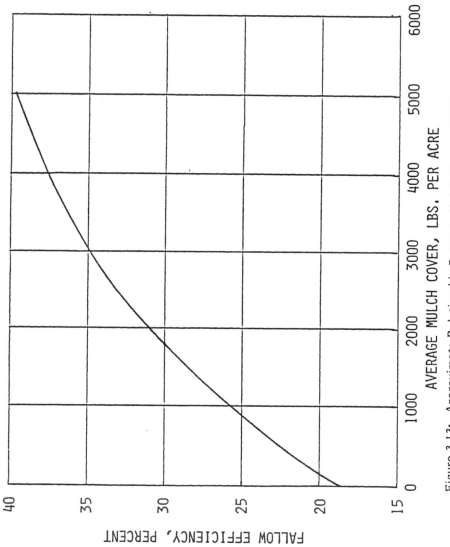

Figure 3.13:   Approximate Relationship Between Mulch Cover and Fallow Efficiency

Table 3.2: Fallow Efficiencies (Greb, 1979)

Tillage System	Fallow Efficiency
Conventional	19% – 24%
Stubble Mulch	27% – 33%
Minimum Till	33% – 40%
No Till	40% – 45%

By knowing precipitation, fallow efficiency, and runoff, we can compute the total available water (TAW) (Koelliker, 1976):

TAW = (precipitation during fallow period x fallow efficiency) + precipitation during plant growth – runoff during plant growth

This represents the amount of water available to a crop from the time it is planted until it is harvested. Total available water can be directly related to potential crop yield as shown by Figure 3.14 (Koelliker, 1976, 1984). Similar curves are available for other crops.   Using these curves, yields can be estimated for various cropping systems.   Potential summer-fallow dryland wheat yield is presently about 71 bu/ac, using average precipitation for the 46 western Kansas counties. Table 3.3 shows the method of computation.

HISTORIC YIELDS

Benefits of land treatment and water conservation techniques adopted in the past can be demonstrated by studying crop yield trends. Wheat is a good example.   Figure 3.15 shows that today the average wheat yield for the 46 western Kansas counties is 36 bu/ac based on analysis of the State Board of Agriculture's county-average yield data. It was only about 10 bu/ac in the late 1930's.

An acre of wheat may yield 26 bu/ac more today than 50 years ago but needs about 5½ inches more moisture to do so. Increased use of fertilizer, improved varieties, better management, and better machinery all contribute to this yield increase; but without the available water gained through conservation, it would not be possible.

FUTURE POTENTIAL

A great potential remains for the future. It is estimated that about half of the cultivated land is now being managed 'for moisture conservation in western Kansas.   On this basis it seems realistic to predict county average wheat yields up to 65 bu/ac by the year 2020, as shown by Figure 3.16. Similar projections can be made for grain sorghum and other crops.

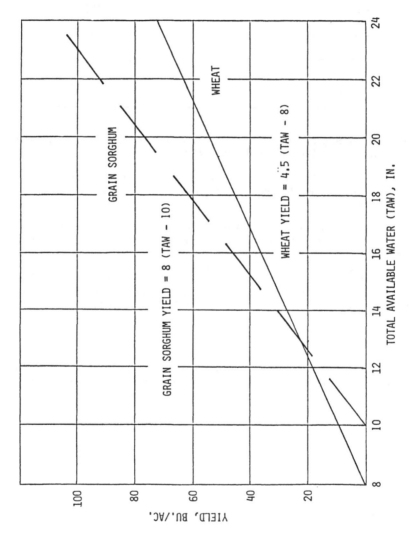

Figure 3.14: Yield Versus TAW

Table 3.3: Potential Dryland Wheat Yield for 46 Western Kansas
Counties

Precipitation during fallow period	= 27.8 inches
Fallow efficiency (min. tillage)	= 40.0 percent
Potential soil water	= 11.1 inches
Typical soil capacity in root zone	= 9.9 inches
Precipitation during wheat growth	= 14.2 inches
Runoff during wheat growth	= 0.2 inches
Total available water (TAW) = 9.9 + 14.2 - 0.2	= 23.9 inches
Yield (from Figure 3.14)	= 71.0 bu/ac

Increased production capability through moisture conservation can offset the expected decline in production from the Ogallala Aquifer. If the water needed for the increased production as a result of conservation measures had to come instead from the Ogallala, it would cause water levels to drop at approximately twice the present rate. The water to be gained through soil and water conservation may be considered a permanent supply to ease the agricultural demand on the Ogallala Aquifer. This will help sustain the Ogallala as a continuing source of municipal and industrial water for the Great Plains.

There is great potential for the future of agriculture on the High Plains, primarily by the conservation and wise use of natural precipitation. It is just a matter of realizing the potential exists, knowing how to take advantage of it, and putting the practices into effect.

SELECTED REFERENCES

Greb, B.W., "Reducing Drought Effects on Croplands in the West-Central Great Plains", Information Bulletin No. 420, 1979, U.S. Department of Agriculture, Washington, D.C.

Koelliker, J.K., "Land Forming Systems to Improve Water Use Efficiency", Kansas Water Resources Research Institute Report, 1976, Kansas State University, Manhattan, Kansas.

Koelliker, J.K., "Choosing Optimum Dryland Moisture Conservation Systems for Western Kansas", Report Developed for the U.S. Soil Conservation Service, 1984, Kansas State University, Manhattan, Kansas.

Rawls, W.J., Onstad, C.A., and Richardson, H.H., "Residue and Tillage Effects on SCS Curve Numbers", Transactions of the American Society of Agriculture Engineers, Vol. 23, No. 2, 1980, pp 357-361.

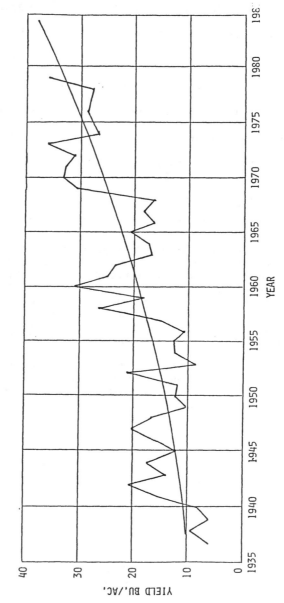

Figure 3.15:   Dryland Wheat Yields for 46 Western Kansas Counties

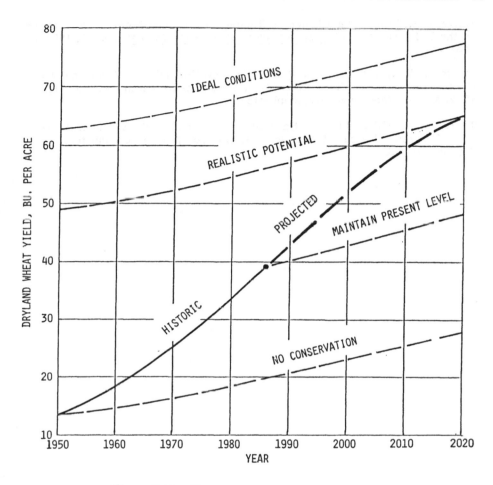

Figure 3.16:   Western Kansas Wheat Yield Projection

Smika, D.E., "Soil Water Change as Related to Position of Wheat Straw Mulch on the Soil Surface", Soil Science Society of America Journal, Vol. 47, 1983, pp 988-991.

Steichen, J.M., and LaForce, R.W., "Conservation Tillage Effects on Water Conservation and Runoff", Kansas Water Resources Research Institute Report, 1982, Kansas State University, Manhattan, Kansas.

Thornthwaite, C.W., "Physical Basis of Water Supply and Its Principal Uses", U.S. Congress, House of Interior and Insular Affairs Committee, 1952, Washington, D.C.

CHAPTER 4

GROUND WATER CONSERVATION TECHNIQUES:
POTENTIAL IMPACTS ON WATER USAGE AND QUALITY

by

Harry P. Mapp

This chapter emphasizes water conservation, focuses on tillage practices and irrigation technology being adopted which will reduce ground water usage, discusses economic and other forces which contribute to the increase in adoption of water conserving technology, and raises some concerns regarding the implications for ground water quality. It is based primarily on research conducted in the southern part of the Great Plains where irrigation water is withdrawn from the Ogallala Formation. The Ogallala is basically a confined aquifer characterized by minimal recharge and extensive ground water mining activity. The Southern High Plains is an area in which a gradual conversion from intensive irrigation to less intensive irrigation to dryland production is occurring. Adoption of water conserving technology tends (other things equal) to slow the conversion to dryland production. Changes in irrigation technology are often accompanied by changes in tillage practices, including substitution of herbicides for machinery operations. This substitution has heightened the concern regarding ground water quality. Little is known regarding the exact relationships among tillage practices and irrigation technology on alternative soils for dryland and irrigated crops over a relatively deep confined aquifer. What are the implications, if any, for water quality in the Ogallala Formation?

IRRIGATION FROM GROUND WATER SOURCES

Rapid development of irrigated production and increased ground water withdrawals throughout the Great Plains and the Western U.S. have been documented. Between 1950 and 1975, ground water use increased from 20 million acre-feet to 58 million acre-feet. Ninety-six percent of the water withdrawals from underground aquifers occurs in 17 western states (Office of Technology Assessment, 1983). The tremendous increase in ground water usage greatly exceeded natural recharge resulting in declining ground water levels. In 1985 approximately 14 million acres, 45 percent of the irrigated acreage in 11 major ground water irrigation states, were located in areas with declining ground water levels (Sloggett and Dickason, 1985).

During the 1970s the physical depletion of aquifers in areas of intensive irrigation development was a major concern. The Ogallala Aquifer which lies under a major portion of six Great Plains states received considerable attention. Total irrigated acreage in the area over the Ogallala expanded from about 3.5 million acres, mostly in Texas and Nebraska in 1950, to more than 15 million acres by 1980. Less than 7 million acre-feet were withdrawn from the Ogallala in 1950. By 1980 more than 21 million acre-feet were pumped annually even though improved irrigation efficiencies had reduced the per acre application of water by about 30 percent from 2 to 1.4 acre-feet per acre (High

Plains Associates, 1982). Acreage irrigated from the Ogallala is small in Oklahoma compared to the other states in the Great Plains, but the pattern of intensive irrigation development and concerns over declining ground water levels are the same as in other states in the region.

SIX-STATE REGIONAL STUDY

Concern over declining ground water levels led the federal government to authorize and fund a study on the feasibility of various ways to insure adequate water supplies for six states in the Great Plains (Colorado, Kansas, Nebraska, New Mexico, Oklahoma, and Texas) overlying the Ogallala. At the time the study was initiated, the importance of the economic (rather than physical) life of the aquifer was not well understood.

ECONOMIC LIFE OF AQUIFER

The number of years over which irrigated production is more profitable than dryland production defines the economic life of the Ogallala Aquifer. Many factors interact to determine the relative profitability of irrigated and dryland production. Some of those factors include the quantities of seed, fertilizer, herbicide, insecticide, fuel, water, borrowed capital, etc. required to produce an acre of alternative commodities, prices of those inputs and prices of the commodities. These factors interact with water availability and cost, and the ability of the manager, to determine the relative profitabilities of irrigated and dryland production. When it becomes more profitable to produce commodities under dryland conditions, the economic life of the underground water supply is exhausted. Because of variations in soil situations, water availability, pumping costs, and management, economic exhaustion will not occur uniformly across the Great Plains or across any single state or county overlying the Ogallala Aquifer.

The importance of agricultural commodity prices in determining the economic life of the water supply for irrigation purposes was emphasized in the Six-State Ogallala Aquifer Area Regional Resources Study. Excess optimism regarding the demand for U.S. agricultural commodities led the U.S. Department of Agriculture to project increases in "real" agricultural commodity prices for the 1980s. That is, commodity prices were projected to increase more rapidly than the rate of inflation. When rising real agricultural commodity prices were incorporated into the High Plains project models, those models projected tremendous increases in irrigated agricultural production throughout most of the Great Plains. Crop yield increases and technological improvements were also built into the models and, in general, those increases have occurred about as projected. The increases in real agricultural commodity prices certainly have not.

In Oklahoma the High Plains project study report contained a baseline analysis projecting increases in irrigated acres, along with a number of sensitivity analyses which varied certain key assumptions relative to the baseline analysis (Figure 4.1). These sensitivity analyses are discussed in detail elsewhere but basically showed that if agricultural commodity prices failed to increase in real terms, irrigated acres in the Oklahoma portion of the study area would fall considerably below the 300,000 acre mark by the 1985-90 period (Warren et al., 1981). Recent irrigation surveys in the area reveal that

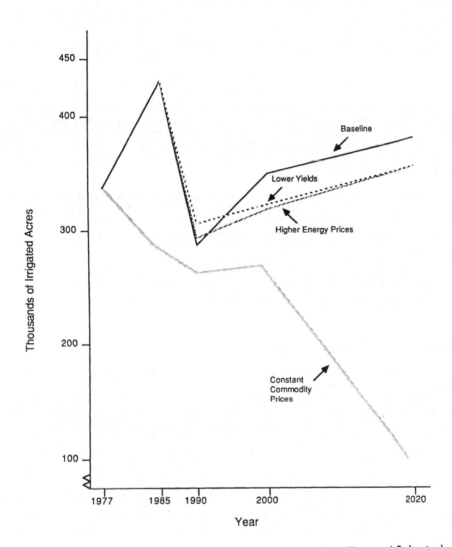

Figure 4.1:   Total Irrigated Acreage Under Oklahoma Baseline and Selected Sensitivity Analyses

irrigated acres in the three Oklahoma Panhandle counties declined from 416,000 acres in 1981 to 338,000 acres in 1983, a 19 percent reduction (Schwab, 1981, 1983) and have continued to decline since 1983. The sensitivity analyses also revealed the results to be more sensitive to prices of agricultural commodities. That is, higher than expected increases in energy costs and lower than projected increases in commodity yields have much less impact on the conversion from irrigated to dryland production than agricultural commodity prices. Results in several other states were very similar to those developed in Oklahoma.

While the High Plains project study was completed in 1982, the results appear to have implications for the recent past, the present, and the future. Agricultural product prices have not increased in real terms nor have they even remained constant in real terms. Precipitous declines in commodity prices have forced producers to make adjustments in irrigation practices, tillage techniques and crop production alternatives. Economic forces are at work to reduce the economic life of the Ogallala Formation. Farmers are responding to these forces in a number of ways.

Intensive irrigation practices result in declining water levels, reduce irrigation well yields and increase pumping costs. With rising pumping costs and falling commodity prices, farmers respond to lower profitability by increasing pumping hours, deepening wells, reconditioning pumps and wells to improve pumping efficiency, or adding new wells. These responses are designed to maintain pumping capacity and do little to reduce water withdrawals. As water levels continue to decline, farmers adopt a combination of water-conserving tillage practices and improved irrigation application techniques. Water conserving techniques utilize more fully the available water by reducing runoff, evaporation, and percolation.

REDUCED TILLAGE TECHNIQUES

A number of water-conserving techniques have been developed and their impacts on water use and crop yields are discussed by Ellis, Lacewell, and Reneau (1984). Runoff from rainfall and irrigation can be reduced by leveling or terrace construction, or by changing tillage practices. In a study of the economic value of terrace construction in the Southern High Plains, parallel terraces were found to be profitable for nearly all moderately sloping soils (Young and Merrick, 1973). Rainfall conservation due to terraces has been estimated to range from 0.02 to 2.74 inches per year, with an additional savings ranging from 2.99 to 5.02 inches of irrigation water on bench terraces. In a study of grain sorghum production on terraced land, improved furrows were found to increase water use efficiency by 34 percent to 40 percent and dryland grain sorghum yields by 51 percent to 59 percent. Mini-bench terraces were even more efficient because of less soil water evaporation due to tillage (Jones, 1981).

Another tillage technique, furrow dams or furrow dikes, has proven effective in reducing runoff, increasing soil water, and increasing crop yields. In dryland grain sorghum trials at Bushland, Texas, additional rainfall of 0.7 to 3.3 inches was retained on the field, resulting in increased yields of 25 percent to 40 percent (Clark, 1979). Dryland cotton yields have been found to increase 11 percent to 25 percent (Runkles, 1980).

Reduced tillage and no-till cropping practices reduce wind and water erosion and increase storage of precipitation and irrigation applications. Stubble-mulch tillage has been shown to increase dryland wheat yields by 17 percent compared to conventional tilled wheat (Johnson and Davis, 1972). Managing wheat residue has increased precipitation storage from 40 percent to 80 percent (Greb, Smika, and Black, 1967; and Unger, Allen and Parker, 1973) and grain yields for subsequent crops by 16 percent to 38 percent. Many of these water-conserving techniques have been used in dryland production for years. Farmers involved in the transition from intensive to less-intensive irrigation and then to dryland production often utilize one or more of these water-conserving practices to improve utilization of scarce rainfall and irrigation water.

IRRIGATION TECHNOLOGY

Accompanying the adoption of water-conserving techniques has been the development of new technology to improve irrigation application efficiency. Considerably more water is withdrawn from underground sources than is utilized by the plant. Losses occur due to evaporation, runoff and deep percolation. Average irrigation water use efficiency is estimated to be 60-70 percent, which means that 30-40 percent more water is typically withdrawn from the aquifer than is used by the plant.

Water use efficiency varies widely by type of irrigation system. Furrow irrigation systems, in general, are much less efficient than sprinkler systems and the efficiency of different types of furrow and sprinkler systems varies widely. Techniques to increase the efficiency of furrow irrigation systems include alternate furrow irrigation, surge flow irrigation, furrow diking and automated furrow systems (Ellis, Lacewell and Reneau, 1984). Alternate furrow irrigation has been tested in Oklahoma and Texas on a variety of soils. Results on hard Texas soils with low porosity have not been encouraging. However, alternate furrow tests on lighter clay and clay loam soils in Oklahoma reduced water use rates on grain sorghum while maintaining crop yields. Similar results were reported on cotton in Oklahoma. Alternate furrow can also be used along with furrow dikes in the nonirrigated rows to reduce rainfall runoff resulting in yield increases for cotton and grain sorghum (Clark, 1979; and Lyle and Dixon, 1977).

Scheduling irrigations in accordance with soil water conditions and the stage of plant development offers an opportunity to reduce water use and maintain grain sorghum yields. An Oklahoma State University study utilized a grain sorghum plant growth computer model validated for Oklahoma panhandle conditions to conduct irrigation scheduling research (Harris, 1981; Harris, Mapp, and Stone, 1983; and Harris and Mapp, 1986). This research revealed that by reducing or eliminating preplant and/or an early season irrigation, and scheduling subsequent irrigations in accordance with critical stages of grain sorghum development, water use could be reduced by 30 to 40 percent, final yields could be maintained, on the average, and producer net returns could be increased.

Another innovation designed to improve the water use efficiency of furrow irrigation is surge flow which creates a surge of water down the furrow on an intermittent cycle. The upper end of the field dries slightly between surges and the water moves further down the row with each surge. As a result, percolation losses are reduced and distribution efficiency is increased. Surge flow has not been widely adopted, but is experiencing increased use in the Texas High Plains.

Improvements in sprinkler irrigation systems have resulted in widespread adoption of sprinklers and replacement of furrow systems with sprinklers on a wide variety of soils. The greater investment costs are partially offset by reduced labor costs and improvements in water application and distribution efficiency. Many farmers have switched from older water-drive systems requiring 95 pounds per square inch (psi) of pressure at the nozzle to self-propelled center pivot systems requiring 65 psi of pressure to low-pressure center pivots requiring 30 to 45 psi of pressure. The reductions in pressure result in corresponding reductions in pumping costs. A recent innovation offers even greater improvements in irrigation efficiency (Lyle and Bordovsky, 1980). The low energy precision application (LEPA) system distributes water at very low pressures (5 psi) through drop tubes and deflectors at plant level directly into a furrow which is usually furrow diked. The LEPA system results in water application efficiencies greater than 98 percent and distribution efficiencies averaging 96 percent, while runoff, both from irrigation and rainfall are essentially eliminated when furrow dikes are included (Lyle et al., 1981).

POTENTIAL IMPACTS ON WATER USAGE

As declining water levels and commodity prices combine to reduce the economic life of underground water supplies, several types of adjustment are likely to continue. Adoption of water-conserving tillage practices, including stubble-mulch and other reduced tillage practices is widespread, particularly on dryland acres. The conversion to improved furrow and/or low pressure sprinkler irrigation systems is being made at a fairly rapid rate. In the Texas High Plains, the conversion from high pressure to low pressure sprinklers has been estimated at 5 to 10 percent of sprinkler acreage per year (Ellis, Lacewell, and Reneau, 1984). The conversion of irrigated acres to the low energy precision application system has occurred more slowly. The investment cost for a new LEPA system is approximately $35,000 while existing sprinkler systems can be converted to LEPA at a cost of $5,000 to $8,000 for the addition of pressure regulators, drop tubes, and nozzles.

What is the likely impact of the adoption of limited tillage and improved irrigation technology on water withdrawals from the Ogallala Formation? This question is the topic of current research at Oklahoma State University, and a similar study was recently completed in Texas (Ellis, Lacewell, and Reneau, 1984). The Texas study projects irrigated and dryland acreage under alternative tillage and irrigation systems from 1980 to 2020. One might expect the improvements in irrigation efficiency to reduce overall water use in future years. However, the Texas study revealed that adoption of improved furrow and LEPA irrigation systems allows spreading the same amount of water across greater acreages while maintaining previous yields, or permits irrigation of more water-intensive crops at less cost. Adoption of the improved irrigation technologies results in increased irrigated acres. Reduced water requirements due to limited tillage also contribute to increased irrigated acres. The Texas study predicts that in the long run, dryland production will replace irrigated production in most areas of the Texas High Plains. However, adoption of improved tillage and irrigation technology slows the transition to dryland, thus helping to sustain irrigated agriculture on the High Plains.

The conclusion of the Texas study is that the adoption of improved furrow and LEPA distribution systems will not greatly extend the economic life of the aquifer for agricultural producers. The reason for this conclusion is that use of these technologies lowers the per-unit cost of obtaining and distributing ground water and results in constant or even greater annual water use. Greater distribution efficiency essentially increases the available water supply within a given time period, allowing more effective and timely application of irrigation water. Both effects encourage greater use of the limited water supply. However, in areas with relatively large pumping lifts, adoption of low pressure distribution systems could extend the economic life of the Ogallala Aquifer. By reducing pumping and application costs, improved technologies increase farmer net returns and contribute to sustaining the economic base of the region.

POTENTIAL IMPACTS ON WATER QUALITY

As indicated earlier, economic forces are at work which encourage widespread adoption of reduced tillage practices and improved irrigation technology. Reduced tillage practices can only be successful if application levels of herbicides are increased. Oklahoma wheat enterprise costs and returns budgets show expenditures for herbicides and insecticides increase from $4.50 per acre for conventional tillage and residual management dryland wheat to $15.98 per acre for a reduced tillage system involving only two tillage operations. Further reductions in tillage to a system involving only one tillage operation on dryland wheat increase per acre expenses for herbicides and insecticides to $27.73 per acre (Table 4.1). Variable and fixed machinery costs, on the other hand, are reduced from $31.73 per acre for dryland wheat under conventional tillage practices to $22.90 for a two-tillage operation production system to $17.55 per acre for a one-tillage operation system. These data are not available for irrigated crops in Oklahoma.

Per acre expenditures for biocides for dryland and irrigated crops under conventional and limited tillage have been estimated for the High Plains of Texas (Ellis, Lacewell, and Reneau, 1984). A shift from conventional to limited tillage practices under irrigation results in per acre increases in biocide expenditures from $28 per acre (conventional) to $42 per acre (reduced) on corn, from $10.50 per acre (conventional) to $17.00 per acre (reduced) on sorghum, from $12.00 per acre (conventional) to $20.00 per acre (reduced) for sunflowers, from $7.00 per acre (conventional) to $12 per acre (reduced) on soybeans, and $5.62 per acre (conventional) to $17.00 per acre (reduced) on wheat. These expenditure increases suggest increases in biocide use ranging from 50 percent on irrigated corn to 200 percent on irrigated wheat as production practices shift from conventional to reduced tillage methods. Little is known regarding the short-run or long-run effects of increased pesticide and herbicide use on ground water quality in underground aquifers. Are there dangers of ground water pollution under fairly intensive irrigation schemes? Does it help that the depth to water in many portions of the Ogallala Aquifer is 100 to 150 feet below the land surface? Is it important that water movement within the Ogallala is only a few feet per year--is there any reason to think that chemicals will move more or less rapidly than water within the aquifer? As the adoption of new irrigation technology greatly reduces runoff and deep percolation, are the concerns over ground water quality similarly reduced? Or are chemical buildups in the soil likely to be a problem for future generations? What are the economics of preventing ground water contamination (if it is likely to be serious problem) relative to the costs of cleanup (if that is even

Table 4.1: Production Inputs for Dryland Wheat Production in Oklahoma

Inputs	Conventional Tillage Moleboard Plow	Residue Management Sweep Plow	One Tillage with Sweep Herbicide	Reduced Tillage 50% One and 5% Two Tillage with Sweep	Two Tillage with Sweep Herbicide
			---$---		
Wheat	4.10	4.10	4.10	4.10	4.10
Fertilizer	22.60	22.60	22.30	22.30	22.30
Insecticide	4.50	4.50	4.36	4.36	4.36
Herbicide	---	---	23.37	17.58	11.62
Custom Combine and Haul	19.48	19.48	19.48	19.48	19.48
Annual Operating Capital	3.58	3.59	4.15	3.94	3.58
Labor Charges	12.15	7.33	2.38	3.05	3.56
Machine Fuel, Lube and Repair	11.21	9.16	6.31	7.71	7.87
Machinery Fixed Costs	20.52	18.50	11.24	13.77	15.03

Source: Oklahoma Crop Enterprise Budgets, 1985.

possible)? Does the likelihood of a return to dryland production, accompanied by a reduction in herbicide, pesticide and fertilizer usage, act to reduce the overall level of concern regarding ground water quality? Certainly these are not questions with easy answers. They are, however, questions worthy of our attention.

SUMMARY AND CONCLUSIONS

From 1950 to 1980 irrigation development expanded rapidly throughout the Great Plains. Water withdrawals from ground water sources greatly exceeded recharge resulting in declining ground water levels. Rising pumping costs have combined with declining agricultural commodity prices to reduce the profitability of irrigated crop production and the economic life of the ground water supplies for irrigation purposes.

Farmers have responded by adopting reduced tillage practices which conserve precipitation and irrigation water. In addition, farmers have adopted new irrigation technology which increases water application and use efficiency. Studies of the conversion from intensive irrigation to water conserving irrigation techniques suggest that overall water use may not be reduced. Thus, the economic life of underground aquifers will be extended only if reduced pumping costs per acre lead to increased profitability from irrigated production.

The adoption of reduced tillage techniques is accompanied by increased use of herbicides. Irrigated production also requires increased levels of fertilizer and pesticides relative to dryland production. The impact of increases in chemical usage during the transition to dryland is uncertain. Adoption of water-conserving techniques will reduce runoff and deep percolation and may reduce the likelihood of ground water contamination. Ground water quality is certainly a concern and many technical and economic relationships must be established before meaningful research in this area can be accomplished.

SELECTED REFERENCES

Clark, R.N. "Furrow Dams for Conserving Rain Water," presented at Crop Production and Utilization Symposium, Texas A&M Research and Extension Center, February 22, 1979, Amarillo, Texas.

Ellis, J.R., Lacewell, R.D., and Reneau, D.R., "Estimated Economic Impact of Irrigation Technology: Texas High Plains," TA-70088, October, 1984, Texas Agricultural Experiment Station, College Station, Texas.

Greb, B.W., Smika, D.E., and Black, A.L., "Effect of Straw Mulch Rates on Soil Water Storage During Summer Fallow in the Great Plains," Soil Science Society of America Proceedings, Vol. 31, 1967, pp 556-559.

Harris, T.R., "Analysis of Irrigation Scheduling for Grain Sorghum in the Oklahoma Panhandle", Ph.D. thesis, 1981, Oklahoma State University, Stillwater, Oklahoma.

Harris, T.R., and Mapp, H.P., "A Stochastic Dominance Comparison of Water Conserving Irrigation Strategies", American Journal of Agricultural Economics, Vol. 68, 1986, forthcoming.

Harris, T.R., Mapp, H.P., and Stone, J.F., "Irrigation Scheduling in the Oklahoma Panhandle: An Application of Stochastic Efficiency and Optimal Control Analyses", Technical Bulletin T-160, September, 1983, Oklahoma Agricultural Experiment Station, Stillwater, Oklahoma.

High Plains Associates, "Six-State High Plains Ogallala Aquifer Regional Resources Study, Summary: A Report to the U.S. Department of Commerce and the High Plains Study Council", July, 1982.

Johnson, W.C., and Davis, R.G., "Research on Stubble-Mulch Farming of Winter Wheat", Conservation Research Report No. 16., 1972, U.S. Department of Agriculture, Washington, D.C.

Jones, O.R., "Land Forming Effects on Dryland Sorghum Production in the Southern Great Plains", Soil Science Society of America Journal, Vol. 45, 1981, pp 606-611.

Lyle, W.M., and Bordovsky, J.P., "New Irrigation System Design for Maximizing Irrigation Efficiency and Increasing Rainfall Utilization," TR-105, May, 1980, Texas Water Resources Institute, Lubbock, Texas.

Lyle, W.M. et al., "Evaluation of Low Energy, Precision Application (LEPA) Irrigation Method at the Texas Agricultural Experiment Station, Halfway, Texas, 1981", Annual Progress Report, 1981, Texas Agricultural Experiment Station at the High Plains Research Foundation, Lubbock, Texas.

Lyle, W.M., and Dixon, D.R., "Basin Tillage for Rainfall Retention", Transactions of the American Society of Agricultural Engineers, Vol. 20, 1977, pp 1013-1021.

Office of Technology Assessment, "Water Related Technologies for Sustainable Agriculture in the U.S. Arid/Semi-Arid Lands", OTA-F-212, October, 1983, U.S. Government Printing Office, Washington, D.C.

Runkles, J.R., "Critical Issues in the Eighties: Water", in Texas Agriculture in the Eighties: The Critical Decade, Texas Agricultural Experiment Station, B-1341, December, 1980, Lubbock, Texas.

Schwab, D., "Irrigation Survey Oklahoma 1981", Department of Agricultural Engineering, 1981, Oklahoma State University, Stillwater, Oklahoma.

Schwab, D., "Irrigation Survey Oklahoma 1983", Department of Agricultural Engineering, 1983, Oklahoma State University, Stillwater, Oklahoma.

Sloggett, G.R., and Dickason, C.L., "Ground Water Mining in the United States", March, 1985, U.S. Department of Agriculture, Economic Research Service, Washington, D.C.

Unger, P.W., Allen, R.W., and Parker, J.J., "Cultural Practices for Irrigated Winter Wheat Production," Soil Science Society of America Proceedings, Vol. 37, 1973, pp 58-62.

Warren, J. et al., "Results of the Oklahoma Agricultural and Farm Level Analysis: Six-State High Plains-Ogallala Aquifer Area Study", AE-8191, Department of Agricultural Economics, 1981, Oklahoma State University, Stillwater, Oklahoma.

Young, K.B., and Merrick, E.B., "Economic Analysis of Parallel and Bench Terrace Use in the Southern High Plains of Texas", Miscellaneous Publication No. 1073, 1973, Texas Agricultural Experiment Station, Lubbock, Texas.

CHAPTER 5

CONJUNCTIVE USE OF SURFACE AND GROUND WATER IN THE SOUTH PLATTE RIVER BASIN: A CASE STUDY OF THE CENTRAL COLORADO WATER CONSERVANCY DISTRICT

by

Thomas V. Cech

Colorado lies within an area of the country that is regarded as semi-arid and, therefore, places great value on its water resources. The South Platte River Basin is one of five major river basins which originate in Colorado and is located within the northeast section of the state; it is bounded by the Rocky Mountains on the west, the Arkansas River Basin to the south, the North Platte River Basin on the north, and the High Plains to the east. The South Platte River originates in South Park, a broad mountain valley approximately eighty miles southwest of Denver and flows over four hundred miles northeast to its confluence with the North Platte River in Nebraska.

The climate of the basin varies as greatly as its terrain. The western area is located along the east slope of the Rocky Mountains and experiences an annual precipitation between fifteen to twenty inches. Summer temperatures seldom exceed 90°F and killing frosts are likely to occur from September 15 through May 15. Concurrently, the annual rainfall on the eastern plains ranges between nine to twelve inches with summertime temperatures frequently reaching 90-95°F. However, since most of the plains are located above 4,000 feet in elevation, nighttime temperatures often fall below 60°F during the summer months. This unique climate allows a wide variety of crops to be grown.

Colorado's population has been growing nearly three times as fast as the national rate. The Front Range is located in the western portion of the South Platte Basin and contains over eighty percent of the state's population. This area also has the major cities of the state including Metro Denver, Boulder, Fort Collins, Greeley, Longmont, and Loveland. The total Front Range includes 2.3 million people, compared to 2.8 million in the entire state. Although irrigated agriculture provides the stable economic base for northeast Colorado, high technology industries such as Kodak of Colorado, Hewlett Packard, IBM, and Anheuser-Busch are rapidly adding to the economic stability of the area.

Total average annual surface water supplies in the South Platte Basin are estimated at 1.8 million acre-feet, with the normal maximum flows occurring from April to June. This includes flows from the major tributaries which include Bear Creek, Boulder Creek, Clear Creek, St. Vrain River, Big Thompson River, and the Cache La Poudre River. Over sixty reservoirs with capacities exceeding 5,000 acre-feet have been constructed in the basin in Colorado and have a total storage capacity in excess of 2,000,000 acre-feet. Primary storage facilities on the mainstream include Antero, Eleven Mile Canyon, and Cheesman Reservoirs. Principle offstream reservoirs include Barr, Standley,

47

Milton, Riverside, Empire, Jackson, North Sterling, Julesburg, and Prewitt Reservoirs.

It is estimated that more than 300,000 acre-feet of water are imported into the South Platte River system annually through sixteen transbasin diversion projects. The Colorado-Big Thompson Project, which began delivering water in 1948, diverts 230,000 acre-feet of water annually from the Colorado River Basin and pumps it across the continental divide to provide supplemental irrigation water to over 700,000 acres in northeast Colorado. A second massive water importation project involves the Moffat Collection System and the Blue River Collection System operated by the City and County of Denver.

Although annual precipitation is just over ten inches across the majority of the basin, irrigation has created a versatile and dynamic agricultural economy. Agriculture ranks second only to manufacturing in retail sales among all of the economic sectors of the state. Diverse crops such as sweet corn, corn and corn silage, sugar beets, blue grass, alfalfa, beans, cabbage, cantaloupe, tomatoes, green peas, onions, carrots, cauliflower, lettuce, barley, asparagus, potatoes, and green peppers are grown extensively along the Front Range in the Denver area. Further east the primary irrigated crops are corn, oats, alfalfa, pinto beans, hay, and sugar beets. Approximately one-half of the state's livestock and crop production is concentrated within the South Platte Basin. Weld County, located just north of Denver, is the fourth leading agricultural producing county in the United States.

The mainstream South Platte River and its tributaries exhibit wide variations in water quality characteristics. Most streams in the upstream part of the basin have small dissolved solids concentrations, providing excellent stream quality for most water uses. However, water quality declines in the South Platte as it flows through the metropolitan Denver area. This is probably due to municipal and industrial wastewater discharges and nonpoint source contributions such as lawn irrigation and urban runoff.

Agricultural return flows and runoff from feed lots have affected water quality downstream of Denver. Large concentrations of nitrogen and phosphorus have been measured in the area in the South Platte River and in the adjacent aquifer. The existence of large amounts of dissolved solids and sulfates have created some problems for downstream municipal and domestic water supplies.

Degradation of ground water has been noted in several areas of the basin. Dissolved solids concentrations in alluvial ground water are consistently greater than average concentrations in the adjacent South Platte River.

WATER LAW IN COLORADO

History of the South Platte Area

The South Platte Valley has a rich heritage of exploration and development. In 1803 the region became property of the United States as part of the Louisiana Purchase. Several expeditions travelled up the South Platte River, including those of Major Stephen H. Long in 1820 and General John C.

Fremont in 1842. Reports of the "Great American Desert" cast an unfavorable light toward settlement in the area.

However, in the 1850's and 1860's, numerous miners came through the basin in search of gold in the Colorado Rockies. This great influx of people created huge demands for food supplies that could not be handled by the freight trains from the east. Therefore, many individuals found another type of gold in the form of irrigated crops to be sold in Denver and the various surrounding mining camps.

The first ditch in northern Colorado was constructed in 1859 on Boulder Creek just east of the City of Boulder. Quickly, other ditches were built near the mouths of the various tributaries of the South Platte. Large scale irrigation development began in the 1870's with the Union Colony organization at the present site of the City of Greeley. Nathan Meeker and Horace Greeley settled a tract of prairie nearly twenty miles east of the Rocky Mountain foothills and created an elaborate irrigation development.

Appropriation Doctrine

In the normally water-rich eastern states, the English or "Common Law" Doctrine of Riparian Rights is the basic law governing surface water allocations. This doctrine allows landowners to divert water from streams, which cross their property, without regard to the effect on downstream diverters. Since this system is unworkable in the arid western states, the Colorado Constitution of 1876 (Article 16, Section 5) declared: "The water of every natural stream, not heretofore appropriated, within the State of Colorado, is hereby declared to be the property of the public, and the same is dedicated to the use of the people of the state, subject to appropriation is hereinafter provided."

The Colorado concept had its beginnings in the mining camps of the Colorado Rockies when gold miners applied the rules of staking mining claims for the utilization of water. It became a common practice for the first diverter of water to have priority over a latter or "junior" diverter. The amount of water miners could use was limited to the amount that they originally appropriated from the stream to operate their mines. The size of the ditch was used to determine the quantity of water that was appropriated. Additionally, failure to use the water constituted abandonment just as failure to work a mine resulted in abandonment of a mining claim. This method of water allocation was placed into legislation in the late 1870's. Also, provisions were made for the establishment of State Water Divisions, Water Districts, and for the Office of the State Engineer.

The Adjudication Act was passed in 1881 by the Colorado Legislature whereby everyone holding surface water rights was required to appear and testify before a court-appointed referee. The person submitting testimony was required to give the date that the right had been put to beneficial use, the location, and the ditch dimensions. The referee then reported to the court and, subsequently, a water decree was issued. An owner of a water right who did not appear before the court lost their appropriation date.

Once approved by the court, a water right became private property that could be bought, sold, and moved geographically, that is, removed from the land, subject to water court approval to prevent injury to senior vested water rights downstream.

Ground Water Legislation in Colorado

The first ground water law in Colorado was passed in 1957 whereby a permitting process was instituted. Additionally, a ground water commission was created and authorized to designate "tentatively critical ground water districts in areas where the withdrawal of ground water appears to have approached, reached, or exceeded the normal rate of replenishment." The law created the basic institutional framework for the administration of ground water in Colorado.

The Ground Water Management Act of 1965 replaced the 1957 Act and authorized the creation of "designated ground water basins" whereby ground water not tributary to any natural stream would be administered by a modified version of the Doctrine of Prior Appropriation. The act authorized the Colorado Ground Water Commission to institute a permitting system for withdrawal of nontributary ground water and to form ground water management districts to administer and manage ground water within designated basins.

The Water Right Determination and Administration Act of 1969 addressed the appropriation of tributary ground water. This act brought diverters, that were pumping from the alluvial aquifer of a stream, within the priority system. To comply with the Water Right Determination and Administration Act of 1969, irrigators who utilized tributary wells were required to cease pumping whenever their appropriation date was not in priority.

The 1969 Act had several other key components which reorganized water administration in the state. To assist the State Engineer, seven water divisions were created, for the nine drainage basins in the state, to eliminate the previous seventy districts. The purpose of the seven districts was to administer and distribute water at the local level. A division engineer, appointed by the State Engineer, was made responsible for the day-to-day operations of the division. In addition to the creation of seven water divisions, the 1969 Act required the Supreme Court of Colorado to designate a "Water Judge" in each division to rule on all water litigation within the division.

Litigation regarding the act was intense in the South Platte River Basin during the early 1970's. Well owners argued that shutting down tributary wells, when a senior call was in effect, would result in a "futile call" and would not immediately result in additional water to the senior diverter. The senior diverters argued that pumping of the tributary well did, in fact, affect the senior vested water right during the course of the irrigation season and, therefore, should be required to follow the new law. On March 15, 1975, the State Engineer issued a set of "Amended Rules and Regulations", based on previous court decisions, and applicable only to underground water in the South Platte Basin. These rules and regulations provided a means for the 10,000 wells of the South Platte River drainage area to continue to pump without significantly harming vested senior water rights downstream. Basically, the State Engineer presented alternatives whereby tributary irrigation wells could pump out of priority. Such an alternative was for the well to be included within a temporary augmentation plan. Such a plan would require that at least five percent of a well's projected annual volume of ground water diversion would be replaced in the stream to offset the stream depletion of the well. A permanent augmentation plan is required to replace the full injury to senior water rights caused by the out-of-priority ground water diversion of the well in question.

The purpose of a plan for augmentation is to attempt to maximize utilization of both surface and ground water by allowing out-of-priority diversions of tributary wells by replacing the resulting depletion to the river so that no injury occurs to senior water rights. This is generally achieved by purchasing replacement water from senior water rights. An augmentation plan is defined in Colorado statute as "A detailed program to increase the supply of water available for beneficial use by the development of new or alternative means or points of diversion, by a policy of water resources, by water exchange projects, by providing substitute supplies of water, by development of new sources of water or by any other appropriate means." This technique provides an alternative, for the State Engineer, to strictly enforce the priority system for tributary wells.

FRAMEWORK OF THE CENTRAL COLORADO WATER CONSERVANCY DISTRICT

The Water Conservancy Act

The Central Colorado Water Conservancy District was formed in 1965 by public petition and vote under the auspices of the Water Conservancy Act of 1937. This legislation enhanced the concept of water development in Colorado by creating public agencies empowered to levy ad valorem taxes on all real and personal property in the district for conserving, developing, stabilizing, and acquiring supplies of water for domestic, irrigation, power, manufacturing, and other beneficial purposes. The water conservancy district concept was adopted to provide for the conservation of water for the benefit of public, industrial, municipal, and irrigation water users. Once a conservancy district has been organized and approved, the District Court can appoint or call for an election of a Board of Directors of not more than fifteen residents from counties within the district. The districts have the power, among others: to acquire and sell water rights, construct and operate facilities, exercise eminent domain, contract with the federal government, and levy taxes.

Formation of the Central Colorado Water Conservancy District

The Central Colorado Water Conservancy District contains over five hundred square miles in the three northeast Colorado counties of Weld, Morgan, and Adams. A board of fifteen directors from these three counties were either appointed by the District Court or designated by an election of the constituents. Since all cities with a population greater than 25,000 elected not to be included within the boundaries of the district, most directors are agriculturalists or have strong farm backgrounds.

Central was originally formed to pursue the transmountain diversion of the Blue River water to the East Slope. The diversion plan subsequently proved to be financially impractical, and the water rights on the western slope were dropped. The organizers of the district, however, agreed to lend support to the Lower South Platte Water Conservancy District in sponsoring the Narrows

Project near Ft. Morgan, Colorado. For several years, the pursuit of this project was the main objective of Central.

Creation of the Ground Water Management Subdistrict

In 1973, again under the authority of the Water Conservancy Act, a second district, or subdistrict, was formed called the Ground Water Management Subdistrict of the Central Colorado Water Conservancy District. The subdistrict creation was in response to the great demand for an entity which could coordinate and administer an augmentation plan for some of the thousands of wells located within the South Platte River Basin. It also covers portions of Adams, Morgan, and Weld Counties in Northeast Colorado and has a geographical area of just over three hundred square miles. The subdistrict utilizes the same Board of Directors, staff, and operational facilities as the parent district. It does, however, have its own ad valorem tax and assesses an annual fee for members of the augmentation plan. As with the parent district, no city with a population greater than 25,000 was included within the subdistrict.

Creation of the Plan for Augmentation

The Ground Water Management Subdistrict has submitted a plan of operation to the State Engineer each year since 1973. This plan includes information on irrigation wells within the plan, sources of water and replacement rates. The purpose of the plan is to show a replacement schedule, based on month and river conditions, that will offset the river depletions caused by pumping of wells out of priority. In 1985 nearly 1,000 wells were served by Central's plan for augmentation which provided supplemental irrigation water for 66,258 acres.

Central's plan includes eight replacement tables which present daily augmentation requirements based on the location of a call for water on the river. For example, a call for water by a senior diverter at the south end of the subdistrict, i.e., higher on the river, will require less replacement water for Central than would a call at the northeast or lower end. This is because Central's member wells have a greater accumulative depletive affect at the downstream or northeast end of the subdistrict. The plan shows the amount of water to be replaced by river reach, based upon the total depletive affect of wells on a particular section of the river. This replacement schedule is based upon the depletion analysis conducted by Central's engineering consultants.

Irrigators who wish to join Central's plan for augmentation are required to fill out a "Class D Water Allotment Contract". This contract requests information regarding irrigated cropping patterns, quantity, and type of surface water utilized with the ground water, method of irrigation, and location of the well. This information, along with additional data, is then used to determine the well's depletive affect on the South Platte River.

Depletion Analysis

In understanding river depletion, a few concepts are presented. Any ground water pumping from an aquifer that is hydraulically connected to surface water will initially cause depletion, or a cone of depression, in the aquifer. After pumping is discontinued, the volume of the dewatered cone of depression will not change, but the shape of the cone will change with time. This hole or void in the aquifer will fill in with water flowing from the perimeter toward the center of the cone. As soon as the perimeter of the cone reaches a river or stream, the cone of depression will begin to fill in from surface water flows. This "refilling of the cone" can take a long time due to the slow velocity of ground water flows. The distance, however, of the well from the stream must also be taken into account.

If a well is drilled on a riverbank, its affect on the river is almost instantaneous and will behave much like a ditch diversion. If a well is drilled several miles from a stream or river, its affect on the river will be delayed much longer. Due to this delay, the total ground water depletion caused by a well may not occur in the same year at the river if the distance between well and river is significant. Part of the depletions will carry over to successive years and will be superimposed on the depletions of past and future years.

After a certain time, the ultimate steady state will be reached where the river depletion will be equal to the on-farm ground water depletion. Due to the seasonal diversion of ground water, there will be a difference between the monthly distribution of the depletion. A well close to the river will have a high peak in its monthly river depletion distribution while a well located away from the river will have a more uniform depletion throughout the year. Wells that are miles from the river may take decades before reaching a uniform depletion rate.

During the period of time when a "call for water by senior water right" is on the river, irrigation wells must replace their depletive affect on the river. Although the "Amended Rules and Regulations of the State Engineer of 1975", which Central follows, requires the replacement of at least five percent of a well's projected annual pumpage, the Directors of Central were not satisfied with the methods used to determine such pumping volumes. Other augmentation groups were utilizing electricity consumption records to determine the extent of ground water pumped, but these figures can be inaccurate due to changes in drawdown, lift, pump efficiency, and application pressure during the course of the irrigation season. Additionally, the proposed use of totalizer flow meters to obtain more accurate readings is as unpopular in Colorado as in other western states.

Therefore, Central utilizes crop consumptive use data through the use of the Blaney-Criddle Formula whereby the consumptive use is determined based on temperature, length of day, and available moisture. Blaney and Criddle found that the amount of water consumptively used by crops during their normal growing season was closely correlated to mean, monthly temperatures and daylight hours. Their method not only considers climatological parameters but also relates to crop types and the growth stage of the given crop. Coefficients were also developed that can be transposed from an area with consumptive use data to an area for which only climatological data are available.

Replacement Procedure

Since its formation in 1973, the Ground Water Management Subdistrict has acquired various types of water for use as augmentation or "replacement" waters. Due to the distributions of Central's member wells along the South Platte River from Denver to Ft. Morgan, depletions arrive at the river throughout the length of this one hundred mile stretch.   Therefore, augmentation waters are located at various places along the river.

Central currently owns or leases 8,000 acre-feet of surface water. Shares of ditches or reservoirs are purchased or leased annually and then are delivered to the South Platte River through augmentation stations consisting of a Thompson partial flume and a Stevens Type F Recorder. Appropriate River Commissioners (Deputies of the State Engineer) read the recorder charts weekly and verify replacement quantities. Credit is only given for water which is delivered to the river when a request for water or "call" is in effect on the river downstream of the point of replacement.

Augmentation water is also obtained in the form of sewage effluent from municipalities, but under certain restrictions. Transmountain water (imported from the West Slope of the Rocky Mountains) can be reused until it is entirely consumed if it is water that historically was not within the South Platte Basin system. Therefore, Central acquires sewage effluent from cities which utilize transmountain water. In 1983, one-thousand acre-feet of effluent were leased from the City of Aurora, a suburb of Denver, and was delivered to the South Platte River through the Denver Metro Sewage discharge site north of Denver.

Two batteries of wells are located at the upper or south end of the subdistrict and are utilized in the latter part of the augmentation season. The ten wells in these fields are pumped beginning approximately forty-five days prior to the end of the annual augmentation season. These wells are located about one-half mile from the South Platte River; and although they pump additional water into the South Platte (up to forty acre-feet per day), they do have a depletion affect on the stream. Therefore, a depletion factor is subtracted from the quantity of water placed in the stream.

This process works because the depletion caused by Central's battery of wells is delayed beyond the call season. Since augmentation is usually not required after mid-September, the remaining depletions currently do not have to be offset.

In 1980 Central developed several recharge sites through cooperation with numerous ditch and reservoir companies in the area. One site north of Denver utilizes the excess carrying capacities of the FRICO and Henrylyn Ditch Systems to deliver approximately twenty acre-feet daily into the dry creek bed of Box Elder Creek. This recharge water infiltrates below ground and has two benefits.   First, recharge water elevates ground water levels and reduces pumping costs for nearby irrigators. Second, since this underground water is utilized by member irrigators of the subdistrict, portions of these flows are deducted from the augmentation requirements of the wells in the area.

The following year, in 1981, a second recharge project was constructed west of Ft. Morgan to store excess water of the South Platte River via the Bijou Ditch System. The subdistrict incorporated a series of earthen/sand dams, approximately ten feet in height, to create recharge infiltration ponds on Kiowa Creek. The ponds are approximately ten acres in size and eight feet deep when filled. A continuous flow of water at a rate of five cubic feet per second will keep the ponds full. Up to 3,000 acre-feet of water have been stored underground annually through the Box Elder and Kiowa recharge systems since their inception in 1980 and 1981.

In 1984 the Platt Valley Irrigation Company, located near Greeley, and Central's Subdistrict began a joint effort on a unique type of recharge project in that area of the South Platte River. Water was diverted from the South Platte, during periods of "no call", into the Platte Valley Ditch at a rate just great enough to allow the water to percolate underground before it reached the end of the ditch system. Approximately 200 acre-feet of water were recharged in a matter of two or three weeks.

In 1986 the Farmer's Independent Ditch, also located near Greeley, began a test project of running recharge water in the spring and fall periods of no demand for water on the South Platte. Over 500 acre-feet of water was recharged In a matter of weeks.

Again in 1986, a series of earthen dams were constructed on Milliron Draw, near Ft. Morgan, to expand the recharge capabilities of the subdistrict. The dams are approximately ten feet in height and hold almost 50 acre-feet of recharge water per pond. A constant flow of approximately five cubic feet per second has been recharged through the three ponds. Two more structures are planned for the near future.

Since the distance to the river through the alluvium is not great, the calculated return flow credits to the South Platte River, from these recharge projects, are reflected annually in Central's plan for augmentation. These return flows are credited as replacement waters during the irrigation season.

THE FUTURE OF GROUND WATER ADMINISTRATION IN COLORADO

Ground water administration has occurred for only twenty-five years in Colorado, but it has developed into a complex network of permits, hearings, and legal doctrine. It does not appear that the utilization of ground water will become less enforced in the future; if anything, it will become more complicated. With the continued competition for water along the Front Range of the state, entities with large water demands and sizable revenues will be best able to continue to develop and acquire valuable water rights. It appears that the cities will slowly but steadily continue to obtain agricultural water rights and convert them to municipal and industrial uses.

The implications for the ground water irrigators are great. Not only will they continue to be required to pay for augmentation water to offset the depletion of the river caused by out-of-priority pumping, but they will also be charged higher and higher fees to acquire such water for augmentation. The physical allocation and administrative tools of ground water in Colorado are in place, but the economic effects of the existing system may be devastating to the ground water irrigator of the future.

REFERENCES

Central Colorado Water Conservancy District, The Central Waterline, Winter, 1983, Greeley, Colorado.

Colorado Front Range Project, "Water: Understanding the Future," September, 1981, State of Colorado, Office of the Governor, Denver, Colorado.

Colorado Water Conservation Board, "South Platte River Basin Assessment Report," August, 1982, Denver, Colorado.

Department of the Interior, Bureau of Reclamation, "Pick-Sloan Missouri Basin Program, Front Range Unit, Longs Peak Division, Colorado, Status Report," October, 1977, U.S. Government Printing Office, Washington, D.C.

Department of the Interior, Bureau of Reclamation, "Water Use and Management in the Upper Platte River Basin," August, 1982, Lower Missouri Region, Denver, Colorado.

Dill, J.M., "A Brief History of the Northern Colorado Water Conservancy District and the Colorado-Big Thompson Project," 1958, Northern Colorado Water Conservancy District, Loveland, Colorado.

Fischer, W.H., and Ray, S.G. et al., "A Guide to Colorado Water Law," September, 1978, Colorado Water Resources Research Institute, Colorado State University, Fort Collins, Colorado.

Hartman, L.M. and Seastone, D., "Water Transfers," 1970, Resources for the Future, Washington, D.C.

Personal Communication, George Palos, Engineer, Resource Consultants, Inc., Fort Collins, Colorado, Memo to the Board of Directors of the Central Colorado Water Conservancy District, Greeley, Colorado, May 24, 1983.

Radosevich, G.E. et al., Evolution and Administration of Colorado Water Law: 1876-1976, 1976, Water Resources Publications, Fort Collins, Colorado.

Trelease, F.J., Water Law - Cases and Materials, Third Edition, May, 1979, West Publishing Company, St. Paul, Minnesota.

U.S. Soil Conservation Service, U.S. Department of Agriculture, "Irrigation Water Requirements," Technical Release Number 21, April, 1967, Washington, D.C.

CHAPTER 6

GROUND WATER RECHARGE FOR OKLAHOMA--
AN ANALYSIS OF PAST AND FUTURE METHODOLOGY

by

Robert K. Thomas, Jr.

Oklahoma is an antithesis of hydrologic contrast. The east half receives fifty plus inches of precipitation a year, rivers abound, lakes are plentiful, and the need for ground water is low. The west half is semi-arid, rivers flow only in their respective alluvium, precipitation is usually less than eighteen inches a year, the evaporation rate is some one hundred plus inches per year, and the aquifer recharge rate is less than one inch per year (Figure 6.1). A critical situation exists in western Oklahoma today. As water levels fall and quality degrades, the need for artificial recharge programs has become a primary concern of the Oklahoma Water Resources Board.

Of the six major aquifers in western Oklahoma, all but the Dog Creek Shale/Blaine Gypsum are showing a steady decline (Figure 6.2). This decline may well be linked to the major user of the available ground water--irrigation. Some seventy-nine percent of the ground water used in Oklahoma (in 1982) could be linked to farming and subsequent irrigation. Without this water the land would quickly become hostile and impossible to farm. Therefore, it is in the best interest of the state, its economy, and its people to instigate recharge projects. The objective of this chapter is to define the parameters of artificial recharge, to observe past recharge successes within the state, and to propose new methods that may best demonstrate how to artificially recharge Oklahoma's aquifers.

Artificial recharge is a means of augmenting the natural infiltration of surface water into a ground water reservoir by means of wells or other specialized construction, by water spreading, or by changing natural conditions (Asano, 1985; and Pettyjohn, 1981). This statement envelopes an idealistic approach to artificial recharge. Where can one get a source of water to be used in recharge in western Oklahoma? At least five potential sources of recharge exist:

(1)   natural stream flow (both surface and alluvium);

(2)   high stream flows diverted from nearby water courses;

(3)   cooling water;

(4)   sewage effluent; and

(5)   irrigation return water.

Each potential source could be used successfully in limited areas. Alluviums could be used instead of the major confined aquifer if the quality and

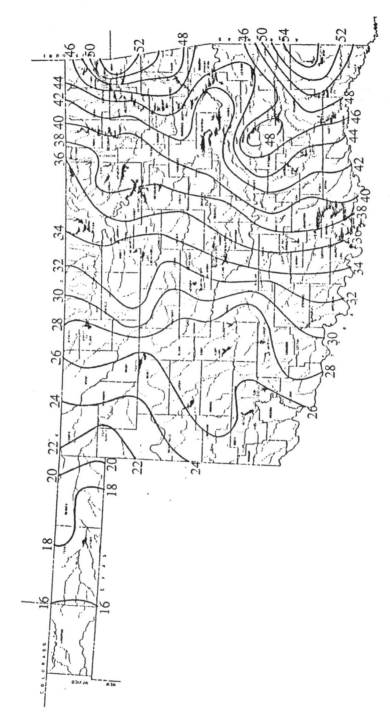

Figure 6.1:   Average Annual Precipitation (in Inches) for 1970-1979 (Data--
National Oceanic and Atmospheric Administration; Mapping--
Oklahoma Water Resources Board)

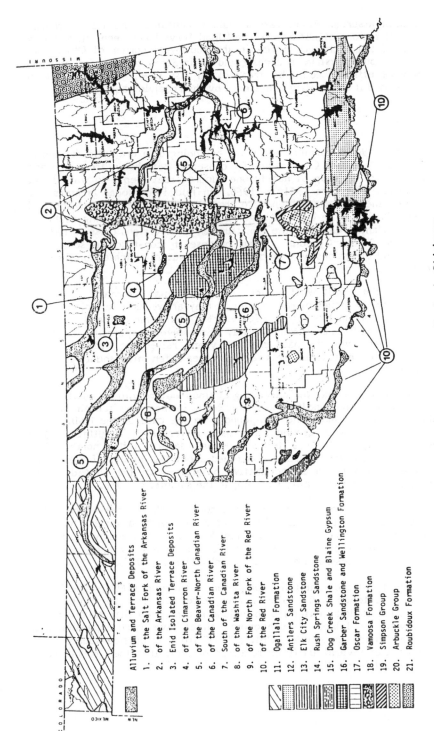

Alluvium and Terrace Deposits

1. of the Salt Fork of the Arkansas River
2. of the Arkansas River
3. Enid Isolated Terrace Deposits
4. of the Cimarron River
5. of the Beaver-North Canadian River
6. of the Canadian River
7. South of the Canadian River
8. of the Washita River
9. of the North Fork of the Red River
10. of the Red River
11. Ogallala Formation
12. Antlers Sandstone
13. Elk City Sandstone
14. Rush Springs Sandstone
15. Dog Creek Shale and Blaine Gypsum
16. Garber Sandstone and Wellington Formation
17. Oscar Formation
18. Vamoosa Formation
19. Simpson Group
20. Arbuckle Group
21. Roubidoux Formation

Figure 6.2:   Major Ground Water Basins in Oklahoma

quantity could be assured. Flood waters could be contained and used to charge the aquifer. Diversions of natural stream/flood flows have been successfully used as an artificial recharge source in the southwest corner of Oklahoma. Cooling waters from industrial sources could be injected to raise an aquifer, locally. Sewage effluent could be treated and also used. Irrigation return water could be used if the quality is matched by the receiving aquifer.

THE HISTORY

In the early 1960's, farmers noted the beginning of a dramatic decline in the Dog Creek Shale/Blaine Gypsum aquifer caused by both heavy irrigation and several "dry" years. A drop of nearly one hundred feet (depth to water) was observed between the years 1961 to 1966. In mid 1965, local farmers and ranchers pooled their time and money to create the Southwest Water and Soil Conservation District. The district began a bold program of drilling recharge wells, in key lowland areas, and the construction of diversions close to natural fractures in the Blaine Gypsum. Their purpose in constructing some two hundred wells and diversions was to capture flood waters which occur three or four times a year.

Although the southwest part of Oklahoma normally receives approximately twenty inches of rain a year, it may arrive as a few cloudbursts inflicting five or more inches within a few short hours. Some of the flood water finds its way into the Blaine Gypsum (by way of the seventy operating wells and diversions) and has, over the past eighteen years, raised the water table to within forty feet of the surface. This recharge area is ten miles square from its hub, the Town of Hollis.

It should be noted that the cascading of flood waters into an aquifer may degrade the quality. The runoff of any field being farmed may contain several undesirable organic and inorganic compounds. In the case of the Hollis area, few, if any, drink the local ground water as it is very high in sulfates. The public water supply is from a terrace deposit several miles north. The Oklahoma Water Resources Board is concerned about the quality of the Blaine Gypsum water and will, in a cooperative program with the United States Geological Survey, conduct quality sampling in the summer of 1986.

The district's success can be seen throughout the Hollis area where fields, once heavily scarred by runoff and flood water, now grow a variety of crops free from potential washout. Roads, often destroyed on a yearly basis, now need only occasional repair. Diversion walls, cut deeply by churning flood waters, denote the successful use of natural sinks (Horton, 1986).

The project was only partially completed in the mid 1960's. The Blaine Gypsum Aquifer extends many miles in every direction from the Hollis area. Today, the project could be extended to include two hundred fifty square miles, if the funds were available. Hard economic times have made expansion with the use of state funds impossible. Only a "gift" from an outside source could help this project or several artificial recharge projects that the Ground Water Division of the Oklahoma Water Resources Board had planned.

THE GIFT

In October, 1985, the United States Bureau of Reclamation challenged several states using ground water from the Ogallala to submit proposals for artificial recharge demonstration projects. Subsequently, the Ogallala Ground Water Recharge Demonstration program was born and soon was expanded to include any major aquifer within a "High-Plains" state (U.S. Bureau of Reclamation, 1985).

The Ground Water Division of the Oklahoma Water Resources Board accepted the challenge and submitted three proposals for artificial ground water recharge funding in early May, 1986. These proposals include:

(1)   the expansion of the Hollis project with more diversions, recharge wells, and several monitoring wells,

(2)   a subsurface dam around a municipal well in a terrace deposit along the North Canadian River for the City of Woodward, and

(3)   effluent treatment and injection facility for the City of Guymon.

THE BLAINE GYPSUM PROJECT

If accepted by the Bureau of Reclamation, the new Hollis Project will follow the success story previously mentioned. The diversion north of Hollis will require two recharge wells and a monitoring well to be drilled. The south diversion will be completed on a ridge, containing one of the first diversions built near the town. Several natural sinks are located there. At least ten new recharge wells will be drilled in key spots of high recharge potential throughout the area. The average depth of a recharge well could be two hundred feet. The impoundments are: "A" five hundred feet long, twenty feet high, thirty feet thick at the base (Figures 6.3a and 6.3b); "B" one thousand feet long, six feet high, and twelve feet thick (Figures 6.4a and 6.4b). All impoundments are made from local soils.

THE WOODWARD PROJECT

The City of Woodward is unique as they pump water from both a terrace deposit of the North Canadian River and the Ogallala Aquifer. The terrace deposits began producing in the mid-1930's, where as the deeper Ogallala wells were drilled just a few years ago (Figure 6.5). Of the seventy wells operating, twenty-nine are terrace, and they produce approximately thirty-nine percent of the required water for the City of Woodward. The terrace wells are pumped only sporadically, as backup for the forty-one Ogallala wells. The decline of the Ogallala water table has been abrupt in the last few years as a result of increased irrigation and Woodward requirements. A water table drop of six feet (from 153 to 159 depth to water) occurred between 1981 and 1982 and an implied decline continues. The City of Woodward mines over two billion gallons of water each year from the Ogallala.

The proposed Woodward recharge project concerns taking more of the abundant terrace water to relieve the strain on the Ogallala. For demonstration purposes, a single municipal well will be isolated by a horse-shoe

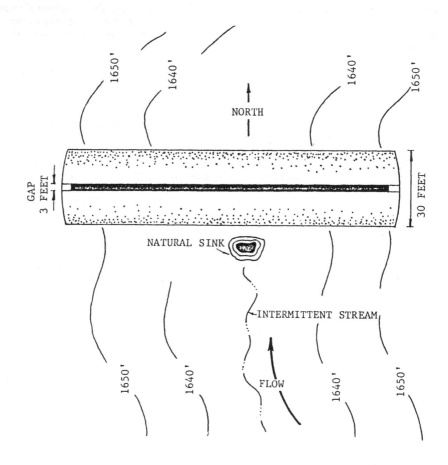

Figure 6.3a: Impoundment "A"--Top View--The Blaine Gypsum Project

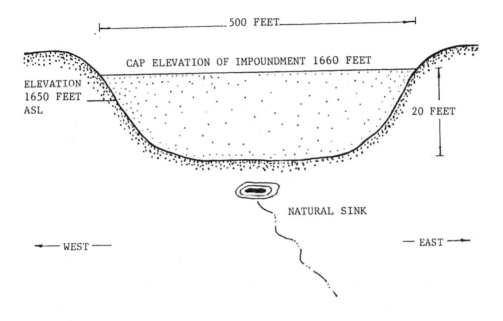

Fibure 6.3b: Impoundment "A"--Side View--The Blaine Gypsum Project

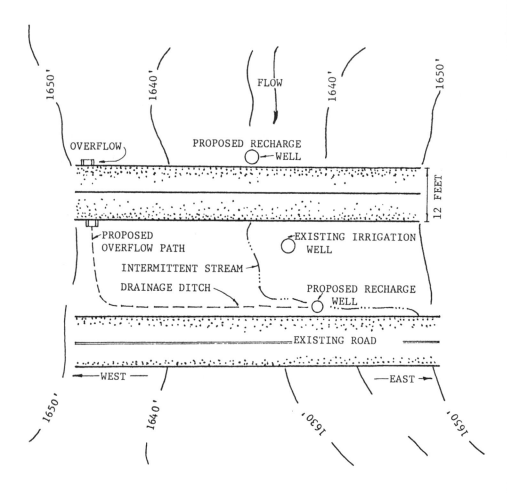

Figure 6.4a: Impoundment "B"--Top View--The Blaine Gypsum Project

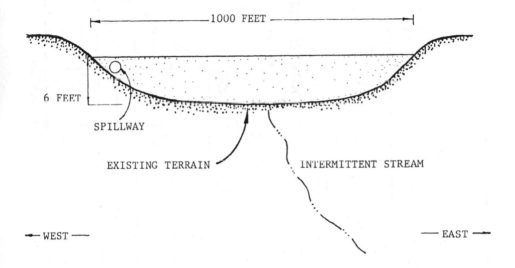

Figure 6.4b: Impoundment "B"--Side View--The Blaine Gypsum Project

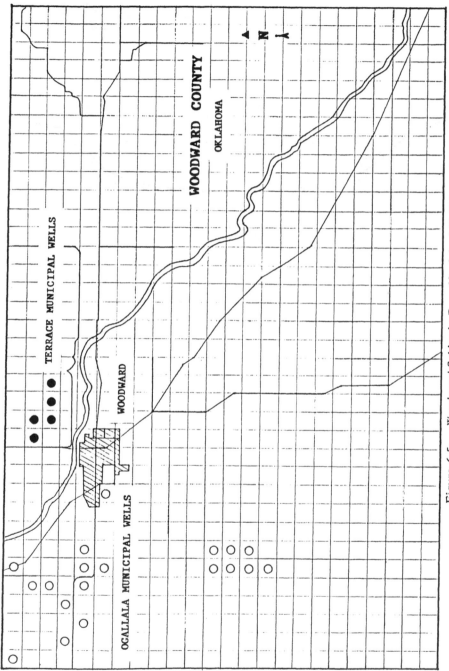

Figure 6.5:   Woodward Subbasin Dam Municipal Well Locations

shaped concrete and clay subsurface dam several hundred feet wide (Figures 6.6a and 6.6b). This dam should keep the ground water level high within the chosen well, permitting the constant use of the well.

Subsurface dams have been constructed many times before. One of the newer successes lies just outside of Glenburn, North Dakota, where the clay dam reached bank to bank in a till bed. The concrete and clay slurry used to build the dam is a Haliburton proprietary method. First used in Italy to create impoundments, the method is now being used for subsurface dam work in Louisiana and California (Figure 6.7). The dam could be constructed in a matter of a few weeks.

THE GUYMON PROJECT

The City of Guymon relies totally upon the Ogallala for its municipal needs. No river or rainfall could be used to recharge this aquifer. The only constant supply of water is from the city effluent. An increase in the number of irrigation wells has steepened the decline of the Ogallala.

The proposal to the Bureau of Reclamation is for the construction and monitoring of a 2.5 million gallon per day processing plant (Figure 6.8). Presently, the city is dumping daily some 1.5 million gallons of partially treated effluent into the North Canadian River. This effluent could be treated and used for recharging the Ogallala. The processing plant consists of five phases:

(1) Settling pond --the majority of solids will precipitate

(2) Filtration--the removal of the majority of minerals

(3) Denitrification--the removal of nitrates and free nitrogen

(4) Chlorination--the removal of bacteria

(5) Injection--injection of the effluent into the Ogallala.

The concept of using wastewater is not new. Plants are in daily operation in such states as New York, Arizona, and California. All equipment to be used in this plant is "off-the-shelf", i.e., readily available locally. Although the cost of the plant is high, the benefits for the community and the Ogallala are enormous.

CONCLUSION

The methods described display a great deal of variance as to region and circumstances. No one method may be the primary instrument of recharge. The answer may lie within the scattered communities of Oklahoma. Perhaps tail water or rain water could be channeled into an area where it would percolate into the aquifer. Perhaps good quality water can be found in alluviums rather than relying on the major aquifer. Perhaps a community would rather pump their treated effluent back down into their aquifer rather than allow it to flow away in a stream. A brighter ground water future lies ahead for Oklahoma if similar artificial recharge methods are instituted.

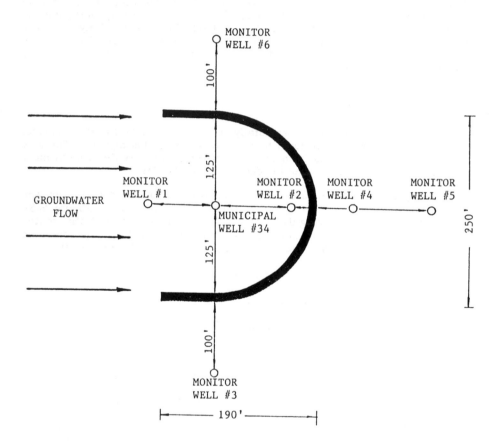

Figure 6.6a: Woodward Subsurface Dam--Top View

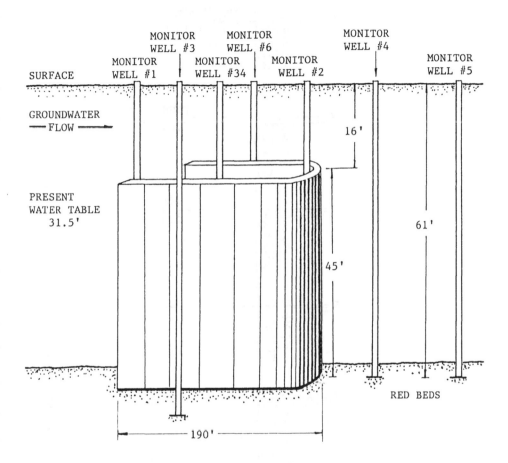

MONITOR
WELL #3

MONITOR
WELL #6

MONITOR
WELL #4

MONITOR
WELL #1

MONITOR
WELL #34

MONITOR
WELL #2

MONITOR
WELL #5

SURFACE

GROUNDWATER
— FLOW —►

16'

PRESENT
WATER TABLE
31.5'

61'

45'

RED BEDS

190'

Figure 6.6b: Woodward Subsurface Dam--Side View

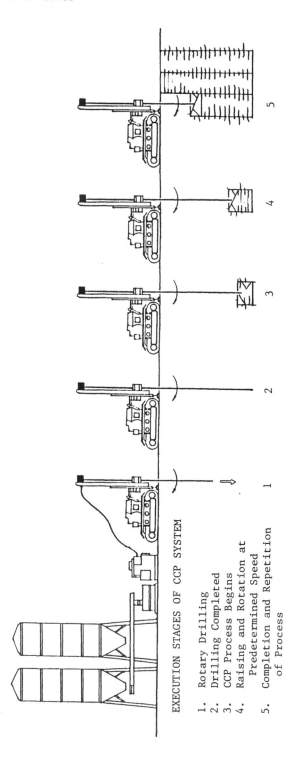

EXECUTION STAGES OF CCP SYSTEM

1. Rotary Drilling
2. Drilling Completed
3. CCP Process Begins
4. Raising and Rotation at
   Predetermined Speed
5. Completion and Repetition
   of Process

Figure 6.7:   Woodward Subsurface Dam Construction Techniques

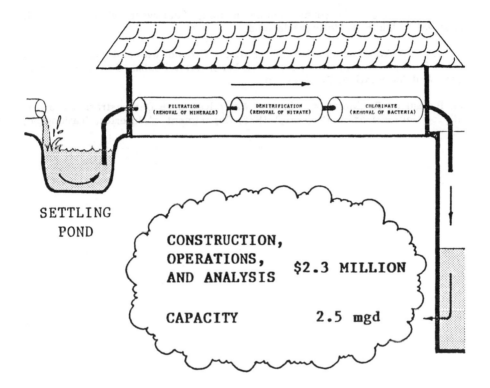

SETTLING
POND

FILTRATION
(REMOVAL OF MINERALS)

DENITRIFICATION
(REMOVAL OF NITRATE)

CHLORINATE
(REMOVAL OF BACTERIA)

CONSTRUCTION,
OPERATIONS,
AND ANALYSIS $2.3 MILLION

CAPACITY   2.5 mgd

Figure 6.8: Guymon Recharge Project--The Recharge Facility

SELECTED REFERENCES

Asano, T., <u>Artificial Recharge of Ground Water</u>, 1985, Butterworth Publishers, Stoneham, Massachusetts.

Horton, P., 1986, Personal Communication, Southwest Water and Soil Conservation District, Hollis, Oklahoma.

Pettyjohn, W.A., "Introduction to Artificial Ground Water Recharge," 1981, Robert S. Kerr Environmental Research Laboratory, Published by the National Water Well Association, Dublin, Ohio.

U.S. Bureau of Reclamation, "Ground Water Recharge Demonstration Project; Call for Demonstration Projects," 1985, Department of Interior, Washington, D.C.

CHAPTER 7

EFFECTS OF IRRIGATION PRACTICES ON STREAM-CONNECTED PHREATIC AQUIFER SYSTEMS

by

S.G. McLin

The development of surface irrigation in arid to semiarid regions of the world has produced substantial changes to stream-connected shallow ground water systems. These changes are typically characterized by long-term water quality degradation for both surface and shallow ground waters. They result when salts are leached from below the crop root zone and enter the phreatic aquifer. If this same aquifer is used to supplement crop water requirements when surface diversions are insufficient, then these leached salts are reintroduced into the ground water system. In addition, saline drainage waters are often returned to conveyance canals or streams for subsequent downstream irrigation reuse. As a result, dissolved salts are continually recirculated through the system, while new salts enter with each new surface diversion.

This surface and subsurface water quality deterioration may affect crop selection, yields, and eventually soil drainage. In severe situations it can threaten continued agricultural practices altogether. While some of these long-term effects may not be completely eliminated, their undesirable impacts can be reduced through appropriate water management practices. However, before any specific management recommendations can be made, a thorough evaluation of the system's historical and existing water and mass balance relationships should be made. This information can then be utilized for an optimization of agricultural water use that ideally improves or maintains crop productivity, while minimizing undesirable effects on downstream water users.

In this chapter evaluation of irrigation related water management practices on stream-connected phreatic aquifer water quality are examined. A systems operation model is presented that can be used to assess these impacts before proposed alternative water management schemes are put into effect. A model sensitivity analysis suggests that the hydraulic and solute response times are critical parameters in determining how a particular irrigation system's water quality will respond to these water conservation measures. Several water utilization schemes that are simulated include: (1) improvements in irrigation efficiency that reduce the leakage fraction of applied waters; (2) lining of conveyance canals; and (3) changes in the ratio of applied surface to ground waters that are used to satisfy crop water requirements. These hypothetical system stress patterns are frequently cited as potential ways to reduce irrigation water requirements; however, each of these conservation techniques will alter the shallow ground water quality in different ways. A field application further demonstrates that the model can accurately simulate changes in stream flow and total dissolved solids (TDS) concentration, in addition to predicting changes in average aquifer TDS.

MODEL STRUCTURE

The simplest type of model that can be used to describe advective transport in a dynamically connected stream-aquifer system is the lumped parameter model. This system is described only to the degree that it relates spatially averaged input-output-storage changes over time (Dooge, 1973; and Gelhar and Wilson, 1974). Hence it is applicable when system-wide temporal responses in ground water levels or water quality resulting from distributed input stresses are of interest. Hydrodynamic dispersion effects are not considered since they have little overall impact on temporal mass output in such systems (Eldor and Dagan, 1972; Konikow and Bredehoeft, 1974a and 1974b; and Konikow and Person, 1985).

For the stream-connected phreatic aquifer system shown in Figure 7.1, a simple water balance statement may be written as:

$$S \frac{dh}{dt} = q_\ell + (1- \alpha)q_a - q_p - q_d \tag{1}$$

where S = average aquifer specific yield (or drainable porosity),

   h = average saturated aquifer thickness,

   t = time,

   q_ℓ = average canal leakage rate per unit surface area,

   q_a = average recharge rate per unit surface area,

   q_p = aquifer pumping rate per unit surface area,

   q_d = natural aquifer drainage rate per unit surface area.

In (1) α is the percentage of total applied irrigation water that goes toward evapotranspiration. It therefore represents a dimensionless irrigation efficiency. In a similar fashion we can define the ground water pumpage (β) and canal leakage (γ) components of total applied waters. Thus,

$$q_{et} = \alpha q, \quad q_p = \beta q_a, \quad \text{and } q_\ell = \gamma q_a \tag{2}$$

where α, β, and γ are all less than one. In general, any number of separately identifiable aquifer water inputs or outputs can be included in (1). Inserting (2) into (1) yields

$$S \frac{dh}{dt} + q_d = q_a(1 - \alpha - \beta + \gamma) \tag{3}$$

The form of the aquifer outflow term (q_d) will determine whether the lumped model aquifer water balance equation is linear or nonlinear. For example, Gelhar and Wilson (1974) approximated q_d as a linear relationship given by

$$q_d = a(h - h_d) \tag{4}$$

where h_d is the stream or drain reference level and the parameter a is a lumped outflow constant having units of inverse time. Substituting (4) into (3) and solving yields

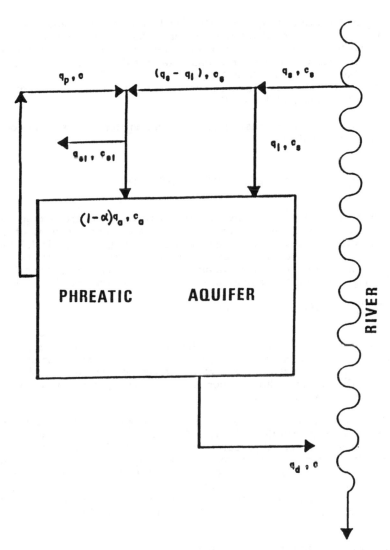

Figure 7.1:   Schematic Diagram for a Lumped Parameter Representation of an Irrigated Stream-connected Phreatic Aquifer System

$$h - h_d = (h_o - h_d) \exp\left(-\frac{at}{S}\right) + \frac{1}{S}\int_0^t E(\tau)\, \exp\left[-\frac{a}{S}(t-\tau)\right] d\tau \quad (5)$$

where h_o is the average aquifer water level at time $t = o$, and τ is a dummy integration variable. In (5) the net aquifer recharge (E) is a function of time and is equal to the right hand side of (3). The term S/a has been referred to as the hydraulic response time since it characterizes the average response time of the water balance equation to net aquifer recharge. Typically S/a has a value of several months. Field applications of both linear and nonlinear lumped models can be found in McLin (1983, 1984).

The corresponding mass balance equation for the stream-connected aquifer system would be

$$n\,\frac{d(hC)}{dt} = (1 - \alpha)\, q_a C_a + q_\ell C_s - q_p C - q_d C \quad (6)$$

where n is the average effective aquifer porosity, C is the average aquifer concentration, and C_a and C_s are the respective concentrations of aquifer recharge waters and surface water diversions. It is implied in (6) that the aquifer is a well-mixed linear or nonlinear reservoir and that any system outflows will carry this average aquifer concentration. Multiplying (3) by C, substituting the resulting expression into (6), and assuming that $S = n$ yields

$$Sh\,\frac{dc}{dt} + Cq_a(1 - \alpha - \beta + \gamma) = q_s C_s \quad (7)$$

A solution to (7) for constant h is similar in form to (5), with appropriate term modifications. The parameter Sh/q_a may be referred to as the solute response time since it characterizes the average response time of (7); physically it represents the average solute residence time in the aquifer. Typically Sh/q_a is on the order of several years.

The water and mass balance statements given by (3) and (7) form a coupled systems description of subsurface water and mass transport in an irrigated stream-connected phreatic aquifer. Thus, h(t) is initially found using (3); this h(t) is then used in (7) so that C(t) may be found. Net stresses are assumed to be uniformly distributed throughout the system, while hydrodynamic dispersion and biodegradation effects are neglected. Input parameter estimation procedures and system characterization for field scale applications can be found in Updegraff and Gelhar (1978) and McLin (1983).

PARAMETER SENSITIVITY ANALYSES

An understanding of the processes governing hydraulic and mass transport in surface-ground water systems can be achieved upon a close inspection of (3) and (7). Because typical irrigated stream-aquifer systems in the western United States have hydraulic response times much less than their corresponding solute response times, and because long-term water quality degradation is of interest, only the latter will be explored here. Hence, it will be assumed that our system is at steady-state with respect to fluid flow. This situation would correspond to a stream-connected phreatic aquifer where average water levels do not fluctuate by more than about ten percent from year to year. A finite difference approximation for (7) can then be written as

$$\frac{C_{i+1} - C_i}{\Delta t} + k_1 \frac{C_{i+1} + C_i}{2} = k_2 \, C_s \qquad (8)$$

where: $k_1 = q_a(1 - \alpha - \beta + \gamma)/Sh$ and $k_2 = q_s/Sh$.

In (8) C_i and C_{i+1} are the average aquifer TDS concentrations at the i-th and i+1 time steps, respectively, and Δt is the time step taken as one year in the following calculations. This approximate solution will be used to predict average aquifer TDS levels in a parameter sensitivity analysis where α, β and γ are systematically varied. In these calculations, C_1 is assumed to be zero, while C_s remains at a constant 100 mg/l; furthermore, S is assumed to be 0.20 while h is either 25 or 100 feet. Table 7.1 summarizes the results of sixteen hypothetical simulations using (8). These cases have been ranked in ascending order according to predicted average aquifer concentration after twenty years. In these calculations, crop consumptive use is held constant at 3.70 ft/yr, the average rate for Socorro, New Mexico, as predicted by the Blaney-Criddle technique (Table 7.2). Once α, β, and γ have been specified, then the remaining input parameters are computed from the relationships in (2) and Figure 7.1 before annual average aquifer TDS levels are computed from (8). Figure 7.2 shows a typical plot of the time history of predicted average aquifer TDS levels from several simulations. A characteristic exponentially increasing TDS level is observed in response to each system input stress pattern.

From these simulations it is obvious that aquifer thickness is one of the most critical parameters examined. In other words, for thick phreatic aquifers more uncontaminated ground water is available to dilute concentrated infiltration waters that percolate through the crop root zone. Put another way, the solute response time (Sh/q_a) in (7) is larger for larger values of h. Variations in α, β, and γ have less influence than h on the predicted aquifer concentration as seen in Table 7.1.

An important point to be made with regard to system management is that we have design control over α, β, and γ, but the system's geometry is largely determined by nature. Hence, one cannot uniformly conclude that higher irrigation efficiencies are always preferable. Indeed, the results in Table 7.1 suggest that for thin aquifers, lower efficiencies may be desirable when applied surface waters generally contain lower TDS levels than ground waters. If one were to simultaneously consider all parameters, then a different case ranking might result, as illustrated in Table 7.3. In actual field simulations, however, the determination of optimal system operation will not be so simple because other factors that were not incorporated into the lumped model should also be considered. Some of these factors might include changes in h and system efficiencies over time, soil root zone salinity, drainage return flow, downstream flows and chemistry, cropping patterns, and economic factors. One example of a field management application can be found in McLin (1984).

FIELD APPLICATION OF MODEL

The area selected for this application lies on a stretch of the Rio Grande in central New Mexico, as seen in Figure 7.3; this area was previously

Table 7.1:   Case Rankings by Predicted Aquifer TDS

PARAMETER	VALUE	UNITS	DESCRIPTION
Alpha	Table	none	Irrigation Efficiency
Beta	Table	none	% of Irrigated Water from Ground Water
Gamma	Table	none	% of Water to Canal Loss
Qet	3.70	ft/yr	Total Crop ET Required
Ci	.00	mg/1	Average Ground Water TDS at Time t = o
Csurf	100.00	mg/1	Average TDS of Applied Surface Water
S	.20	none	Aquifer Drainable Porosity
h	Table	ft	Average Aquifer Thickness
Time	20.00	yrs	Time of Caq Prediction
Caq	Table	mg/1	Prediction Average Aquifer TDS

NUMBER	ALPHA	BETA	GAMMA	h	Caq
Case 1	.50	.25	.20	100.00	203.64
Case 9	.50	.25	.20	25.00	211.11
Case 5	.50	.25	.05	100.00	237.84
Case 13	.50	.25	.05	25.00	266.64
Case 2	.50	.50	.20	100.00	270.52
Case 8	.75	.50	.05	100.00	271.65
Case 3	.75	.25	.20	100.00	298.15
Case 6	.50	.50	.05	100.00	340.53
Case 4	.75	.50	.20	100.00	345.73
Case 10	.50	.50	.20	25.00	349.11
Case 7	.75	.25	.05	100.00	350.12
Case 11	.75	.25	.20	25.00	465.98
Case 14	.50	.50	.05	25.00	850.20
Case 15	.75	.25	.05	25.00	1004.29
Case 16	.75	.50	.05	25.00	1086.59
Case 12	.75	.50	.20	25.00	1382.93

investigated by Bakr and Gelhar (1974). This region extends about 45 miles from San Acacia, where river flows are diverted for irrigation, downstream to San Marcial. USGS gaging stations measure surface flows at these locations, in addition to providing surface water quality data. These data are graphically illustrated in Figures 7.4 (Qsa and Qsm vs. time) and 7.5 (CSACAC and CSMARC vs. time), respectively; Table 7.4 lists these and other available input data for the lumped model that was previously developed. Individual term definitions in Table 7.4 are given below, or were previously described.

Table 7.2:   Computed Blaney-Criddle Evapotranspiration

MONTH	JAN	FEB	MARCH	APRIL	MAY	JUNE	TOTAL
T	36.30	42.50	49.40	57.70	65.80	74.80	
P	7.05	6.88	8.35	8.83	9.76	9.77	
ET (ft)	.16	.18	.26	.32	.40	.46	1.78

MONTH	JULY	AUG	SEPT	OCT	NOV	DEC	TOTAL
T	77.90	75.00	69.10	58.10	45.50	37.20	
P	9.93	9.37	8.36	7.87	6.97	6.86	
ET (ft)	.48	.44	.36	.29	.20	.16	1.93

Total Average Annual Crop Consumptive Use (feet/year) =   3.70

Location: Socorro, New Mexico, U.S.A.
Equation: ET (inches) = K * Sum (T * P)
          K = Average Annual Crop Coefficient = 0.75
          T = Average Monthly Temperature (degrees F)
          P = Percentage of Monthly Daylight Hours (%)
Reference: Chow, 1964, p. 11.27

Some of the Rio Grande flow is diverted at San Acacia for irrigation within this study reach. According to Bakr and Gelhar (1974), Lansford and Sorensen (1972) reported that the irrigated land area in the Socorro region fluctuated between 13,000 and 16,500 acres between 1940 and 1970; hence an average value of 14,750 irrigated acres was used in this study. Total irrigation water requirements in 1970 were 61,000 acre-feet, yielding a crop evapotranspiration rate of 3.70 ft/yr (Qa = 61000/16500); this value was previously computed in Table 7.2. As was done by Bakr and Gelhar (1974), an initial Qa value corresponding to year 1940 was assumed to be 2.0 ft/yr per irrigated acre. Annual Qa values between 1940 and 1970 were obtained by a linear interpolation between the assumed 1940 value and the known 1970 value; these results are summarized in Table 7.4.

The percentage of irrigation water derived from ground water pumpage (β) was reported to be 0.41 in 1970 (Lansford et al., 1973); Bakr and Gelhar (1974) assumed that the 1940 β value was 0.10, and that value was used here. Table 7.4 lists the linearly interpolated β values between the 1940 and 1970 values. The average irrigation efficiency (α) was estimated to be 0.50, while conveyance canal losses (γ) were assumed to be 0.20. In Table 7.4 yearly values for Qs and Qd were computed from the relationships

$$Qs = Qa (1 - \beta + \gamma) \quad \text{and} \quad Q_d = Qa (1 - \alpha - \beta + \gamma) \qquad (9)$$

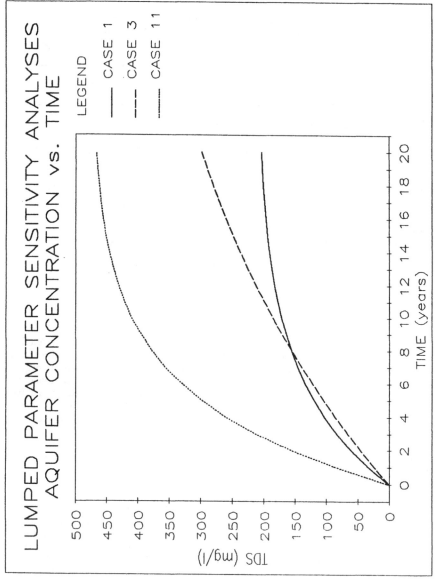

Figure 7.2:   Average Aquifer Outflow Concentration for Several Hypothetical Situations

Table 7.3: Case Rankings Using Multiple Criteria

SYMBOL	VALUE	UNITS	DESCRIPTION
Qa	Table	Points	Position Ranking from Table 7.1; see NOTE
Qs	Table	Points	Position Ranking from Table 7.1; see NOTE
Ql	Table	Points	Position Ranking from Table 7.1; see NOTE
Qp	Table	Points	Position Ranking from Table 7.1; see NOTE
Caq	Table	Points	Position Ranking from Table 7.1 for Predicted AQ TDS
Total	Table	Pt Sum	Total Pts = 10 * Caq + 5 * Qa + 3 * Qp + Ql + Qs

NUMBER	ALPHA	BETA	GAMMA	h(ft)	Qa	Qs	Ql	Qp	Caq	TOTAL
Case 8	.75	.50	.05	100.00	1.00	1.00	1.00	9.00	6.00	99.00
Case 3	.75	.50	.20	100.00	1.00	9.00	9.00	1.00	7.00	133.00
Case 7	.75	.25	.05	100.00	1.00	5.00	1.00	1.00	11.00	137.00
Case 5	.50	.25	.05	100.00	9.00	13.00	5.00	5.00	3.00	153.00
Case 4	.75	.50	.20	100.00	1.00	3.00	9.00	9.00	9.00	159.00
Case 1	.50	.25	.20	100.00	9.00	15.00	13.00	5.00	1.00	163.00
Case 13	.50	.25	.05	25.00	9.00	13.00	5.00	5.00	4.00	163.00
Case 15	.75	.25	.05	25.00	1.00	5.00	1.00	1.00	14.00	167.00
Case 9	.50	.25	.20	25.00	9.00	15.00	13.00	5.00	2.00	173.00
Case 11	.75	.25	.20	25.00	1.00	9.00	9.00	1.00	12.00	183.00
Case 16	.75	.50	.05	25.00	1.00	1.00	1.00	9.00	15.00	189.00
Case 6	.50	.50	.05	100.00	9.00	7.00	5.00	13.00	8.00	209.00
Case 2	.50	.50	.20	100.00	9.00	11.00	13.00	13.00	5.00	215.00
Case 12	.75	.50	.20	25.00	1.00	3.00	9.00	9.00	16.00	229.00
Case 14	.50	.50	.05	25.00	9.00	7.00	5.00	13.00	13.00	259.00
Case 10	.50	.50	.20	25.00	9.00	11.00	13.00	13.00	10.00	265.00

NOTE: Computed values initially arranged in ascending order; one point given for each succeeding position.

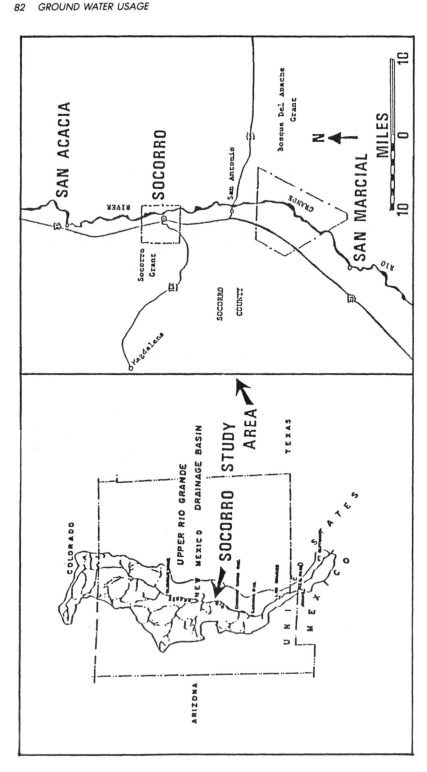

Figure 7.3: Location of the Socorro Region Study Area

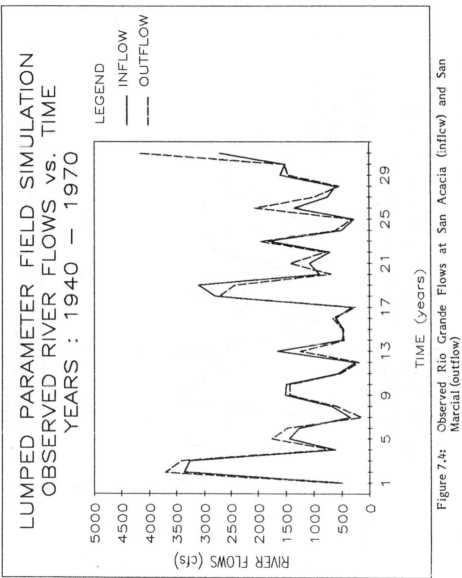

Figure 7.4: Observed Rio Grande Flows at San Acacia (inflcw) and San Marcial (outflow)

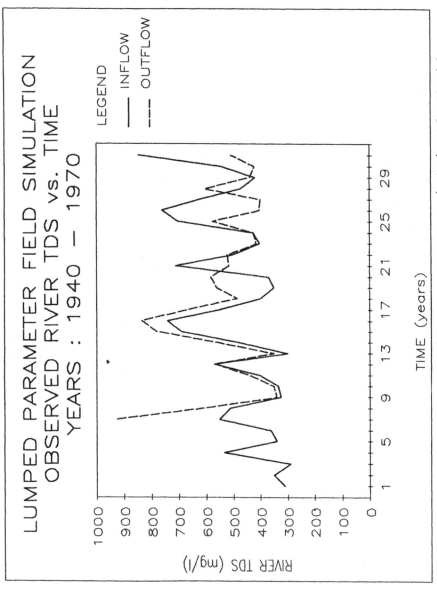

Figure 7.5: Observed Rio Grande TDS at San Acacia (inflow) and San Marcial (outflow).

Table 7.4: Lumped Parameter Model Field Simulation Study: Socorro
Region of the Rio Grande, Central New Mexico

YEAR	BETA (%)	Qa (ft/yr)	Qs (ft/yr)	Qd (ft/yr)	CONCGW (mg/l)	CONCSM (mg/l)
1940	.10	2.00	2.20	1.20	310.00	291.09
	.11	2.06	2.24	1.21	329.82	319.47
	.12	2.11	2.28	1.22	342.32	280.14
	.13	2.17	2.32	1.23	381.79	501.99
	.14	2.23	2.36	1.24	397.63	285.92
1945	.15	2.28	2.39	1.25	415.27	317.40
	.16	2.34	2.43	1.26	454.68	1481.40
	.17	2.40	2.46	1.26	487.69	611.51
	.18	2.45	2.50	1.27	497.01	351.77
	.19	2.51	2.53	1.27	507.39	360.22
1950	.20	2.57	2.56	1.27	525.63	474.35
	.21	2.62	2.59	1.28	564.60	851.12
	.22	2.68	2.62	1.28	567.72	410.79
	.23	2.74	2.64	1.27	594.13	544.64
	.24	2.79	2.67	1.27	646.74	694.72
1955	.26	2.85	2.69	1.27	703.58	704.54
	.27	2.91	2.72	1.26	732.92	757.18
	.28	2.96	2.74	1.26	741.38	420.14
	.29	3.02	2.76	1.25	743.25	454.29
	.30	3.08	2.78	1.24	748.38	506.16
1960	.31	3.13	2.80	1.23	800.04	548.76
	.32	3.19	2.82	1.22	820.94	472.67
	.33	3.25	2.83	1.21	828.01	382.13
	.34	3.30	2.85	1.20	838.67	383.78
	.35	3.36	2.86	1.18	888.10	693.88
1965	.36	3.42	2.88	1.17	943.91	510.69
	.37	3.47	2.89	1.15	978.40	487.98
	.38	3.53	2.90	1.13	991.15	566.88
	.39	3.59	2.91	1.11	996.55	474.27
	.40	3.64	2.92	1.09	1020.40	554.13
1970	.41	3.70	2.92	1.07	1087.35	552.96

These equations were obtained directly from (2) and (3) for steady hydraulic
flow.

The saturated thickness of the alluvial aquifer adjacent to the Rio Grande
in the Socorro region varies between 78 and 141 feet (Bushman, 1963); an
average aquifer thickness of 100 feet and a porosity of 0.20 were used in this
study.   Locations of wells used in establishing aquifer depth and electrical
conductivity of shallow ground water were previously shown in Bakr and Gelhar
(1974).

The finite difference approximation of the mass transport relationship
given by (8) was solved for the ground water TDS concentration (CONCGW)
using the input data of Table 7.4. These results are graphically shown in Figure

7.6. Equation (8) predicted that the average aquifer TDS level would gradually increase from its approximate 1940 value of 310 mg/l to 1087 mg/l in 1970. Bakr and Gelhar (1974) indicated a similar increasing trend from 310mg/l in 1940 to over 3000 mg/l in 1970; however, their approach did not consider any canal leakage effect. The average aquifer TDS level in 1960 predicted by (8) was 800 mg/l. In 1960 an average value of about 1090 mg/l for shallow ground water TDS from 22 monitor wells was reported by Bakr and Gelhar (1974). This single comparison of model output to observed average water quality is insufficient for any system generalizations, but may suggest that the assumed model leakage fraction is incorrect. No other observed ground water data are available for model comparisons.

Figure 7.7 shows the predicted and observed Rio Grande TDS at San Marcial. This predicted TDS concentration at San Marcial was computed from a simple volumetric mass balance relationship written as

$$Q_{sa} \, C_{sa} + Aq_d \, C_d = Q_{sm} \, C_{sm} \tag{10}$$

where Q_{sa} and Q_{sm} are the observed stream flows for San Acacia and San Marcial, respectively; C_{sa} and C_{sm} are the observed and predicted TDS concentrations at San Acacia and San Marcial, respectively; A is the average irrigated acreage of 14,750 acres; q_d is the computed aquifer drain flow defined previously; and C_d is the predicted average aquifer TDS from (8). Note that in (10) care must be taken so that dimensionally homogeneous units are used.

Considering the approximations made for some of the input data listed in Table 7.4, the simulations represented in Figures 7.6 and 7.7 are quite encouraging. These results demonstrate the effectiveness of the lumped parameter approach to modeling stream-connected phreatic aquifer systems. Other examples of field applications (McLin and Gelhar, 1979; and McLin, 1983, 1984) further demonstrate the model's flexability to variations in types of observed input data. Finally, it should be noted that access to a mainframe computer is not required with this approach. The simulations made here were performed on a popular electronic spreadsheet with graphics capability and was run on a popular microcomputer with a floppy disk operating system.

CONCLUSIONS

Both the model parameter sensitivity analysis and field simulation indicate that a definite trend of increasing TDS levels in shallow ground water over time can be expected from irrigated agriculture in semiarid settings. This trend is especially significant when the stream-connected phreatic aquifer is shallow, or when a significant portion of applied irrigation waters are obtained from ground water pumpage. Furthermore, changes in irrigation efficiency or conveyance canal losses can affect shallow ground water quality. In effect, these higher quality surface waters which are lost to the shallow aquifer actually serve to dilute the concentrated irrigation waters that penetrate the crop root zone and recharge the aquifer.

The effects of a transient hydraulic flow system have been neglected in this analysis. These effects will often be important, especially when they result in large fluctuations in average aquifer water levels. In these situations the drain flow predicted by equation (4) will be directly affected. Furthermore, when surface diversions fluctuate over time, then fluctuations in irrigation

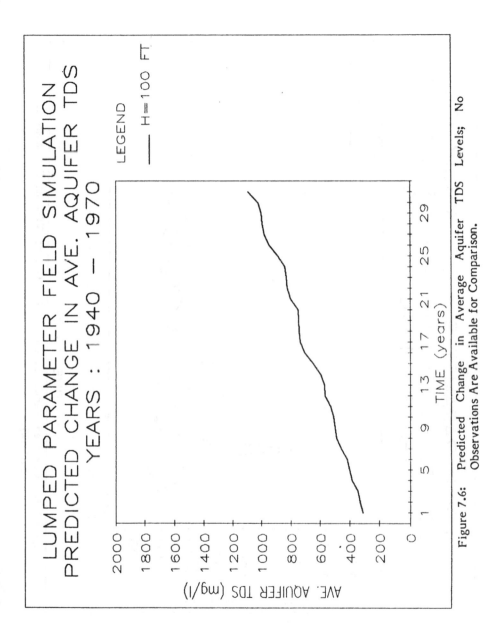

Figure 7.6:   Predicted Change in Average Aquifer TDS Levels; No Observations Are Available for Comparison.

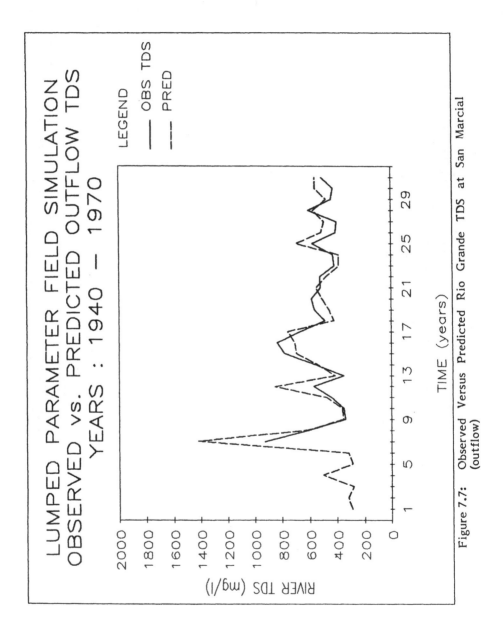

Figure 7.7: Observed Versus Predicted Rio Grande TDS at San Marcial (outflow)

efficiency, canal leakage, and ground water pumpage can be expected. These parameters have a direct affect on predicted aquifer concentration. Hence, in actual field simulations a coupled flow dynamics and mass transport description will provide a more complete predictive management tool. While this approach does not yield a system-wide spatial distribution of simulated aquifer water levels and contamination, it does provide a means to estimate system averaged behavior with limited input data requirements.

SELECTED REFERENCES

Bakr, A.A., and Gelhar, L.W., "Groundwater Quality Simulation for the Socorro Region of the Rio Grande in New Mexico," 1974, Hydrology Section, Department of Geoscience, New Mexico Institute of Mining and Technology, Socorro, New Mexico.

Bushman, F.X., "Groundwater in the Socorro Valley in New Mexico," Geologic Society, 14th Field Conference, 1963, Bureau of Mines and Mineral Resources, Socorro, New Mexico.

Chow, V.E., Handbook of Applied Hydrology, McGraw Hill, New York, 1964, Chapter 11.

Dooge, J.C.I., "Linear Theory of Hydrologic Systems," Technical Bulletin 1468, 1973, U.S. Department of Agriculture, Washington, D.C.

Eldor, M., and Dagan, G., "Solutions of Hydrodynamic Dispersion in Porous Media ," Water Resources Research, Vol. 8, No. 5, 1972, pp. 1316-1331.

Gelhar, L.W., and Wilson, J.L., "Ground-water Quality Modeling," Ground Water, Vol. 12, No. 6, 1974, pp 399-408.

Konikow, L.F., and Bredehoeft, J.D., "A Water Quality Model to Evaluate Water Management Practices in an Irrigated Stream-aquifer System," in Proceedings of the 15th Annual Water Resources Conference, edited by J.E. Flack and C.W. Howe, 1974a, Merriman Publishing Company, Boulder, Colorado, pp 36-59.

Konikow, L.W., and Bredehoeft, J.D., "Modeling Flow and Chemical Quality Changes in an Irrigated Stream-aquifer System," Water Resources Research, Vol. 8, No. 4, 1974b, pp 546-562.

Konikow, L.W. and Person, M., 'Assessment of Long-term Salinity Changes in an Irrigated Stream-aquifer System", Water Resources Research, Vol. 21, No. 11, 1985, pp 1611-1624.

Lansford, R.R., and Sorenson, E.F., "Trends in Irrigated Agriculture 1940-1972," in New Mexico Agriculture-1972, Research Report 260, Agricultural Experiment Station, New Mexico State University, Las Cruces, New Mexico, 1972.

Lansford, R.R. et al., "An Analytical Interdisciplinary Evaluation of the Utilization of the Water Resources of the Rio Grande in New Mexico," Water Resources Research Institute Report No. 020, 1973, New Mexico State University, Las Cruces, New Mexico.

McLin, S.G., and Gelhar, L.W., "A Field Comparison Between the USBR-EPA Hydrosalinity and Generalized Lumped Parameter Models, in The Hydrology of Areas of Low Precipitation, IAHS Publication No. 128, 1979, pp 339-348.

McLin, S.G., "Evaluation of Irrigated Related Water Management Practices with the Lumped Parameter Hydrosalinity Model," Proceedings of the 3rd AGU Front Range Branch Conference, Colorado State University, Ft. Collins, Colorado, 1983, pp 104-138.

McLin, S.G., "A Managerial Technique for Groundwater Quality Modeling," in Proceedings of Conference on Practical Applications of Ground Water Models, 1984, National Water Well Association, Worthington, Ohio.

Updegraff, C.D., and Gelhar, L.W., "Parameter Estimation for a Lumped Parameter Groundwater Model of the Mesilla Valley, New Mexico," New Mexico Water Resources Research Institute Report 097, 1978, New Mexico State University, Las Cruces, New Mexico.

CHAPTER 8

GROUND WATER CONTAMINATION FROM SALTWATER INTRUSION AND LIMITATIONS ON AGRICULTURAL ACTIVITIES

by

Samuel F. Atkinson

Recognition of the trend of increasing ground water usage, especially for agricultural purposes, has brought with it the realization that, in many areas of the United States, ground water quality is related to agricultural activities. One type of ground water quality impact related to agricultural activities which has received increasing concern is saltwater intrusion. Agricultural activities promote saltwater intrusion in three different ways. One involves overpumping of fresh water in coastal and tidal areas, or from aquifers which overlay, underlay, or occur adjacent to saline water. The second way is through inadequate water well construction which results in leaking wells that promote mixing of fresh and saline water. Finally, agricultural practices which change the composition of local root systems from expanded throughout the soil profile to only shallow roots allow increased percolation of water through soluble salts in the soil.

Saltwater intrusion can impact agricultural activities by decreasing crop yield, completely eliminating the capability of growing certain crops, and impacting the health of domestic livestock. Additionally, the salt content of soils has been shown to be related to the potential for erosion.

Fortunately, there are numerous ways in which saltwater intrusion can be controlled or managed. Some solutions are simple, such as controlling pumpage rates, strategically locating withdrawal wells, or the sealing of leaking wells. Other solutions are more complicated and costly. For example, artificial recharge, extraction wells, or desalination represent techniques which are applicable primarily on a regional basis.

IMPACTS OF SALTWATER INTRUSION ON AGRICULTURAL ACTIVITIES

Plant Response

As ground water salinity increases, plant growth usually decreases. Most attempts to explain this phenomenon include the concept of 'physiological drought'. This view, as depicted by Figure 8.1, relates to the process of osmosis. Osmosis involves the movement of water across a semipermeable membrane until the concentration of dissolved substances on both sides of the membrane are equal. The factors which influence the movement of ground water into a cell are: (1) osmotic pressure, (2) hydrostatic pressure, (3) the permeability of the cell wall and cell membrane to water, and (4) a species specific ability for the cell wall/membrane to transport various ions across the barrier.

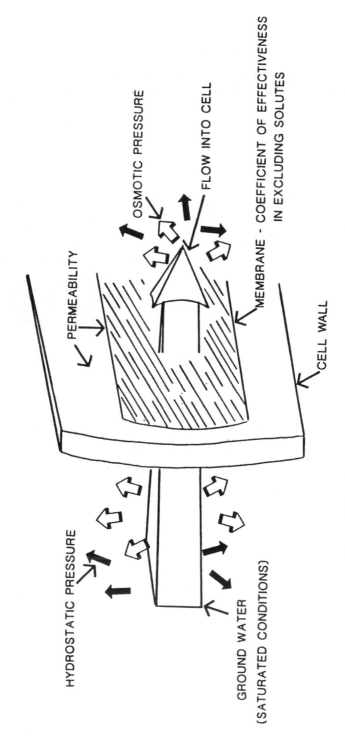

Figure 8.1:   Factors Involved in Ground Water Movement into Plant Cell

A typical situation involves a transpiring plant and, therefore, a lower hydrostatic pressure inside the root cell than in the ground water. Additionally, the solute concentration inside the root cell is higher than the ground water outside the cell due to the biological process of photosynthesis and respiration. Finally, the root-cell/membrane is permeable to water and has a fairly high tendency to prevent ion transport into and out of the cell. Thus, all factors promote water movement from the ground water into the cell.

However, if ground water salinity is higher than root cell solute concentration, osmosis tends to move water out rather than into the root cell. If ground water salinity is high enough, the plant may experience reduced transpiration and, ultimately, not enough water movement into the root cells. This situation has been referred to as physiological drought since the plant cannot receive enough water for normal biological processes.

However, some studies have shown that factors other than water movement can generate the symptoms of drought. The evidence is that some salts are more detrimental to plants than other salts, even under equal concentrations. In response to the differing views on how salinity affects plant growth, O'Leary (1970) presented a mathematical relationship which related water movement into and out of root cells. The equation relates water flow to the permeability of the plant, the pressure inside and outside of the plant, the osmotic pressure of the water inside and outside of the root cell, and how effective the plant is at excluding solutes. O'Leary's equation is:

$$J_V = L_P ((P_o - P_i) - \Omega (\pi_o - \pi_i))$$

where: J_V = volume flow

L_P = permeability coefficient

P = hydrostatic pressure

π = osmotic pressure

Ω = coefficient which expresses the effectiveness
of the roots in preventing ion transport

The subscripts "o" and "i" indicate the outside and inside of the root, respectively.

O'Leary (1970) looked at root permeability under saline conditions and found that it was reduced considerably by small increases in salinity. This reduction in root permeability also reduced the amount of water entering the plant; and, ultimately, there was not enough water for sustained stomatal opening. The reduced stomatal aperture resulted in less transpiration water loss in an attempt by the plant to reestablish a water balance equilibrium.

O'Leary, knowing that there is some minimal turgor pressure necessary for leaf expansion (i.e., growth), hypothesized that as long as the leaves are suffering from water deficit, leaf expansion will be suppressed. Furthermore, if the evaporative demand of the atmosphere was high enough, partial stomatal closure would not prevent a water deficit from occurring and the plant would give the typical appearance of a plant suffering from drought. However, if the evaporative demand of the air was low, the plant should be able to maintain a favorable water balance.

The hypothesis was tested by O'Leary in closed microcosms where atmospheric humidity was maintained at an extremely high level. The high humidity resulted in low evaporative demand. However, plant growth was still inhibited, indicating a direct effect of salinity on growth. What O'Leary finally showed was that the symptoms of water and salt stress often resemble the symptoms of senescence (old age). Experimental evidence found that less cytokinin, a substance which seems to control leaf senescence, was recovered from root systems subjected to water stress, and that the amount of cytokinin reaching the shoots under normal conditions drops off significantly when leaf senescence begins.

With the flow chart shown in Figure 8.2, O'Leary hypothesized that a significant effect of increasing salinity was decreased delivery of cytokinins to the leaves. The indication is that increasing ground water salinity results in two separate but related effects on plant growth. One, as salinity increases, the osmotic pressure within the plant becomes higher and water tends to move out of the root system inducing physiological drought. Two, as the plant becomes stressed, cytokinin production is reduced and the plant's leaves begin senescence.

Crop Tolerance

The U.S. Department of Agriculture Salinity Laboratory in Riverside, California, has been conducting laboratory and field tests on the salt tolerance of crops since its establishment in 1937. Salt tolerance tests are usually conducted in small experimental plots, where commercial practices are followed as closely as possible, with adequate moisture and fertility. To ensure an acceptable stand, researchers plant seed in a nonsaline seedbed and impose salinity by adding calcium and sodium chloride salts to the irrigation water after the seedlings have emerged. They test several salinity levels to determine both the threshold levels that begins to decrease yield and the rate of yield reduction caused by higher salinity levels. They have found, in general, that, if the threshold level of a crop is high, a less severe decrease in yield can be expected.

The following discussion of the effects of salinity on crops is based on the work of E.V. Maas (1984), a plant physiologist at the U.S. Department of Agriculture Salinity Laboratory. In the agricultural environment, salinity levels often change throughout the season. Although most crops become more tolerant at later stages of growth, there are some exceptions. For example, salt seems to affect rice during pollination and may decrease seed set and grain yield. Plants are generally more sensitive during the seedling and early vegetative stages of growth. In sweet corn, for example, it was found that although seedling growth was reduced by salinity, the salt level of the irrigation water could be increased up to about 5,800 mg/l during the tasseling and grain-filling states without affecting yield.

Research at the U.S. Department of Agriculture Salinity Laboratory has allowed the grouping of agricultural crops into four sensitivity groups. Sensitive crops are those that have a salinity threshold less than 1200 mg/l. For example, carrots and strawberries have threshold levels at approximately 640

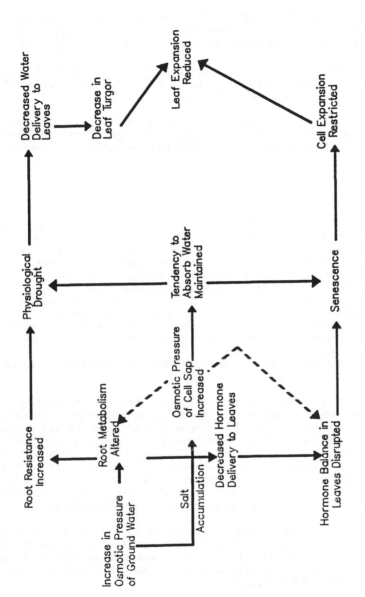

Figure 8.2:   Effect of Salinity on Plant Growth (Based on O'Leary, 1970)

mg/l. Oranges, apricots, peaches, and grapefruit are also considered sensitive crops. Once the threshold has been reached, decreases in yield will range from 10 to 33 percent.

Moderately sensitive crops can tolerate salinity up to about 2,000 mg/l. Corn, tomatoes, and rice are representative of this category. Above the threshold, yield can be expected to decrease by between 6 and 19 percent.

Moderately tolerant crops, including squash, soybeans, wheat, and sorghum, have threshold levels at less than 4,300 mg/l. These crops will have a decrease in yield by between 4 and 20 percent.

Finally, tolerant crops, such as sugar beet, cotton, and barley, have been found to have threshold levels up to about 5,100 mg/l. Above the threshold, these crops can be expected to have yields decreased by up to approximately 7 percent.

Erosion

Jenkins and Moore (1984) described the erosion of soils as a surface phenomenon involving the removal and entrainment of particles due to fluid flow-induced shearing stresses. They point out that the nature of the forces resisting the stress is different for two general categories of soils. "Cohesionless" soils are coarse-textured materials comprised of sands and gravels. "Cohesive" soils are fine-textured materials predominately comprised of silts and clays with various amounts of fine sand. Due to the relatively large size and inert surface potential of cohesionless soil particles, erosion is predominately resisted by gravitation forces. These forces also contribute to cohesive soil resistance but to a much lesser extent. Instead, the erosion resistant forces of cohesive soils are attributed to the internal cohesion developed by net attraction-repulsion potentials between electrically-charged clay particle surfaces and edges.

Unlike the erosion mechanism of cohesionless soils controlled purely by physical parameters of the system, the cohesive soil eroding mechanism is complex, involving physio-chemical interactions between the soil matrix and environment. Jenkins and Moore (1984) conducted experiments to identify the factors affecting cohesive soil stability. They devised a laboratory procedure which utilized a small-scale, straight-channel, recirculating flume to study the effects of salinity on cohesive soil erosion. Jenkins and Moore found that the flume measured the gross effect of increasing total salt concentration on cohesive soil erodibility with a high degree of statistical significance.

The cohesive soils were demonstrated to have increased stability and resistance to erosion as the total salt concentration in the eroding water increased. The stabilizing effect of increased salinity was attributed to suppression of interparticle repulsive forces and the existence of constant attractive forces in the fine grained soils.

Jenkins and Moore pointed out that cohesive soils are typical of marshland areas. These marshland environments are frequently altered by an influx of fresh water which tends to increase erodibility. An increased erodibility may have serious effects on the stability of unlined ditches, drainage canals, embankments and other typical agricultural features. Since the designed usage of these features would be hindered by siltation and erosion, maintenance costs could be substantial to clean and dredge canals and channels. Therefore, in areas where saltwater intrusion control plans are considered, a decrease in salinity may result in increased erosion. Careful assessment is required to determine if the benefits of saltwater intrusion control would outweigh the impacts of increased erosion.

IMPACTS OF AGRICULTURAL ACTIVITIES ON SALTWATER INTRUSION

Sea Water Intrusion

Before fresh ground water is pumped from aquifers that are in hydraulic connection with the sea, a dynamic equilibrium is reached between the sea water and the fresh ground water. Figure 8.3 indicates this equilibrium for an unconfined aquifer, and the situation is very much the same with a confined aquifer. The piezometric head of the ground water is higher than that of the sea and, therefore, fresh water tends to flow into the sea in an attempt to equilibrate the two different piezometric heads. However, sea water is more dense than fresh water and subsequently it tends to flow underneath the fresh water. In dynamic equilibrium, a wedge of sea water is established underneath the freshwater, with a zone of transition in between.

When fresh water is pumped from the aquifer for agricultural (or other) use, the piezometric head of the fresh water is lowered, and sea water will flow inland in an attempt to reestablish the dynamic equilibrium. This situation, shown in Figure 8.4, is known as sea water intrusion. This situation has been known to occur in every coastal state in the nation (Atkinson et al., 1985). Sea water intrusion can also occur without actually pumping ground water: if surface water is captured and not allowed to recharge the fresh water aquifer when under normal conditions it would have done so, the piezometric head of the aquifer may drop below the sea water piezometric head, thus inducing sea water intrusion.

Leaking Wells

Saltwater intrusion via leaking wells has been documented in many areas of the United States. Although the problems have been primarily associated with injection wells, production wells (petroleum, gas, and water) have been known to have been poorly constructed or even deteriorate to the point where saltwater intrusion has occurred. The following mechanisms have been listed as routes for saltwater contamination via injection wells (Miller, 1980):

(1)   direct emplacement into fresh water zone;

(2)   escape into a fresh water zone by well-bore failure;

(3)   upward migration from receiving zone along the outside of casing;

(4)   leakage through inadequate confining beds;

(5)   leakage through confining beds due to unplanned hydraulic fracturing;

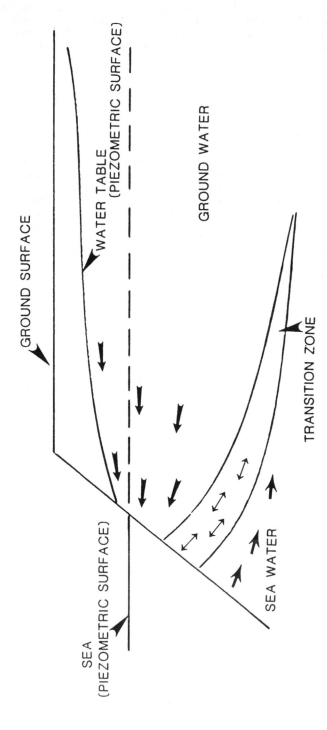

Figure 8.3:   Dynamic Equilibrium Between Sea Water and Ground Water (After Todd, 1974)

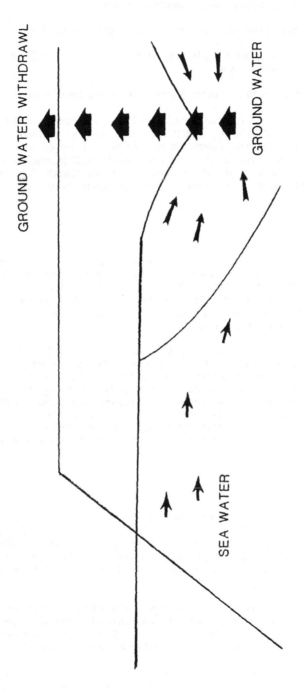

Figure 8.4: Sea Water Intrusion (After Todd, 1974)

(6)   leakage through deep abandoned wells;

(7)   displacement of saline water into a potable aquifer;

(8)   injection into a saline aquifer eventually classified as a potable water 'source; and

(9)   migration to a potable water zone of the same aquifer.

The utilization of injection wells has been practiced in many parts of the U.S. as a means of disposing industrial waste, sewage effluent, spent cooling water, irrigation drainage, storm water, and brine from oil and gas activities. If the structural integrity of the injection well fails, it is possible for salt water to find its way to a fresh water zone and contaminate that aquifer. If a production well loses its structural integrity, saline water can also find this conduit.

Upconing

This term describes the situation where a producing well is drawing fresh water at a rate and from an area which is located close enough to saline water to cause salt water to be drawn into the producing well. Although generally referred to as 'upconing', this situation can actually be 'downconing' or 'lateral intrusion' ('sideconing', if you will). Figure 8.5 represents the process of upconing in an unconfined aquifer.

An estimated two thirds of the continental United States is underlain by saline ground water that could intrude into fresh water supplies as a result of upconing. In many areas, where water tables have been declining from mining ground water for agricultural use, water wells are being drilled deeper and deeper. As the depth of water wells increases, the potential for salt-water intrusion from upconing also increases. The patterns and magnitude of pumping, the presence of interconnected zones of preferential permeability, and the presence of fault zones in the aquifer are factors which influence upconing.

Natural Salinity

The term 'natural salinity' can be defined in many ways. Researchers have classified mineralized waters into at least 40 chemical types. In almost all waters containing more than 20,000 ppm, the dominant anion is chloride (Feth, 1981). The major exceptions are waters associated with saline seep or the oxidation of sulfide minerals in mines in which sulfate is the dominant anion.

The origins of natural salinity in ground water have been considered to be (Craig, 1970): (1) juvenile waters magmatically derived from fluids; (2) meteoric waters--derived from rainfall; (3) marine waters--derived from ancient seas; and (4) evaporites--salts left when water evaporates.

"Salt sieving" is a process which has been used to describe how some ground waters have become saline. The theory, first described by deSitter (1947) and updated by Bredehoeft et al., (1963) states that as fine silts and clays

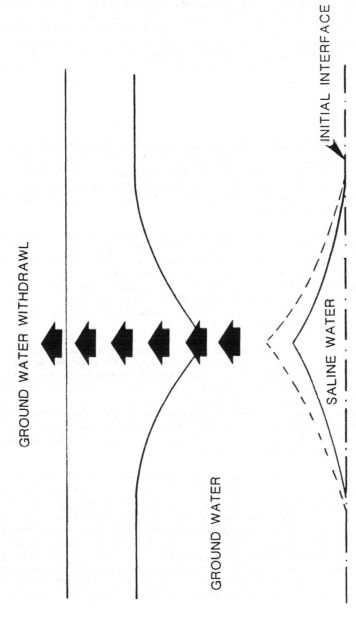

Figure 8.5: Upconing (After U.S. Environmental Protection Agency, 1973)

are deposited, the water and aqueous ions that are associated with them are free to move unimpeded. As compaction occurs and the clay minerals are placed very close together, the fixed negative charges on the clays come close enough to repel the aqueous anions (e.g. chloride) while permitting the passage of uncharged water molecules. The hydraulic head in the aquifer can be considered one of the driving forces, moving water and ions through the clay membrane. This theory can explain many of the phenomena observed in subsurface waters: increasing acidity and salinity with depth; changes in ion ratios that coincide with their relative mobility through clays; and the occurrence of brines in areas away from the sea or salt deposits.

Saline Seep

Miller et al., (1980) defined saline seep as recently developed saline soils in nonirrigated areas that are wet some or all of the time, often having white salt crusts. They occur in places where once productive land became non-productive due to the seepage of saline ground water. They usually are found in areas of thin alluvial or glacial aquifers, where the ground water systems are shallow, locally derived, and perched.

The process develops in response to the elevation of the water table. As water moves beneath the root zone but above shallow impermeable layers, a local ground water flow system is formed. Salts in the soil are dissolved and move with the ground water downslope to discharge areas (seeps), where the water evaporates and salt is deposited on the surface. The primary cause appears to be farming practices in general, and the alternate crop-fallow system in particular (Thompson and Custer, 1976; Miller et al., 1980; and Donovan, Sonderegger, and Miller, 1981).

In nature, the grasses, forbes, and shrubs of the prairie make efficient use of water and restrict excessive vertical percolation. This efficient water use implies a permanently expanded root system at all depths in the soil profile. Annual grasses and forbes, like small grain crops, obtain their water only from the upper soil levels; their root systems shallow, expanding and functioning only during the growing season. Perennial grasses and shrubs, absent from most cropping systems, use water that escapes from shallow-rooted plants into the subsoil; their root systems are deeper, permanently expanded, and functional over a longer period (Bahls and Miller, 1975).

However, for agricultural purposes, during a summer fallow period in the crop fallow rotation, the soil is kept bare of all vegetation to enhance the soil moisture status for the next crop. Only about 27 percent of the precipitation is stored in the soil during the fallow period; the remainder is lost to evaporation and to deep percolation (Thompson and Custer, 1976). Indications are that the fallow areas can undergo a water-table rise of 1 to 15 feet during years of average or above-average spring precipitation. The water levels gradually decline during the rest of the year but normally do not reach the low of the previous year. As a result, excess water accumulates through the years, causing expansion of the saline seeps during each succeeding wet cycle (Miller et al., 1980).

COMBATTING SALTWATER INTRUSION

Planning Control Programs

Saltwater intrusion is widespread and continually becoming an important consideration for agricultural activities. Locating and defining the magnitude of the problems is the first step to controlling the problem. Each control strategy must be designed for the specific type(s) of intrusion, the source(s) of intrusion, the hydrology of the area, and the areal extent of the problem. A generalized planning approach to saltwater intrusion control consists of (Atkinson et al., 1985):

(1)   problem definition;

(2)   inventory and analysis;

(3)   formulation of alternative control plans;

(4)   comparative evaluation of control plans; and

(5)   selection and implementation of controls.

The problem definition involves defining exactly what type of salt-water intrusion is occurring. Initial estimates must be verified by monitoring ground water quality in locations which will confirm or reject the probable sources.

Once the sources have been confirmed, an inventory and analysis of the magnitude and extent of the problem can be estimated through monitoring and modeling programs. An inventory of water users and patterns of pumpage must be completed to understand and predict the movement of saline water under given conditions. Predicting the movement of saltwater intrusion can be accomplished by mathematical and physical modeling techniques.

The formulation of alternative control plans can be based upon results of the inventory and analysis step. Numerous control techniques are available to combat saltwater intrusion. Several suggestions follow in the next section. Finally, a comparative evaluation of control plans, examining costs and benefits, should result in a preferred control option. Once the preferred plan has been selected, the program can be implemented.

Control Techniques

There are two approaches to controlling saltwater intrusion problems. One involves engineering solutions and one involves management solutions. Some suggested techniques are in reality a combination of the two. Effectively controlling saltwater intrusion from an agricultural perspective will undoubtedly involve both approaches. The following are summaries of control techniques, some much more applicable for agricultural programs than others.

Pumping Patterns

Based on the difference in density between sea water and fresh water, proper pumping rates can maintain the piezometric head of the fresh water at a level high enough to prevent intrusion. By reducing the pumping rate so that the fresh water piezometric head does not significantly decline, sea water intrusion can be controlled.

Well Location

For noncoastal problems (i.e., upconing) several wells may be used to pump the same amount of fresh water as one well did before. If the spacing of the wells is appropriate, the cone of depression may be larger in areal extent than a single withdrawal well, but not as deep. This reduction in the depth of the cone of depression may preclude upconing. In coastal areas, water wells can be moved farther inland so that the cone of depression does not reach sea water, thus preventing intrusion.

Artificial Recharge

Sea water intrusion can be controlled by artificially recharging an intruded aquifer by the use of surface spreading or recharge wells. The artificial recharge produces a hydraulic barrier which has the effect of a physical barrier to sea water intrusion. The barrier is formed by creating a fresh water ridge adjacent and parallel to the shoreline. The ridge raises the piezometric surface above sea level and thereby prevents sea water from intruding.

Extraction Wells

The reverse of artificial recharge, this method involves a line of wells adjacent and parallel to the coast which extract sea water from the sea water wedge. The extraction of sea water forms a barrier to intrusion by creating a stationary wedge where the sea water's piezometric head is maintained at a lower level than the fresh water head.

Recharge/Extraction Combination

By combining an extraction trough and a fresh water ridge, a more effective barrier can be created. Generating an extraction trough near the shoreline and a fresh water ridge slightly inland allows a significant difference in the sea water and fresh water piezometric head. This difference provides the barrier to sea water intrusion.

Subsurface Barriers

By placing a physical wall between salt water and fresh water, intrusion can be prevented. A subsurface barrier is an impermeable, vertical wall which is usually in one of three forms: slurry walls, grout cutoff, or steel sheet piles. Slurry walls involve pumping water and bentonite clay/cement into a trench. The slurry forms an impermeable wall. Grout cutoffs involve the injection of a liquid emulsion under pressure in order to occupy the available pore space. Eventually the grout hardens and forms an impermeable barrier. Finally, sheet piling involves driving lengths of steel that connect together into the ground to form a thin impermeable barrier.

Seal Wells

When the structural integrity of a well has failed, and saltwater intrusion occurs through this conduit, the channel must be closed. Sealing the well typically involves some method of plugging the well hole. Today, the most common type of plugging material is a cement slurry.

Air Injection

As the percent saturation of granular material decreases, so does its relative permeability. Therefore, the injection of air can be utilized to retard the movement of saline water. Much like artificial recharge, air injection forms a barrier adjacent and parallel to a shoreline. By displacing a zone of water ahead of the saltwater interface and replenishing the air as it migrates or becomes absorbed in the water, sea water intrusion can be effectively controlled.

Diversion of Surface Water

Saline seeps are caused by water percolation through soils with high salt content. By simply diverting excess water from the areas of high salt and not allowing excess soil moisture to develop, seeps can be controlled. Diversion can be achieved by routing surface water around susceptible areas, or using farming practices such as continuous cropping (or at least planting crops every year instead of every other year). This is probably the only practical means of controlling saline seep.

Desalination

Once salt water has intruded into an area, it may be more technically or economically feasible to desalinate the water rather than turn to alternative sources or restore the ground water to its original quality. The most often mentioned methods for desalination are distillation, electrodialysis, and reverse osmosis.

SUMMARY OF SALTWATER INTRUSION SURVEY

As mentioned earlier, a survey of the extent of ground water contamination from saltwater intrusion was completed in 1985 (Atkinson et al., 1985). This section contains a synopsis of that survey. The Atkinson et al. survey was an effort which involved extensive literature searches, a nation-wide telephone survey, and a compilation of the results to define the national status of limitations to agricultural activities from saltwater intrusion.

Figure 8.6 (Atkinson et al., 1985) represents the states which have identified saltwater intrusion problems. Forty of the forty-eight contiguous states have reported problems of saltwater intrusion. Only Vermont, West

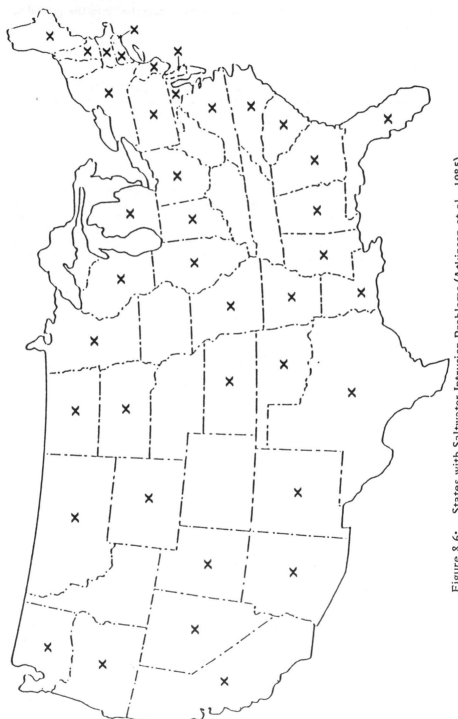

Figure 8.6:   States with Saltwater Intrusion Problems (Atkinson et al., 1985)

Virginia, Kentucky, Tennessee, Iowa, Nebraska, Colorado, and Idaho have been free of saltwater intrusion problems. It is significant to note that every coastal state has experienced problems. Although the problems within each state may or may not be isolated incidents, most of the problems of saltwater intrusion can be considered regional issues.

Limitations to agricultural uses of ground water as a result of increasing salinity from saltwater intrusion can be represented by analyzing areas known as Aggregated Subareas (ASAs). The ASAs were delineated by the U.S. Water Resources Council and are generally definable with hydraulic parameters which separate the areas. There are a total of 99 ASAs in the contiguous U.S. Figure 8.7 (Atkinson et al., 1985) represents the status of limitations to agricultural activities from saltwater intrusion for each of the 99 ASAs. Four categories are shown:

(1) Continuing Problems--if an ASA is known to have had or to be experiencing saltwater intrusion, and it is expected to continue in the future.

(2) High Potential for Future Problems--if an ASA has not been known to have had saltwater intrusion problems, but problems have been predicted by researchers in the area.

(3) Moderate Potential for Future Problems--if an ASA has not been known to have saltwater intrusion problems, and problems have not been predicted, but there is shallow saline ground water (less than 500 feet depth to saline water).

(4) Low Potential for Future Problems--if an ASA has not experienced salt-water intrusion, no problems are predicted, and there is no shallow saline ground water.

Of the 99 ASAs, 55 are considered to have "Continuing Problems". These areas are currently experiencing saltwater intrusion or are in close proximity to deteriorated water quality. These 55 ASAs represent approximately 57 percent of the surface area of the continental U.S. A total of 9 ASAs fall into the "High Potential" category. These 9 ASAs are not currently experiencing salt-water intrusion problems but are very susceptible to problems and careful planning should precede extensive ground water development. These 9 ASAs represent approximately 5 percent of the surface area of the continental U.S. There are 29 ASAs not experiencing saltwater intrusion and considered only to have a "Moderate Potential" for future problems. These Moderate Potential ASAs represent approximately 29 percent of the surface area of the continental U.S. Finally, only 6 ASAs are considered to have a "Low Potential" for salt-water intrusion. These 6 Low Potential ASAs represent only 9 percent of the continental U.S.

It is very evident that saltwater intrusion is a serious threat for agricultural activities. In many situations a "Catch 22" exists: more water is needed to continue agricultural activities, but pumping more water means more saltwater intrusion. As saltwater intrusion increases, crop yield is reduced and the cycle continues. Fortunately, there are many possible techniques for combatting saltwater intrusion. However, many techniques are costly. Therefore, in areas not currently experiencing saltwater intrusion it is

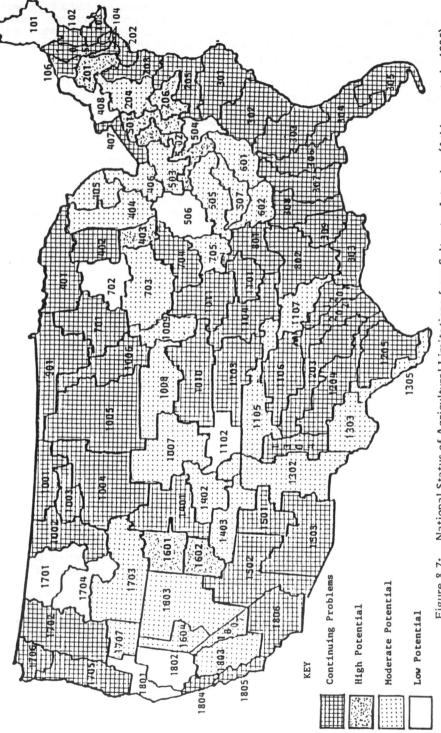

Figure 8.7:   National Status of Agricultural Limitations from Saltwater Intrusion (Atkinson et al., 1985)

KEY

Continuing Problems

High Potential

Moderate Potential

Low Potential

imperative to properly plan ground water use so additional problems can be avoided.

SELECTED REFERENCES

Atkinson, S.F. et al., "Salt-Water Intrusion of Ground Water in the Contiguous United States," June, 1985, Environmental and Ground Water Institute, University of Oklahoma, Norman, Oklahoma.

Bahls, L.L., and Miller, M.R., "Ground Water Seepage and Its Effects on Saline Soils-Completion Report," MUJWRRC Report No. 66, October, 1975, Montana Universities Joint Water Resources Research Center, Montana State University, Bozeman, Montana.

Bredehoeft, J.D. et al., "Possible Mechanisms for Concentration of Brines in Subsurface Formations," American Association of Petroleum Geologists' Bulletin, Vol. 47, 1963, pp 257-269.

Craig, J.R., "Saline Waters: Genesis and Relationship to Sediments and Host Rocks," in Mattox, R.B. (ed), Proceedings of the Symposium on Groundwater Salinity, 46th Annual Meeting of the Southwestern and Rocky Mountain Division of the American Association for the Advancement of Science, Contribution No. 13 of the Committee on Desert and Arid Zones Research, Las Vegas, Nevada, 1970.

deSitter, L.U., "Diagenesis of Oil Field Brines", American Association of Petroleum Geologists' Bulletin, Vol. 31, 1947, pp. 2030-2040.

Donovan, J.J., Sonderegger, J.L., and Miller, M.R., "Investigations of Soluble Salt Loads, Controlling Mineralogy, and Factors Affecting Rates and Amounts of Leached Salts," MWRRC Report No. 120, September, 1981, Montana Water Resources Research Center, Montana State University, Bozeman, Montana.

Feth, J.H., "Chloride in Natural Continental Water--A Review," Water-Supply Paper 2176, 1981, U.S. Geological Survey, Washington, D.C.

Jenkins, S.R., and Moore, R.K., "Effect of Saltwater Intrusion on Soil Erodibility of Alabama Marshlands", July, 1984, Water Resources Research Institute, Auburn University, Auburn, Alabama.

Maas, E.V., "Crop Tolerance", California Agriculture--Special Issue: Salinity in California, Vol. 38, No. 10, October, 1984, pp 20-22.

Miller, D.W., (ed), Waste Disposal Effects on Ground Water: A Comprehensive Survey of the Occurrence and Control of Ground Water Contamination Resulting from Waste Disposal Practices, 1980, Premier Press, Berkeley, California.

Miller, M.R. et al., "Regional Assessment of the Saline-Seep Problem and a Water-Quality Inventory of the Montana Plains," MWRRC Report No. 107, April, 1980, Montana Water Resources Research Center, Montana State University, Bozeman, Montana.

O'Leary, J.W., "The Influence of Ground Water Salinity on Plant Growth", in Mattox, R.B.,(ed) Proceedings of the Symposium on Ground Water Salinity, 46th Annual Meeting of the Southwestern and Rocky Mountain Division of the American Association for the Advancement of Science, Las Vegas, Nevada, 1970.

Thompson, G.R., and Custer, S.G., "Shallow Ground-Water Salinization in Dryland-Farm Areas of Montana," MUJWRRC Report No. 79, September, 1976, Montana Universities Joint Water Resources Center, Bozeman, Montana.

Todd, D.K., "Salt-Water Intrusion and Its Control", Journal of American Water Works Association, Vol. 66, No. 3, March, 1974, pp 181-187.

U.S. Environmental Protection Agency, Office of Air and Water, "Identification and Control of Pollution from Salt Water Intrusion," EPA-430/9-73-013, 1973, Washington, D.C.

CHAPTER 9

ECONOMIC AND ENVIRONMENTAL IMPACTS OF USING MUNICIPAL SEWAGE EFFLUENT FOR AGRICULTURAL PRODUCTION

by

Daniel D. Badger
and
Donald E. Thomason, Jr.

Land application of municipal sewage effluent, a practice which dates back to 1872 in the United States, is a treatment alternative which is gaining in importance due to two factors: the necessity to dispose of sewage effluent in a proper manner, and the increasing demand for water, including irrigation water for agricultural purposes. However, concerns have been expressed over the potential for pollution of both ground waters and surface waters from improper treatment and/or disposal of municipal effluent.

GENERAL PROBLEM

Municipal governments are faced with increasing institutional restrictions on conventional treatment and handling of sewage effluent. Federal and state environmental quality regulations encourage innovative and alternative (I & A) methods of sewage effluent treatment, such as land application. The Federal Water Pollution Control Act Amendments of 1972 (PL 92-500) set a goal of zero discharge of pollutants into navigable waters by 1985. This act, along with the Clean Water Act of 1977 (PL 95-217), provides for construction cost-share grants for wastewater treatment facilities.

Under the U.S. Environmental Protection Agency (EPA) funding policy, I & A methods of treatment have had a comparative advantage in funding over conventional treatment processes in the percentage of cost borne by the federal government (85 percent federal—15 percent local versus 75 percent federal—25 percent local). Beginning October 1, 1984, funding for I & A methods was reduced to 75 percent federal—25 percent local, and funding for conventional methods was reduced to 55 percent federal—45 percent local.

Although funding for both categories decreased, I & A treatment processes such as land application still have a large comparative advantage over conventional treatment processes. On the state level, the Oklahoma State Department of Health (OSDH) recommends land application as the most effective method of wastewater treatment for most towns in Oklahoma.

Regulations regarding conventional treatment and disposal methods have become more stringent relative to I & A methods. Examples include: the increasing restrictions on issuance of National Pollution Discharge Elimination System (NPDES) permits for disposal of pollutants into navigable waters and oceans; the eventual phase-out of the NPDES permits; termination of funding for conventional treatment processes beyond secondary levels by EPA under the

Reagan administration; and requirements by EPA of applicants for construction cost-share grants to thoroughly justify rejecting land application in the facilities plan if land application is not included in the recommended plan.

In many agricultural regions farmers depend on irrigation water as a sole or supplemental source of water for agricultural production during the summer growing season. Many of those farmers also face increasing economic scarcity, and in some cases physical scarcity, of conventional sources of irrigation water. Municipal sewage effluent can be utilized as a partial or total water source for crop or forage production, and may also provide some benefits as an additional supply of plant nutrients (nitrogen, phosphorus, and potassium) when used in a slow rate application system.

The costs of constructing, operating, and maintaining a land treatment facility are generally considered to be lower than corresponding costs of conventional wastewater treatment systems (Williams, Connor, and Libby, 1977). As municipal governments face increasing costs and limitations with conventional sewage effluent treatment and disposal methods, and as farmers are confronted with increasingly costly conventional irrigation water resources, the benefits of land application become more significant to both the farmer and the municipality.

Despite these factors conducive to the utilization of municipal sewage effluent in a land treatment system with agricultural production as a component, the full potential of municipal wastewaters is not being realized. Reluctance exists on the part of farmers to use sewage effluent in their crop and forage-producing enterprises. An apparent reluctance by municipalities to adopt land application as a component in a municipal sewage treatment system exists as well; in the seven year period following the passage of PL 92-500 in 1972, less than 10 percent of all new systems included land application (Jewell and Seabrook, 1979).

A factor which may contribute to economic underutilization of municipal sewage effluent is the conflict of objectives between the supplier and the user. The municipality's goal is to minimize the costs of the wastewater treatment system. Since storage lagoons represent a substantial cost, a municipality with a land application facility often will maximize its wastewater loading rate on the site while still providing an acceptable quality of effluent treatment. This goal is often reflected in the contractual agreement between the municipality and the farmer. Thus, the farmer is often obligated to take a fixed amount of effluent.

The farmer's goal is profit maximization. Implicit in such a goal is efficient utilization of irrigation water and economic optimization of crop yields. The sewage effluent loading rate that he faces in such instances is higher than that which is optimal for crop or forage production. The producer may be required to irrigate with a volume of wastewater which is detrimental to the crop or pasture, or to the land, unless he/she can spread the effluent over more acres than the city's consulting engineer designed into the system.

Another contributing factor may be inflexibility in the volume of sewage effluent supplied to the farmer. Often in the contractual agreement, the farmer is obligated to receive a constant volume of wastewater on the irrigation site. However, the demand for irrigation water for crops or forage is seasonal. Water use by crops in Oklahoma during the summer may be 0.3-0.4

in/day, while winter water usage may be less than 0.1 in/day (Schwab and Stiegler, 1984). The farmer considering using municipal sewage effluent as irrigation water may be reluctant to enter into an agreement with the municipality under a requirement to take a constant amount of water all twelve months of the year.

The costs of storage facilities for effluent, borne by the municipality, are another factor in wastewater underutilization, and they also underlie the previous two factors. Though largely defrayed by construction cost-share grants, municipalities put forth great effort to minimize the expense of storage lagoons in a land application system. This behavior tends to limit consideration of lowering the irrigation site loading rate since, given no change in the acreage to be irrigated, more storage would be needed.

Negative attitudes toward reuse of municipal sewage effluent on the part of municipalities, farmers, or communities may limit the extent to which sewage effluent can be utilized in crop or forage production. If a municipality's officials, or its consulting engineers, view land application in a perspective of disposal of wastes rather than one of reclamation of wastes, efficient wastewater reuse is less likely to be considered. Farmers may object to using effluent for irrigation water, fearing the possibility of pathogens, salts, or heavy metals which may be detrimental to the crop, the soil, the underlying ground water, or to animal or human health. The citizens of a given community may resist land application technology as a result of mistrust, misinformation, health concerns, individual value judgments, or concern over reduced real estate values.

SPECIFIC PROBLEM

A major concern of this study is the impact of excess application of municipal sewage effluent, during part or all of a given period of time, on land producing crops, forage, or pasture. Applying municipal wastewater at times when plants cannot utilize the water has several consequences: plant growth, possible pollution of ground water, wastewater treatment effectiveness, pricing of wastewater as an irrigation source, and crop or forage selection for profit maximization.

At application rates which exceed the capacity of plant usage, the treatment effectiveness of irrigating with municipal effluent diminishes. Plants do not remove all of the nitrogen and phosphorus from the effluent. Higher loading rates of effluent increase the percolation rate through the soil profile. Absorption of chemicals contained in the effluent by soil particles decreases, and leaching of the constituents of the wastewater into the ground water becomes more likely. If the loading rate is sufficiently high to saturate the soil, the resultant anaerobic conditions would reduce the effectiveness of removal of BOD and suspended solids; therefore, the likelihood of buildup of these constituents is enhanced.

Ground water pollution is salient in its importance relative to surface water pollution. Ground water generally mixes poorly and flows slowly, typically less than 1 ft/day. Consequently, the dispersion of a pollutant through ground water is slow, and the pollutant is capable of being transported long distances over a long period of time. Because the capacity of dispersion in ground water is low, a ground water supply, once contaminated, tends to stay contaminated for a long time.

The cropping mix and the irrigation schedule are affected by high wastewater application rates. A constant, inflexible application schedule does not complement the seasonal water demand of crops or forage. The opportunity for water surplus or shortage during certain periods exists as a result. Such loading rate inflexibility may cause the farmer to irrigate in winter months, when crops cannot benefit from the water, and to irrigate conservatively in the summer months to make the water last through the growing season. Crops which are of lesser economic value but which use or tolerate large volumes of water may be selected for production to maintain system effectiveness and workability of the soil. An example of such crop and land use is found in Wells, Sweazy, and Whetstone (1979).

Satisfaction of only the municipality's goal of cost minimization tends to encourage high, inflexible application rates of sewage effluent on the site. The potential of utilizing the effluent in the most efficient manner diminishes when the loading rate exceeds plant intake capacity, and certain effluent treatment problems increase in significance. Demand by farmers for effluent as irrigation water is decreased, distorting the relationship, as reflected by the price of the effluent, between the resource (effluent) and its use in producing the product (crops or forage). If the value of the marginal product of the effluent differs from the effluent price, then a more efficient combination of municipal sewage effluent and other inputs of agricultural production exists for a given production enterprise.

OBJECTIVES OF STUDY

The general objective was to analyze the treatment method of land application of municipal sewage effluent in terms of economic impacts on farmers and municipalities and environmental consequences on communities and farms. Specific objectives were to: (1) examine the economic and environmental impacts of the land application method of handling municipal sewage effluent; and (2) analyze the economics of land application of municipal sewage effluent in a crop or pasture irrigation system designed to increase agricultural production while providing acceptable levels of treatment of the effluent.

PREVIOUS RESEARCH

Land application of sewage effluent is a treatment process that has been practiced for many years. Wastewater irrigation with crop production as a component has been used in San Antonio, Texas, since 1900 and in Lubbock, Texas, since 1925; sewage irrigation was practiced in Augusta, Maine, in 1872, Cheyenne, Wyoming, in 1881, and in Los Angeles, California, in 1883 (Christensen, 1982; and Wells, Sweazy, and Whetstone, 1979).

A study by Bishop et al., (1973) at Utah State University examined social, economic, environmental, technical and legal aspects of various forms of water reuse such as industrial recycling, land application and other forms of sequential reuse, and irrigation return flow. A framework within which water reuse policies would be formulated on several levels was developed. The

Wasatch Front Area in north central Utah was analyzed from the standpoint of economic efficiency. Supply functions for water resources in the area were developed from mathematical programming models of local hydrology and cost structures; demand functions for water resources were derived from empirical studies and mathematical programming models.

Kardos (1970) reported on a land application study conducted at Pennsylvania State University. Initiated in 1963, the study focused on application of the sewage effluent from both the university and the Town of State College on agricultural and forest land. The effectiveness of removal of plant nutrients, bacteria, detergent residues, and mineral salts from the effluent was examined, as was the impact of crop and forest growth and ground water and soil water quality. Application of sewage effluent increased crop yields over those of the control plot, but wastewater irrigation retarded growth in some species of trees. Removal of phosphorus, nitrogen, bacteria, and detergent residues from the effluent was highly efficient while mineral salt removal was somewhat less efficient. The quality of both the ground water and soil water was improved. The author stressed that the success of the land application method was contingent upon maintenance of certain desirable aspects of the soil condition such as adequate infiltration, percolation, and a high exchange capacity.

Walker (1976), in an EPA publication, recounted the history of the municipal wastewater treatment facility in Muskegon County, Michigan, with respect to the land application/agricultural production component of the works. The conditions which prompted the inclusion of land treatment in the works, the costs of constructing the systems, the design of the system, the operation and management of the system, and the performance and cost reductions of the facility with the land application system were discussed.

A report by Christensen et al., (1976) examined the technical, legal, and institutional components of land application in Michigan. Land acquisition options and impacts on farmers' goals (income generation, wealth accumulation, firm growth, autonomy in decision making, and a sense of community participation) were discussed. The experiences of several Michigan communities with municipal land application treatment systems were reported and analyzed. The legal environment of farmers, municipalities, and Michigan communities, in general, was discussed.

In 1976, Webb and Badger analyzed the operation of a land treatment site with an agricultural production component in Pauls Valley, Oklahoma. Efficiency of operation of the treatment system was examined within a multiobjective framework including both the municipality's goal of disposal of all effluent in a safe manner and the farmer-manager's goal of maximizing agricultural production. System performance was measured by a water balance schedule for a twelve month period. For the year analyzed, the volume of wastewater applied was found to exceed normal application volume due to lack of holding pond capacity, and application timing was determined to be nonoptimal as a result. The authors concluded that system efficiency from an economic and environmental standpoint would be improved by expanding effluent storage capacity and, in general, that proper management of the treatment system is essential for the full benefits of a land treatment/agricultural production system to materialize.

Sullivan, Cohn, and Baxter (1973), in an EPA publication, reported the results of a 1972 American Public Works Association nationwide on-site field survey of approximately 100 land treatment facilities, as well as surveys of officials of land application treatment facilities, health and water pollution regulations at the state level, and experiences with land application in several countries. Several general and site-specific factors influencing adoption and implementation of land application technology were evaluated.

A five-year study by Cabbiness and Badger (1980) at Oklahoma State University estimated the potential fertilizer value of municipal sewage effluent. The nutrients supplied by the effluent from Pauls Valley, Oklahoma, were evaluated in the forms of anhydrous ammonia, ammonium nitrate, superphosphate, and potash. The nutrient value per acre of the effluent as applied to various forage corps was calculated. No attempt was made to evaluate the value of the water component of the wastewater.

Wells, Sweazy, and Whetstone (1979) examined the value of municipal sewage effluent as a water resource. They argued that, despite the nutrients present, wastewater is worth less than conventional irrigation water because the sewage effluent is supplied at a constant rate while demand for irrigation water is variable, depending on temperature, wind, and other conditions affecting soil moisture.

Some studies have indicated that sewage effluent has been used in a suboptimum manner. Wells, Sweazy, and Whetstone (1979) reported that as the volume of wastewater produced by the City of Lubbock, Texas, increased over time, the farmer who owned the rights to the effluent could not increase his farming operation at a similar rate. This caused part of his land to be utilized only as a dumping site and a significant portion of his acreage to be devoted to forage crops which utilize high volumes of water but are not necessarily high profit crops.

A study by Cabbiness and Badger (1981) analyzed the impacts of land application of municipal sewage effluent previously returned to the North Canadian River basin on farmers holding water rights downstream. Ground water to surface water transfers and surface water to surface water transfers were examined. Farmers were surveyed along the North Canadian River with respect to attitudes toward using municipal sewage effluent as their source of irrigation water. The study showed that if 100 percent of the wastewater in the study area was applied to land, no water would be available to downstream water rights holders in certain seasons.

Carlson and Young (1975), in a study of factors affecting the adoption of land treatment of municipal sewage effluent, analyzed the effect of quality of effluent on demand for the land application treatment option. Data were collected on 125 U.S. cities, and both log linear and nonlinear demand curves for land application were derived as well as elasticities of demand for the various factors. Profit maximization was assumed. All variables had the expected signs. Factors which were determined to have a significant impact on adoption or nonadoption of land application were: (1) the local cost share on construction grants, (2) the price of wastewater for agricultural irrigation, (3) required degree of treatment, (4) capital costs for nonland treatment options, (5) rainfall, (6) volume of stream flow, and (7) daily volume of effluent. The price of land was deemed to be insignificant. The authors concluded that the municipalities in the study responded to economic incentives (construction grant share, value of irrigation water, stream discharge regulations, required treatment level, and prices for labor and capital) when selecting sewage effluent treatment technologies for their wastewater systems.

Young (1976) developed a simulation model, named the Cost of Land Application of Wastewater (CLAW) model, to estimate costs of land application systems under varying structural conditions. Cost estimates for six land application methods were derived. Young (1978) used the CLAW model to compare the cost effectiveness of land application with that of conventional wastewater treatment. Factors affecting the cost effectiveness of wastewater treatment facilities, with land application facilities highlighted, were examined. Crop selection in a land application/agricultural production system was found to have the most significant impact on cost effectiveness; other factors with significant impact were buffer zones, chlorination, and application rate.

PROCEDURE

The methodology involves the construction of a linear programming model for a land application treatment system in Oklahoma which includes production of crops or forage. Basic data and parameters of the model were obtained from surveys of farmers and municipal officials, land treatment system design manuals, and Oklahoma State University Enterprise Budgets.

Most of the matrix coefficients came from the Oklahoma State University Enterprise Budgets. The budgets provided information on requirements for labor, nitrogen, phosphorus, water, operating capital, and other inputs into crop and forage production enterprises, as well as information on productivity of alternative enterprises.

The coefficients used for nitrogen and phosphorus concentrations in municipal sewage effluent were average city effluent figures for Oklahoma municipalities (Schwab and Stiegler, 1984). Information on dry matter content of forages and daily minimum dry matter consumption by beef cattle came from NRC tables in Animal Feeding and Nutrition. Dry matter data were used to convert animal unit months (AUM) of forage to units for which selling prices were obtainable.

Information on the sewage effluent system design data was obtained from Process Design Manual for Land Treatment of Municipal Wastewater (published jointly by EPA, USDA, and the U.S. Army Corps of Engineers) and the Oklahoma State Department of Health's Design Guidelines for Land Application of Municipal Wastewater. Included in the data are annual rate of precipitation, evapotranspiration rates, operation and management (O & M) costs, and water usage.

The empirical model used for analyzing economic impacts of land application of municpal wastewater was a polyperiod linear programming model. A representative farm budgeting approach was used. The model considers 240 application periods and 20 production periods, covering a span of 20 years. The scope of the model corresponds with the minimum time period required by EPA for lease agreements between a municipality and a farmer or landowner.

Two models--one irrigated and one dryland--were used to consider the economic impacts of irrigation of cropland with municipal sewage effluent. The irrigation model is that of a farmer who is a profit maximizer granting the municipality easement to his/her land for application of sewage effluent in a treatment system with agricultural production as a component. The municipal government, with EPA cost sharing, provides the center pivot sprayer irrigation equipment and bears the costs of storage lagoons; the city also bears all O & M costs.

The model is constrained to either apply the municipal sewage effluent to the land or store it; all effluent supplied by the city in a given month must be accounted for. Effluent placed in storage in a given period is available for application in future periods. The dryland model is also that of a farmer who is maximizing profit from his/her production enterprises. Agricultural production in this situation takes place under dryland conditions. The producer has full control and responsibility over his/her production activities; there is no partnership with a municipality or any other party.

A representative farm approach was used in the modeling process to make the results of the model optimizations as realistic to Oklahoma farm producers as possible. The acreage of both farms was set at 240 acres, subdivided into three tracts of 80 acres. All of the 240 acres of the irrigation model can be irrigated with municipal sewage effluent. The farmer has 2,500 hours of labor in each production period. Initial operating capital of $15,000 was available to the farmer in year 1. The farmer possessed no initial stocks of nitrogen or phosphorus. Additional units of labor, nitrogen, phosphorus, and operating capital may be purchased by the farmer in both models.

Fixed amounts of nitrogen and phosphorus are applied as commercial fertilizer in the irrigation model. Such an assumption is made to insure that seasonal peak demands for plant nutrients are met, since production periods are in years and effluent application periods are in months. For each acre, 60 pounds of phosphorus are applied to alfalfa, 35 pounds of nitrogen and 20 pounds of phosphorus are applied to bermuda grass, 35 pounds of nitrogen are applied to grain sorghum, and 30 pounds of nitrogen and 30 pounds of phosphorus are applied to wheat. Variable and fixed costs of fertilizer and application are included in the production costs. Crop nutrient requirements above what is met by this fertilization are met by either the nutrients in sewage effluent or the purchase of additional fertilizer. Not all plant nutrients in the effluent are available for crop use due to soil fixation and leaching; 50 percent availability in a given year is assumed.

The cropping pattern is identical for both the irrigated and dryland models. Alfalfa and grain sorghum are rotated on tract A; alfalfa is raised five years and grain sorghum is raised the sixth year. Bermuda grass is grown on tract B all 20 years as a permanent pasture. Wheat is produced on tract C for all 20 years; such constant production, while presenting increased difficulty in weed control and in maintaining the productivity level, is a common practice by Oklahoma wheat producers. Production on tracts A and B start in spring of year 1, while wheat production on tract C begins in July of year 1.

Municipal sewage effluent is produced by a hypothetical Oklahoma town with a population of 6,000, approxiately the mean population of the cities in the study area. Production of 100 gallons per capita per day is assumed, making the daily flow of the municipality 0.60 million gallons per day. Neither the population of the town nor the per capita production of effluent are assumed to change during the model time frame. The net evapotranspiration rate, in keeping with the regional orientation of the enterprise budgets used, is assumed to be 35 inches per year; that rate corresponds with the rate of net evapotranspiration found in much of western Oklahoma.

Three oxidation lagoons and one holding lagoon are assumed. Monthly evapotranspiration was calculated using the surface acreage of the lagoons. The oxidation cells cover a total of 14 surface acres and have an average depth of four feet for each cell. The holding cell covers 20 surface acres and has an average depth of 10 feet. Municipal sewage effluent flows into the oxidation lagoons from the primary treatment plant and remains in those lagoons for a period of 30 days before being released into the holding lagoon. Wastewater in the holding lagoon is then used when required for irrigation. Startup of the land application facilities occurs in December of year 0.

No attempt was made to model how municipal sewage effluent and its constituents are affected by application to, and percolation through, the soil. Effluent is applied in the irrigated model at a rate which does not exceed crop and forage water needs. An implicit assumption of such an application rate is that environmental hazards, such as buildup of harmful wastewater constituents or leaching of nitrogen into ground water supplies, will not be significant.

Although the structure of each model varies significantly from the other model, the theoretical differences between the two models are slight. The objective function of the irrigation model can be stated as:

$$(1) \quad \max \quad \sum_{j=1}^{n} c_j x_j$$

subject to:

$$(2) \quad \sum_{j=1}^{n} a_{ij} x_j \leq b_i \qquad i = 1, \ldots, r$$

$$(3) \quad \sum_{j-1}^{n} a_{ij} x_j = b_j \qquad i = r+1, \ldots, m$$

$$(4) \quad x_j \geq 0 \qquad j = 1, \ldots, m$$

where c_j is the return per unit to unpaid resources for the jth activity, x_j is the quantity produced by the jth activity, a_{ij} is the amount of the ith resource required per unit of the jth activity, and b_i is the level of the ith unpaid resource available. Equation (2) states that m number of equations of primal variable are constrained to be less than or equal to the corresponding resource level. Equation (3) states that (m - r) number of equations are constrained to equal the corresponding resource level. These (m - r) equations represent the municipal sewage effluent produced by the city each month. Equation (4) states that all quantities produced are nonnegative. The objective function of the dryland model is identical to that of the irrigation model; the only difference in the two models theoretically is that equation (3), which represents the municipal wastewater supply constraints for each application period, is omitted in the dryland model.

To analyze the value of municipal sewage effluent in optimal agricultural production, the dual solution of the irrigation model was generated, with the optimal values for the dual variables obtained. The theoretical validity of this approach is presented below. Given the objective function in equation (1), the objective function of the dual is:

$$(5) \quad \text{Min} \quad \sum_{i=1}^{n} b_i y_i$$

subject to:

$$(6) \quad \sum_{i=1}^{n} a_{ij} y_i \geq C_j \quad\quad j = 1, \ldots, s$$

$$(7) \quad y_i \text{ unrestricted} \quad\quad i = s + 1, \ldots, n$$

$$(8) \quad y_i \geq 0 \quad\quad i = 1, \ldots, s$$

where y_i is the shadow price, or imputed value, of the ith resource and all other variables have the same meaning as in the primal. There are some significant differences between the dual and the primal with respect to the constraints. Equation (6) contains a reversed sign compared to its counterpart, equation (2), in the primal. Where the primal contains an equality constraint in equation (3), the dual specifics in equation (7) that y_i is unrestricted in sign over the same (n - s) number of equations. Equation (7), like equation (3) in the primal, does not exist in the dryland model.

RESOURCE REQUIREMENTS, LEVELS, AND COSTS

The requirements for irrigation water per acre by month, year, and crop are shown in Table 9.1. Demand for irrigation water in general by plants is greatest during the period of April through September. Alfalfa and bermuda grass require the most water during the year (33 and 32 acre inches per year, respectively). The rotation of grain sorghum with alfalfa every sixth year decreases the irrigation requirements by nine acre inches during those years when grain sorghum is produced. Wheat requires the lowest volume of irrigation water per year. Also, since it is a winter crop, wheat uses water in periods when other crops and forages are either dormant or not planted.

The total monthly water demand for a farm operation at full capacity (240 acres) is calculated by multiplying the monthly requirements of water for each crop (Table 9.1) by 80 acres and summing the products for the month. The monthly demand for irrigation water in acre inches is presented in Table 9.2.

The net inflow of effluent from the hypothetical town of 6,000 population is shown in Table 9.2. Starting with a daily flow of 0.60 mgd, evapotranspiration is assumed in each month at a rate of 62 inches annually, distributed appropriately among the months of the year. Annual precipitation of 30 inches adds to the water stock in the lagoons, so an annual net evapotranspiration rate of 35 inches is used to derive the net inflow into the

Table 9.1: Monthly Irrigation Water Requirements, by Crop, in Inches per Acre, for Oklahoma, 1984 Conditions (Oklahoma State University Enterprise Budgets)

MONTH	ALFALFA	BERMUDA GRASS	WHEAT	GRAIN SORGHUM[a]
January	--	--	--	--
February	--	--	--	--
March	--	--	3.0	--
April	6.0	--	3.0	--
May	3.0	4.0	6.0	6.0
June	4.0	8.0	--	3.6
July	8.0	8.0	--	7.2
August	8.0	8.0	--	7.2
September	4.0	4.0	3.0	--
October	--	--	--	--
November	--	--	3.0	--
December	--	--	--	--
TOTAL	33.0	32.0	18.0	24.0

[a]Water requirements for years 6, 12, and 18

holding lagoon. Evapotranspiration is assumed to occur both in the holding lagoon for a given month and in the oxidation lagoons for the previous month. The net inflow in Table 9.2, therefore, is a measure of new effluent available for irrigation for a given month.

Using the net inflow figures from Table 9.2 and the irrigation water requirements from Table 9.1, the monthly net balance of municipal sewage effluent also was calculated and is shown in Table 9.2. The water requirements used are those for production of alfalfa, bermuda grass, and wheat. From October through March, municipal sewage effluent is accumulating in storage. From April through September, a net volume of effluent is being drawn from storage, due to both high crop demand and increased evapotranspiration. In this particular system, a surplus of 234 acre inches, or 20 acre feet, occurs each year. Other basic assumptions on inputs, yields and prices are presented in the thesis by Thomason (1985).

Table 9:2:   Monthly Net Inflow of Municipal Sewage Effluent in Acre Inches, Crop Water Use, and Model Design Net Balance, Oklahoma Representative Farm, 1984 Conditions

Month	Gross Inflow	Net Evapotranspiration	Net Inflow	Crop Water Use[a]	Balance
January	685	36	649	0	649
February	619	43	576	0	576
March	685	69	616	240	376
April	663	104	559	720	-161
May	684	126	559	1040	-481
June	663	138	525	960	-435
July	685	157	528	1280	-752
August	685	160	525	1280	-755
September	663	134	529	880	-351
October	685	105	580	0	580
November	663	74	589	240	349
December	685	46	639	0	639
TOTAL	8066	1192	6874	6640	234

[a]The representative farm has 80 acres each of alfalfa, bermuda grass and wheat.

RETURNS TO IRRIGATED AND DRYLAND PRODUCTION

The net present value of returns to land, risk, and management was much greater for the irrigated model than that for the dryland model. The discounted net returns in each year for both models are shown in Table 9.3. Net returns to fixed assets and managements were $441,186.87 for the irrigated model and $82,512.00 for the dryland model. Income for each year is much greater in the irrigated model than in the dryland model. Both models produced all 240 acres available in every year.

Nitrogen and phosphorus are applied in the irrigated model in excess of crop requirements. The extra plant nutrient application comes from nitrogen and phosphorus in the sewage effluent, given the base level of commercial

Table 9.3: Present Value of Net Returns to Selected Factors[a] for Irrigation and Dryland Models, 1984 Oklahoma Conditions

Year	Irrigated Model	Dryland Model
1	$ 21,394.67	$ 3,410.49
2	37,428.12	7,086.51
3	35,047.42	6,685.38
4	32,627.19	6,306.97
5	30,753,23	5,575.39
6	22,848.05	5,172.08
7	27,272.71	4,918.41
8	25,770.62	4,637.62
9	24,298.56	4,372.61
10	22,910.85	4,122.91
11	21,603.04	3,887.56
12	16,045.94	3,633.54
13	19,178.35	3,456.47
14	18,113.57	3,259.65
15	17,081.61	3,073.90
16	16,108.63	2,898.81
17	15,191.40	2,733.74
18	11,272.99	2,555.43
19	13,496.32	2,431.33
20	12,743.60	2,293.20
	$441,186.87	$82,512.00

[a]Land, risk, and management
Net present value is based on a 6.0 percent discount rate.

fertilizer application assumed in the irrigated model. Phosphorus was always at excess levels. Nitrogen exceeded crop requirements in 17 of the 20 years; nitrogen fertilizer was purchased in years 6, 12, and 18, when grain sorghum replaced alfalfa in the crop rotation. In those years, approximately 29 pounds of nitrogen fertilizer per acre was purchased.

SUMMARY AND CONCLUSIONS

Land application treatment systems in Oklahoma generated benefits to both the farmer and the municipality. Farmers involved in land application systems obtained a more stable water source compared to dependence on rainfall. Municipal sewage effluent gave farmers another source of nitrogen, phosphorus, and potassium, and farmers were well aware of the production value of these nutrients. Both farmers and city officials responded that, with the addition of municipal wastewater, yields increased over those of dryland production (the previous method of production on nearly all the application sites), sometimes twofold. City officials in general felt that land application was a low-cost handling method relative to available alternatives. A problem of handling a waste which needed to be disposed of was turned into a situation where a waste product was used and reclaimed. Concern with meeting federal and state regulations on handling municipal sewage effluent was a significant factor with many officials, so a feeling of operating the city's sewage effluent system "within the law" was a benefit to the city.

Certain problems occur with land application systems and proposed systems. The severity of the problems range from slight to severe. Seasonal supply fluctuations cause many land application systems to apply wastewater in winter when plants do not need it and to apply levels below plant needs in summer months due to short supply of effluent. Seepage and erosion in cell walls and dikes are significant concerns of city officials; either problem can seriously detract from successful operation of land treatment facilities and pose a risk of soil and ground water contamination.

A few cities have experienced the consequences of design errors which caused the land application system to be mismatched with the application site in any of a number of ways. Irrigation equipment has not functioned as intended in some systems. Bogging of irrigation equipment and tracking of the application site have caused problems. Some land application systems have experienced delays in startup due to equipment breakdown, seepage of effluent from lagoons, or miscalculations by the designers of the system.

A problem faced by many Oklahoma land application systems was exhibited in the pattern of effluent storage in the 240 acre irrigated model. The volume of wastewater to be stored usually exceeded the capacity of the holding cell from February through May in each year. This corresponded with the period of low demand for irrigation water and low net evapotranspiration. Holding pond overflow would result unless wastewater was applied during late fall and winter months, even though little or no crop demand for water existed from October through February.

Two of the most important factors in the success of land application treatment systems with agricultural production components are system management and cooperation between city and farmer. Because the objectives of the city and the farmer in land application systems are often conflicting, the level of cooperation between the two parties can dictate how successful the system is. Each party needs to be aware of the objectives of the other to increase understanding on both sides.

The potential for operational and environmental problems in a given system illustrates the significance of proper management of the land application system. Timing of wastewater application is one of the most important aspects of system management, depending upon what site-specific factors are present. A host of potential hazards, such as surface runoff, insufficient infiltration and/or percolation, bogging or tracking of equipment, and pond overflow compel the city and the farmer to take an active role in planning the operation of the land application system. Under conditions conducive to agricultural production, the potential exists for increased farmer willingness, or demand, to utilize municipal sewage in production enterprises.

This study was funded by Hatch Project 1906 of the Oklahoma Agricultural Experiment Station.

SELECTED REFERENCES

Bishop, A.B. et al., "Social, Economic, Environmental, and Technical Factors Influencing Water Reuse," December, 1973, Utah Water Research Laboratory, College of Engineering, Utah State University, Logan, Utah.

Cabbiness, S.G., and Badger, D.D., "Estimating the Economic Value of Nutrients in Municipal Sewage Effluent Used for Irrigation at Pauls Valley, Oklahoma," Oklahoma Current Farm Economics, Vol. 53, No. 4, December, 1980.

Cabbiness, S.G., and Badger, D.D., "Implications for Water Rights Users of Land Application of Treated Sewage Effluent: The North Canadian River Case," AE-8167, June, 1981, Department of Agricultural Economics, Oklahoma Agricultural Experiment Station, and Oklahoma Water Resources Research Institute, Oklahoma State University, Stillwater, Oklahoma.

Carlson, G.B., and Young, C.E., "Factors Affecting Adoption of Land Treatment of Municipal Wastewater," Water Resources Research, Vol. 11, No. 5, October, 1975.

Christensen, L.A., "Irrigating with Municipal Effluent: A Socioeconomic Study of Community Experience," ERS-672, January, 1982, Natural Resource Economics Divison, Economic Research Service, U.S. Department of Agriculture, Washington, D.C.

Christensen, L.A. et al., "Land Treatment of Municipal Wastewater: A Water Quality Option for Michigan Communities," Report No. 41, January, 1976, Center for Rural Manpower and Public Affairs, Michigan State University, East Lansing, Michigan.

Jewell, W.J., and Seabrook, B.L., "A History of Land Application as a Treatment Alternative," MCD-40, April, 1979, U.S. Environmental Protection Agency, Washington, D.C.

Kardos, L.T., "A New Prospect," Environment, Vol. 12, No. 2, 1970.

Schwab, D., and Stiegler, J., "Agricultural Use of Municipal/Industrial Wastewaters," OSU Extension Facts No. 1510, July, 1984, Cooperative Extension Service, Oklahoma State University, Stillwater, Oklahoma.

Sullivan, R.H., Cohn, M.M. and Baxter, S.S., "Survey of Facilities Using Land Application of Wastewater," EPA 430/9-73-006, July, 1973, U.S. Environmental Protection Agency, Washington, D.C.

Thomason, Jr., D.E., "Economic and Environmental Impacts of Using Municipal

Sewage Effluent for Agricultural Production," Unpublished M.S. Thesis, May, 1985, Oklahoma State University, Stillwater, Oklahoma.

Walker, O.M., "Wastewater: Is Muskegon County's Solution Your Solution?", MCD-34, November, 1976, U.S. Environmental Protection Agency, Oklahoma State University, Stillwater, Oklahoma.

Webb, A., and Badger, D.D., "Environmental and Economic Aspects of Land Application of Municipal Effluents,"Oklahoma Agricultural Experiment Station No. 3252, November, 1976, Oklahoma State University, Stillwater, Oklahoma.

Wells, D.M., Sweazy, R.M., and Whetstone, G.A., "Long-term Experiences with Effluent Reuse," Journal of the Water Pollution Control Federation, Vol. 51, No. 11, November, 1979.

Williams, J.R., Connor, L.J., and Libby, L.W., "Case Studies and Comparative Cost Analyses of Land and Conventional Treatment of Wastewater by Small Municipalities in Michigan," Department of Agricultural Economics Report No. 329, October, 1977, Michigan State University, East Lansing, Michigan.

Young, C.E., "The Cost of Land Application of Wastewater: A Simulation Analysis," U.S. Department of Agriculture Technical Bulletin No. 1555, November, 1976, Natural Resource Economics Division, Economics Research Service, Washington, D.C.

Young, C.E., "Land Application of Wastewater: A Cost Analysis," U.S. Department of Agriculture Technical Bulletin No. 1594, September, 1978, Natural Resource Economics Division, Economics, Statistics, and Cooperatives Service, Washington, D.C.

CHAPTER 10

SOIL TESTING AS A GUIDE TO PRUDENT USE OF
NITROGEN FERTILIZERS IN OKLAHOMA AGRICULTURE

by

Gordon V. Johnson

Oklahoma agricultural producers purposefully add approximately 570 million pounds of nitrogen in the form of chemical fertilizers to their fields each year. To the casual student of environmental concerns, this would seem to be an horrendously excessive use of chemicals and an obvious source for nitrogen contamination of ground water.

Ground water chemical composition is a result of chemical reactions from water passing through strata above the aquifer and within the aquifer. Many factors will influence the final chemical composition of the ground water. Among the major factors will be the rate of water movement through the overlying material and the degree to which that material, as well as the aquifer itself, are soluble in water. For example, ground water rich in calcium, carbonate, and sulfate may be expected from a limestone aquifer overlain by gypsum as diagrammed below.

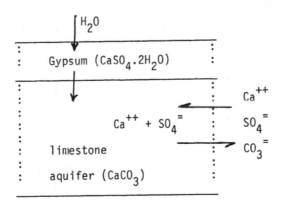

Expanding on this simple approach to estimating ground water composition, if a water soluble material such as ammonium nitrate (NH_4NO_3) fertilizer were added to the earth's surface, one might expect it to strongly affect ground water composition. Simple logic would lead to an expansion of the previous illustration as shown in the next diagram.

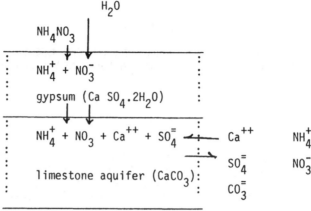

Fortunately, the chemical, physical, and biological dynamics occurring at the earth's surface negate this hypothetical effect of using nitrogen fertilizers.

NITROGEN IN SOILS

Nitrogen exists in soils primarily as organically bound nitrogen in the form of humus and inorganic nitrogen present as ammonium (NH_4^+) and nitrate (NO_3^-). Soils usually contain about 99 percent of the total soil nitrogen in the organic form. Plants cannot absorb significant amounts of organic nitrogen. Except for legumes which utilize N_2 from the atmosphere, crops obtain the nitrogen needed for protein synthesis by absorption of mineral nitrogen through their roots. Most of the mineral nitrogen is present in soil as NO_3^-, and this is the chemical form absorbed in greatest quantities by plants.

Soil microorganisms are responsible for the existence of a dynamic equilibrium between organic and mineral nitrogen. While the balance between mineral and organic nitrogen in soil is influenced by many complex factors, soil organic matter generally serves as a buffer against large changes in the quantity of mineral nitrogen brought about externally. Hence, if large amounts of mineral nitrogen are added to soil as fertilizer just before planting or while a crop is growing, the total amount added will not persist as mineral nitrogen in the soil for more than a few days. Some will be absorbed by the growing crop and some of the mineral nitrogen will be utilized by the growth of microbial biomass. When all the mineral nitrogen is utilized before the crop matures, the deficiency will cause a decrease in yield and/or protein content of the crop. Microbial activity will also decrease, and as the organic energy source is depleted, mineral nitrogen may eventually reappear from the decay of microbial tissue.

When sufficient mineral nitrogen exists in the soil to support crop and microbial needs, the excess may be present as NO_3^-. This form is highly soluble and is subject to movement as water moves through soils. In order for NO_3^- from fertilizer sources to reach ground water, the following criteria must be met.

(1) NO_3^- must be present in the soil in excess of that which can be utilized by soil microorganisms and/or the crop in the existing environment.

(2)   The amount of rainfall and irrigation must exceed evapotranspiration to provide excess soil water for downward percolation.

(3)   The soil and substrata must be sufficiently porous to allow water and NO_3^- to percolate to the ground water.

Variables associated with the second and third criteria are defined to a large degree by climate, weather, and geologic formations of the geographic area under consideration. Whether or not the first criteria for leaching NO_3^- to the ground water is met will depend on how the use of nitrogen fertilizers is managed in crop production systems.

FIELD NITROGEN FERTILIZER REQUIREMENTS

Nitrate Absorption by Plants

The potential for accumulation of NO_3^- in the root zone of crops and its movement into ground water depends on: (1) how well farmers can predict the nitrogen required by their crops and; (2) how well they can account for available soil nitrogen and apply fertilizer nitrogen to satisfy this requirement but still avoid excessive NO_3^- in the soil.

Growing crops are able to extract NO_3^- from the soil near their roots. As this occurs and as the soil water near roots is absorbed, water moves toward the roots from soil some distance away. Because NO_3^- is a very soluble anion (it does not form insoluble compounds with iron, aluminum, or calcium like phosphate ions) and it is not readily adsorbed as cations are by the soil colloids, NO_3^- present in soil water can be moved to the roots from considerable distance. Consequently, unless plants are spaced quite far apart (about 10 feet), adjacent plants will compete for NO_3^- present in soil between them. The amount of competition among plants will depend on plant population density and the total amount of available NO_3^-. Since crop yield is directly influenced by plant population density, yield and available nitrogen are directly related.

Crop Nitrogen Requirement

The crop nitrogen requirement can be easily calculated from yield and nitrogen content information. For example, for a 40 bushel/acre yield (60 pounds of grain per bushel) of wheat containing 13 percent protein (17.5 percent N in protein) the calculations would be as follows:

Nitrogen requirement/acre = (40 bu/ac) (60 lbs/bu)

(0.13 crude protein) (0.175 N/crude protein)

Nitrogen requirement = 55 lbs N/ac

As a first step in estimating fertilizer nitrogen needs of nonlegume crops, the amount of nitrogen removed in the harvesting of the crop would have to be replaced.   If this is not provided by addition of fertilizer or other outside nitrogen sources, there will be a depletion of soil organic nitrogen and the attendant loss of soil organic matter.

Research in many states has shown that crops use only about 50 to 70 percent of the fertilizer nitrogen applied. Most of what is not used directly by the crop is absorbed by the soil biomass. Unused nitrogen does not accumulate as a large pool of NO_3^- subject to leaching.   Hence, in order for a 40 bu/ac wheat crop to store 55 lbs of nitrogen in the harvested grain, about 80 lbs must be available during the growing season (70 percent of 80 lbs = 55 lbs). This 80 lbs requirement of available nitrogen minus the available nitrogen (usually NO_3^-) present in the soil at planting will be the fertilizer nitrogen requirement.

The nitrogen requirement of common Oklahoma crops has been calculated using the above approach.   An example for wheat is shown in Table 10.1. Obviously, farmers must have an idea of how much their crop will yield and how much available nitrogen is present in the soil before they plant in order to arrive at how much fertilizer nitrogen should be used.

Table 10.1:   Available   Soil   N   Requirement   in Relation to Wheat Grain Yield

Yield	Crop N* Removal	Available N Requirement**
20	27	39
30	41	59
40	55	79
50	68	97
60	82	117
70	96	137
80	109	156
90	123	176
100	137	196

*Assumes 13 percent crude protein
**Assumes 70 percent N use efficiency

Yield Goals

Yield is most often limited in dryland agriculture by lack of rainfall. Predicting crop yields in advance of planting assumes the same risks and uncertainties as trying to predict rainfall several weeks in advance. However, to assure that rainfall and not available nitrogen will be the yield limiting factor, farmers must use yield goals or projections assuming good (better than average) rainfall. In this way they will have an improved chance of high yields when rainfall permits.   During poor years, when limited rainfall prevents adequate crop development to use all the nitrogen available, the excess nitrogen is unlikely to leach out of the root zone.

Realistic yield goals are based on experience or the yield history of each field. Two approaches are often used. One is to set the yield goal equal to the best yield of the last five years. The second approach is to set the yield goal at 50 percent above the long-term yield average for the field.

Available Soil Nitrogen

Measuring available soil nitrogen is the final task necessary to determine the fertilizer nitrogen requirement. This involves obtaining a reliable soil sample and submitting it to a soil testing laboratory which tests for available soil nitrogen. Since NO_3^- is mobile in the soil, it is important to sample the subsoil (7-24 inch depth) of each field as well as the surface soil. At the Oklahoma State University (OSU) Cooperative Extension Service Laboratory, soils are tested for available nitrogen using a NO_3^- specific ion electrode.

An example of how the soil test information is used to determine fertilizer nitrogen requirement for a field of wheat follows.

Yield Goal = 40 bu/ac

    Available nitrogen requirement   = 80 lbs/ac

    Available soil nitrogen

        Surface soil test = 12 lbs/ac
        Subsoil soil test = 8 lbs/ac

        Total available nitrogen = 20

    Fertilizer nitrogen requirement   = 60 lbs/ac

The above example illustrates the educational approach used by the OSU Cooperative Extension Service to help farmers apply nitrogen fertilizer for profitable yields and minimum environmental hazard.

OKLAHOMA NITROGEN NEEDS AND USAGE

A Nitrogen Balance Sheet

A broad assessment of how Oklahoma farmers use nitrogen fertilizer was made by comparing the nitrogen crop requirement to the nitrogen fertilizer sales for the entire state. This provided an estimate of whether or not there was a net gain or loss in soil nitrogen. A review of recent soil test information from an extensive soil testing program in the state added insight to fertilizer nitrogen use.

Table 10.2 shows the amount of nitrogen required for production in 1984 calculated from statistics on crop yields for that year. Compared to the approximately 570 million pounds of fertilizer sold for the same period, there appears to be about 100 million pounds more nitrogen required than was applied. This nitrogen shortage could have been made up by nitrogen mineralized from soil organic matter or crop residue. While some nitrogen could have been supplied from mineralization of legume residue which fixed atmospheric nitrogen the previous year, this is not very likely since most of the crops listed in Table 10.2 were not preceded by a legume crop.

The nitrogen requirement shown in Table 10.2 assumes a 70 percent efficiency for use of available soil nitrogen relative to nitrogen in harvested plant material. If only harvested nitrogen (usually removed from the field) is considered, the amount would be about 470 million pounds. The 100 million pounds of fertilizer nitrogen not removed may easily be accounted for in the nonharvested portion (residue) of crops. Obviously the protein nitrogen in leaves, stems and roots of wheat plants must be accounted for even though only the grain is commonly removed from the field. This kind of comparison further documents the importance of soil organic matter as a reservoir of nitrogen.

Table 10.2:   Estimated Nitrogen Requirement for Oklahoma Cropland[*]

Crop	Production	Nitrogen Required
	----thousands----	
Winter wheat	190,800 bu	381,600 lbs
Oats	3,680 bu	4,089
Barley	2,050 bu	3,280
Corn	5,250 bu	6,825
Grain Sorghum	18,000 bu	16,000
Cotton	183 bales	10,980
Hay	2,400 tons	120,000
Pasture[**]	2,500 tons	125,000
TOTAL		667,774 lbs

[*]1984 Oklahoma Agricultural Statistics

[**]Estimated from the planted wheat acreage minus harvested grain wheat acreage.

Nationwide estimates place the composition of nitrogen in soil organic matter at about 5 percent. Many productive Oklahoma soils contain 1 percent organic matter and, therefore, would hold about 1,000 pounds of nitrogen in this form. Generally, only about one to two percent of this nitrogen is biologically transformed to NO_3^- each year. Thus, organic nitrogen in soils contributes little to crop requirements or ground water contamination.

Residual Nitrates

The soil test for available nitrogen measures NO_3^-. This is one of the predominant forms of nitrogen expected to be found in the soil when crop production does not require all the fertilizer nitrogen applied. The other form of nitrogen which may be a storage place for leftover nitrogen is, of course, soil organic matter.

During June and July, 1985, OSU offered a free soil test to Oklahoma wheat farmers as part of an educational effort to help them improve their enterprises economically. A summary of the NO_3^- test results of this program is shown in Table 10.3. Most of the fields tested (67%) had less than 15 pounds (7.5 ppm) of available nitrogen. This small amount of leftover, or residual, nitrogen, in a majority of the fields tested, is good evidence that Oklahoma wheat farmers are not purposefully using amounts of fertilizer nitrogen in large excess of what the subsequent wheat crop will utilize. Some fields (6 percent) had residual NO_3^--N levels of 50 to 70 pounds per acre. This may have resulted from the application of more fertilizer nitrogen than could realistically be expected to be used by the crop. However, it will also occur when realistic amounts of nitrogen are applied and there is less than normal utilization of nitrogen by the crop due to adverse weather, disease, insects, or weeds.

A final interpretation of the summary in Table 10.3 is that soil testing in 1985 reduced the use of nitrogen fertilizers on wheat by about 29 million pounds of nitrogen. Had the soils not been tested, farmers may have assumed the nitrogen applied to previous crops was completely used by those crops and that the total requirement for the anticipated yield in 1986 would have to be met from fertilizer nitrogen. By using the soil test, they were advised to reduce their fertilizer nitrogen rate according to what was found by the test. This, of course, helps prevent buildup of large excesses of NO_3^- which could potentially reach the ground water as well as reducing the farmers production costs.

SUMMARY

Oklahoma farmers and ranchers apply about 570 million pounds of nitrogen to their fields each year as a part of normal crop production practices. Calculations on the amount of nitrogen required by these crops show that more nitrogen is required by the crop plants than is applied as fertilizer. The difference is apparently accounted for by recycled nitrogen contained in the nonharvested portion (crop residue) of the plants and soil organic matter. Based on these calculations and comparisons, Oklahoma farmers seem to be using nitrogen fertilizers in a prudent manner.

A second evaluation of the risk of ground water contamination from fertilizer use is provided by the result of a recent soil testing campaign. The soil test measures how much NO_3^--N is leftover from the past cropping season. In 1985, 65 percent of over 17,000 wheat fields tested had less than 15 lbs of NO_3^--N per acre (7.5 ppm) left in the soil. The fact that the remaining fields showed higher levels should not be interpreted as misuse of fertilizers or a threat to ground water. High levels of residual nitrogen are expected whenever crop failures due to weather and/or pests occur. Nitrogen determinations from the soil test helps farmers reduce their fertilizer use to avoid unnecessary carryover from one year to the next. Since unused nitrogen is found by the soil test as NO_3^--N and is more concentrated in the surface soils, the conditions

Table 10.3: Available Nitrogen in 1985 Soil Tests for Wheat

lbs N/AC in Test	5	15	30	50	70
Surface* samples (%)	4,726 (27)	6,940 (40)	4,650 (27)	828 (5)	228 (1)
Pounds N** (X 1000)	1,890	8,328	11,160	3,312	1,276
		Total Samples = 17,402			
		Total Surface N = 25,966,000 pounds			
Subsoil*** samples		487	531	284	76
Pounds N** (X 1000)		390	1,274	1,136	426
		Total Subsoil N = 3,226,000 pounds			
		Total N = 29,192,000 pounds			

*0-6 inch depth of soil; lbs N/Ac ÷ 2 = ppm NO_3-N
**Assumes 80 acre fields
***7-24 inch depth of soil; lbs N/Ac ÷ 5.4 = ppm NO_3-N

test as NO_3^--N and is more concentrated in the surface soils, the conditions necessary for transporting it to ground water are either absent or very unfavorable.

These findings support the conclusion that use of fertilizer nitrogen is appropriate with what is required for current crop production levels. Furthermore, unused nitrogen accumulates in the surface soil for lack of suitable conditions to transport it to ground water.

REFERENCES

Hauck, R.D. (ed), Nitrogen in Crop Production, American Society of Agronomy, 1984, Madison, Wisconsin.

Johnson, G.V., "Available Nitrogen and Small Grain Production," Facts No. 2232, 1982, Oklahoma State University Extension, Stillwater, Oklahoma.

Johnson G.V., and Tucker, B.B., "OSU Soil Test Calibrations," Facts No. 2225, 1982, Oklahoma State University Extension, Stillwater, Oklahoma.

Oklahoma State Department of Agriculture, "Tonnage Distribution of Fertilizer in Oklahoma Counties by Grades and Material," Annual Report, 1985, Oklahoma City, Oklahoma.

CHAPTER 11

EFFICIENT NITROGEN FERTILIZATION IN
AGRICULTURAL PRODUCTION SYSTEMS

by

Robert L. Westerman

The United States is the world's largest producer, exporter, and importer of nitrogen (N) fertilizer. The United States produces 18.8 percent of the world total N capacity (~110.5 million metric tons), consumes 17.7 percent, exports 21.2 percent, and imports 16.8 percent (Stangel, 1984). The United States is second in consumption only to the Peoples Republic of China which consumes 19.5 percent of the total world production. The major N consuming areas in the United States are the East North Central and West North Central Regions. States included in the East and West North Central Regions are Illinois, Indiana, Ohio, Michigan, and Wisconsin; and Kansas, Missouri, Nebraska, Iowa, South Dakota, North Dakota, and Minnesota, respectively. Most of the N that is used in these areas is applied to corn (Zea mays L.), wheat (Triticum aestivum L.), grain sorghum (Sorghum bicolor (L.) Moench) and forages. Soybeans (Glycine max (L.) Merr.) are grown in significant hectares in these regions; but because they are a legume, they are not frequently thought to need N fertilizer.

Oklahoma lies in the West South Central Region which includes Texas, Arkansas, and Louisiana. The Southern Great Plains comprises approximately 75 percent of the State of Oklahoma and only excludes the Coastal Plains and the Ozark and Quachita Highlands (Tucker and Murdock, 1984). The climate in Oklahoma varies from semiarid to humid and summers are generally, hot, dry, and windy. Winter temperatures vary greatly resulting in growing seasons of 180 to 220 days. Soils are quite variable with large extremes in pH, texture, and depth. Approximately 90 percent of the area is dryland and in areas where irrigation is available, crop yields are markedly increased because of limited rainfall.

Nitrogen is the fertilizer nutrient needed in the largest quantity for crop production in Oklahoma. Usage of N has ranged from a low of 203,770 metric tons in 1975-76 to a high of 273,520 metric tons in 1979-80 (Figure 11.1). In the most recent years, N usage has declined to 252,870 and 249,260 metric tons for 1983-84, and 1984-85, respectively. Phosphorus fertilizer usage has followed the same general trend as N, but the average quantity of plant food consumed in the past 10 years is only 25 percent of the total (361,490 metric tons) compared to 66 percent and 9 percent, respectively, for N and K_2O.

Nitrogen fertilizer is used mainly for production of wheat, cotton (Gossypium hirsutum L.), grain sorghum, corn, bermuda grass (Cynodon dactylon (L.)), and other nonlegume hay forages. Legumes such as alfalfa (Medicago sativa L.), soybeans, and peanuts (Arachis hypogaea L.) are grown but do not require N fertilization.

Typical N fertilization rates for dryland wheat production range from 75 to 115 kg N ha⁻¹ for grain production and an additional 40 kg N ha⁻¹ if the farmer wishes to graze the wheat pasture plus harvest the grain. Dryland cotton production requires 40 to 75 kg N ha⁻¹ in most areas for optimum production. Crop yields are increased with irrigation, likewise a concomitant increase of N is required since yield times percent composition of N in the plant represents the soil nutrient requirement. This amount of N must be supplied by the soil and/or soil plus fertilizer to obtain the yield goal. Dryland production of grain sorghum requires 20 to 60 kg of fertilizer N ha⁻¹ for normal yields of 2250 to 4500 kg ha⁻¹. Under irrigation where yields may reach 9000 kg ha⁻¹, application of 240 kg N ha⁻¹ may be required. The N requirement for corn is slightly higher than for grain sorghum. Typical N fertilization rates for bermuda grass production ranges from 100 to 200 kg N ha⁻¹ and is generally applied in split applications depending on available soil moisture.

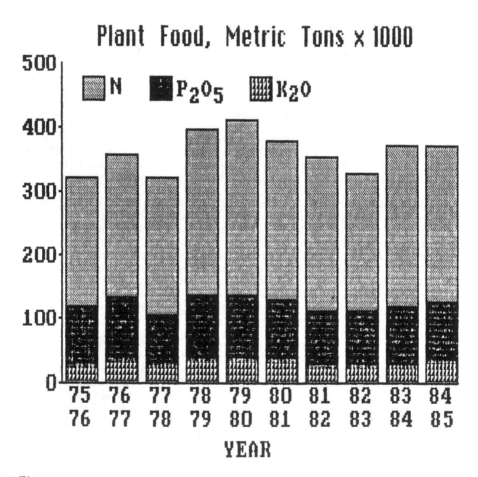

Figure 11.1: Fertilizer N, P_2O_5, and K_2O (thousand metric tons) Consumed in Oklahoma Each Fiscal Year from July 1 to June 30, 1975-76 to 1984-85 (Oklahoma State Department of Agriculture, 1985)

NITROGEN DYNAMICS

The dynamic nature of N reactions in agricultural soil-plant production systems is illustrated in Figure 11.2. Nitrogen input sources for soil nitrate and ammonium include fertilizers, a very small amount of N_2 oxidized during electrical storms that falls with rain, N that is fixed by both symbiotic and nonsymbiotic microorganisms, crop residues, and animal manures. Once either mineral N as fertilizer or N organically bound in crop residue or manures is added to soil, a series of dynamic reactions occur.

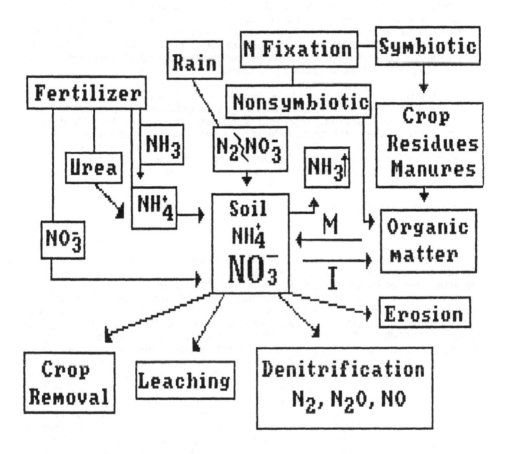

Figure 11.2: Nitrogen Dynamics in Agricultural Soil-plant Production Systems

Since plants take up N in soil, both as ammonium (NH_4^+) and nitrate (NO_3^-), these are the two mineral forms that are of most interest. Generally, NO_3^- is taken up in larger quantities than NH_4^+ by most plants because it is present in larger quantities in aerobic systems than NH_4^+ and is mobile. Two major soil transformations (mineralization and immobilization) determine the net quantity of NH_4^+ and NO_3^-. Mineralization (M) is defined as the conversion of organic N to mineral N, whereas immobilization (I) is the conversion of mineral N to organic N. Both of these transformation processes occur simultaneously in soil and the net effect is dependent on the organic C:N ratio, temperature, moisture, pH, and available oxygen supply. The addition of crop residue to soils favors immobilization until the organic C source for heterotrophic microorganisms is depleted. Once the organic C:N ratio has been narrowed and there is not enough C for consumption by the heterotrophic microorganisms, the population of these organisms declines with a subsequent release of N to mineral forms. Generally, if the organic C:N ratio is >40:1, the net effect is considered to be immobilization whereas if the organic C:N ratio is <20:1, the net effect is considered to be mineralization. In addition to supplying N requirements for the crop, some of the mineral N supplied as fertilizer is used by microorganisms in decomposing processes.

Crop removal of NH_4^+ and NO_3^-, leaching of NO_3^-, erosion of soil containing both organic and mineral N, and gaseous loss processes such as denitrification and ammonia volatilization, represent significant losses from the soil system. Increasing N uptake in the crop, reducing the potential for leaching, minimizing gaseous loss of N as volatilization and denitrification, and controlling soil erosion would lead to more efficient use of fertilizer N and minimize the potential for NO_3^- contamination of ground water and streams.

NITROGEN MANAGEMENT

Soil NO_3^--N Test

The best management tool that the producer has available for making N fertilizer recommendations is the soil NO_3^--N test. Although this test is a measurement of concentration of the mobile NO_3^- ion in soil at a specific point in time, the test is remarkably accurate in predicting response to applied fertilizer N. Data in Table 11.1 was collected from an experiment that was conducted on a farmer's wheat field west of Cordell, Oklahoma, in the cropping year of 1979-80. The farmer collected soil samples from his field and submitted them to the Soil, Water, Plant Analysis Laboratory at Oklahoma State University for routine analysis (pH, NO_3^--N, P, and K). Analyses showed that 116 kg N ha^{-1} was present in his soil which was enough to produce approximately 3,000 kg ha^{-1} of grain. Based on the potential yield goal in that area, no N fertilizer was recommended. The farmer was unsure of our recommendation and resampled his field for further analyses. Again, the field tested high in soil NO_3^--N and no N fertilizer was recommended. In order to further verify the validity of soil NO_3^- test for N fertilization recommendations, a field experiment was conducted that consisted of multiple rates of N replicated four times in a randomized complete block design. There was no statistically significant response in grain yield due to N fertilization. Grain yield and N concentration in wheat forage was not increased significantly statistically and only the high N rate increased the N concentration of the grain. Numerous other experiments (data not reported) have shown the lack of crop response to N fertilization when the soil NO_3^--N test was high. Conversely, data in Tables 11.2 and 11.3 show marked responses in grain yield and N uptake in wheat and yield of grain sorghum when the soil NO_3^--N tests were low and N fertilizer was applied.

Table 11.1: The Effect of Soil NO_3^--N on Yield and N Concentration in Winter Wheat, Cordell, Oklahoma (Tally, 1981)

Soil Test		Grain	N Concentration	
NO_3^--N	N Rate	Yield	Grain	Forage
----------------kg ha^{-1}-----------------			-------mg g^{-1}-------	
116	0	2910	24.7	52.5
	28	3110	25.0	52.4
	56	3170	25.0	50.3
	112	3190	26.4	53.3
LSD* (0.05)		419	1.6	4.7

*Least significant difference

The soil NO_3^--N tests provide the basis for judicious use of N fertilizer while supplying adequate but not excessive amounts of N for the crop yet minimizing the potential for NO_3^- contamination of ground water and streams.

Table 11.2: The Effect of Soil NO_3^--N on Yield and N Uptake in Winter Wheat, Altus, Oklahoma, 1982 (Westerman and Edlund, 1985)

Soil Test		Grain	N
NO_3^--N	N Rate	Yield	Uptake
-----------------------------kg ha^{-1}-----------------------------			
14	0	2320	45
	112	3300*	88*

*Indicates significance at P = 0.001 level of probability using nonorthogonal single df contrasts.

Inhibitors

Nitrogen in the form of NO_3^- in the soil is mobile and moves with water as it passes through the soil. However, N in the form of NH_4^+ is immobile in soil due to the net negative charge on the clay particles. In sands without a significant exchange capacity, NH_4^+ as well as other cations may become relatively mobile.

The process whereby NH_4^+ is oxidized to NO_3^- is called nitrification and is illustrated as follows:

Table 11.3:   The Effect of Soil NO_3^--N on Yield of Grain Sorghum at Tipton, Oklahoma, 1977, and Altus, Oklahoma, 1979 (Westerman, Edlund, and Minter, 1981)

Soil Test NO_3^--N		N Rate	Yield	
Tipton	Altus		Tipton	Altus
--kg ha^{-1}-------------------------------------				
11	34	0	2380	3690
		67	3630	4590
		134	3760	4960
		201	2860	4250
FLSD* (0.05)			814	651

*Fisher's least significant difference

$$NH_4^+ + 3/2\ O_2 \xrightarrow[6\ e^-]{Nitrosomonas} NO_2^- + H_2O + 2H^+ + 66\ Kcal$$

$$NO_2^- + \tfrac{1}{2}\ O_2 \xrightarrow[2\ e^-]{Nitrobacter} NO_3^- + 17\ Kcal$$

This is a naturally occurring oxidizing process in soils and is accomplished by Nitrosomonas and Nitrobacter bacteria. Although there are other microorganisms in soil that may oxidize NH_4^+, Nitrosomonas and Nitrobacter are considered to be the most significant genera. Nitrification requires available oxygen and is an autotrophic process.

The use of nitrification inhibitors such as those listed in Table 11.4 may be used to keep fertilizer N in the NH_4^+ form longer, thus reducing the potential for NO_3^- leaching.

The use of nitrification inhibitors in more humid areas of the United States, and in areas under irrigation, has had some success in improving N utilization by the crop. However, in dryland agriculture where moisture often limits yields, the use of nitrification inhibitors have met with limited success.

In Oklahoma numerous experiments have been conducted to determine if nitrification inhibitors have the potential to improve yields by minimizing NO_3^- leaching. Nitrapyrin and etradiazole had no effect on the yield of winter wheat grown in a wide variety of soils in north central and northwestern Oklahoma (Table 11.5) even though statistically significant increases in yield due to N fertilization were observed at three of the four locations. Numerous other experiments have also shown no effect due to inhibitor application (Westerman and Edlund, 1985).

Table 11.4:   Nitrification and Urease Inhibitors to Suppress Nitrification

---

1.   Nitrification $(NH_4^+ -- X -- NO_2^- \longrightarrow NO_3^-)$

   Inhibitors

   a.   Nitrapyrin--(2-chloro, 6-(trichloromethyl) pyridine)

   b.   Etradiazole--(5-ethoxy, 3-(trichloromethyl)-- 1,2,4 thiazole)

   c.   Dicyandiamide (DCD)

   d.   Thiourea

2.   Urease $(NH_2CONH_2 ---- X \longrightarrow 2NH_3 + CO_2 + H_2O)$

   Inhibitors

   a.   Phenylphosphorodiamidate (PPDA)

   b.   Thiourea

   Yield of grain sorghum on a sandy soil at Tipton and a clay loam soil with limited supplemental irrigation at Altus was not improved by using nitrification inhibitors (Table 11.6). However, yields were increased with N fertilization at both locations in the three years of the study.

   Fertilizers containing urea are decomposed in soil enzymatically to CO_2 and NH_3 as follows:

$$NH_2CONH_2 + 2H_2O \xrightarrow{\text{urease}} CO_2 + 2NH_3 + H_2O$$

The use of urease inhibitors (Table 11.4) with urea fertilizers blocks the formation of ammonia and offers the potential to minimize loss of NH_3 by volatilization.

   The use of urease inhibitors is relatively new compared to the use of nitrification inhibitors and little field data regarding their effectiveness are available. However, losses of N by NH_3 volatilization under climatic conditions in Oklahoma are thought to be minimal, except in cases where soil pH and temperatures are high and urea is surface applied to these areas. Losses are minimized when urea is incorporated into the soil.

Table 11.5:   Effect of Nitrification Inhibitors on Grain Yield of Winter Wheat
(Westerman, unpublished data)

N rate	Inhibitor*	Yield, kg ha^{-1}			
kg ha^{-1}		Helena	Pond Creek	Orienta (W)	Orienta (E)
0	0	1120	2420	3520	1650
56	0	1600	2980	3880	1900
56	N	1450	3170	3870	2020
56	E	1550	2910	3980	1960
112	0	1700	3450	4030	1990
112	N	1610	3360	3910	2130
112	E	1410	3410	3750	1880
LSD**(0.05)		420	550	370	260

*N-nitrapyrin and E-etradiazole applied (0.56 kg ha^{-1}) with
anhydrous ammonia.
**Least significant difference

Nitrogen Use Efficiency

    Efficient use of N in agricultural production systems can be viewed from
agronomic, economic, and environmental perspectives. The overall objective of
fertilizer N research is to maximize the efficiency of plant use of applied N.
Any increase in this efficiency will increase the agronomic and economic value
of the fertilizer as a means of increasing crop production, conserve energy and
the raw materials needed to make N fertilizers, and minimize possible adverse
effects on the environment that may result from inefficient N use (Bremner and
Hauck, 1974).

    The most common definitions of N fertilizer efficiency are based on plant
uptake of N, expressed either as the amount of fertilizer N in the entire plant
or in the harvestable crop components, or as the percentage recovery of applied
N (Hauck, 1984).

    The difference method (Westerman and Kurtz, 1974) is widely used to
estimate N recovery in the crop. It is calculated as follows:

$$((TN_i - TN_0/(N_i \text{ Rate})) \times 100 = \% \text{ Fertilizer N Recovery}$$

where TN_i is the total N in the crop at N fertilizer rate i, TN_0 is total N in the
crop in the unfertilized plot, and N_i is the rate of N fertilizer applied. Since N
reactions in soils are dynamic and plants take up N from both soil and fertilizer
sources, the difference method for estimating N recovery in the crop is not

without error. A more accurate method for determining fertilizer N recovery is the ^{15}N method, although it is much more expensive and time consuming. Fertilizer N labeled with the heavy isotope, ^{15}N, is added to soil and uptake of ^{15}N in the crop is measured. Fertilizer N recovery by the ^{15}N method is calculated as follows (Westerman, Kurtz, and Hauck, 1972):

$$\left[TN_i \; (A\% \; ^{15}N_{TN_i} - A\% \; ^{15}N_{TN_0})/(\text{excess } A\% \; ^{15}N_{N_i}) \right] \times 100 =$$

% Fertilizer N Recovery

where TN_i is the total N in the crop at the N_i rate, $A \% \; ^{15}N_{TN_i}$ is the atom percent ^{15}N in the unfertilized crop and excess $A \% \; ^{15}N_{N_i}$ is the atom percent ^{15}N in the fertilizer source minus normal abundance atom percent ^{15}N. In addition to determining fertilizer N recovery by the crop, the ^{15}N method is used to trace N atoms derived from fertilizer through the soil-water-crop production system. This method also allows the partitioning of native soil N atoms from fertilizer N atoms in the many complex N transformations in the soil-water-plant system.

Table 11.6:   Effect of Nitrification Inhibitors on Yield of Grain Sorghum (Westerman, Edlund, and Minter, 1981)

Nitrogen[*]		Inhibitor[**]		Yield, kg/ha		
Source	Rate	Source	Rate	Tipton	Altus	
	kg/ha		kg/ha	1977	1978	1979
	0		0	2380	3940	3960
UAN	67		0	3630	5340	4590
UAN	67	N	0.56	3480	5200	5110
UAN	67	E	0.56	3000	4890	4910
UAN	134		0	3760	4760	4960
UAN	134	N	0.56	2700	5330	4710
UAN	134	E	0.56	3400	5340	4260
UAN	201		0	2860	5700	4250
UAN	201	N	0.56	2880	5140	5130
UAN	201	E	0.56	2640	4980	4910
FLSD[***](0.10)				810	800	650

[*]UAN--urea ammonium nitrate $[(NH_2)_2CO \cdot NH_4NO_3]$
[**]N-nitrapyrin and E-etradiazole applied (0.56 kg ha^{-1}) with UAN
[***]Fisher's least significant difference

Due to the expense involved in conducting ^{15}N experiments, most estimates of nitrogen use efficiency (NUE) are made using the difference method.

The use of nitrapyrin and etradiazole nitrification inhibitors did not increase NUE significantly in winter wheat production (Tables 11.7 and 11.8). The mineralizable N in soil at Haskell was high and no yield response to applied N was observed; therefore, NUE was exceptionally low.

In bermuda grass forage experiments, slight differences in NUE were observed due to nitrogen sources (Table 11.9). Lower NUE for anhydrous ammonia and anhydrous ammonia-cold-flo than with other N sources were due to the method in which anhydrous ammonia was applied to sod. In sod-type crops, anhydrous ammonia was applied behind thin shanked knives with 45 cm spacings. When soils are dry, it is difficult to cover the soil behind the knives, which generally results in lower NUE. Generally, there is little difference in NUE when nitrogen sources are applied properly.

Table 11.7: Effect of Nitrification Inhibitors on Nitrogen Use Efficiency in Winter Wheat (Westerman, unpublished data)

N rate	Inhibitor*	% Nitrogen Use Efficiency**			
kg ha^{-1}		Helena	Pond Creek	Orienta (W)	Orienta (E)
0	0				
56	0	25	18	40	13
56	N	22	28	15	21
56	E	26	14	43	21
112	0	18	20	24	11
112	N	16	23	22	19
112	E	10	23	18	11

*N-nitrapyrin and E-etradiazole applied (0.56 kg ha^{-1}) with anhydrous ammonia.

**NUE based on N removal in the grain.

Split N Applications

Splitting N applications offers the potential to improve fertilizer N use in areas where significant losses of N occur by leaching, erosion, or gaseous loss processes. The effect of split N applications on N uptake in bermuda grass is shown in Figure 11.3. Split N applications did not improve N uptake in bermuda grass studies conducted over a three-year period at Haskell (Westerman and McMurphy, 1986). However, in forage production there is considerable advantage in splitting N applications because farmers can hedge on the favorableness of the growing season, since moisture is not often a limiting factor for high yields.

Table 11.8:   Nitrapyrin Effect on Nitrogen Use Efficiency in Winter Wheat (Westerman and Edlund, 1985)

Year	% Nitrogen Use Efficiency*			
	Altus		Haskell	
N rate, kg ha^{-1}	112	112 + N	112	112 + N
1980	---	---	21	25
1981	37	38	---	---
1982	35	37	5	10
1983	27	23	6	7

*N - nitrapyrin applied (0.56 kg ha^{-1}) with anhydrous ammonia.
**NUE based on N removal in the grain.

Table 11.9:   Percent Nitrogen Use Efficiencies of N Sources by Bermuda Grass at the Eastern Research Station at Haskell, Oklahoma (Westerman et al., 1983).

N Source*	1978	1979	1980
AA	33.6 ab**	19.3 a	35.2 ab
ACF	26.0 a	19.6 a	28.5 a
UAN	38.1 bc	35.7 b	44.9 b
UREA	30.7 ab	36.6 b	32.1 a
AS	43.0 c	48.3 c	37.2 ab

Means in columns followed by the same letter are not significantly different by Duncan's New Multiple Range Test at P = 0.05.

*AA - anhydrous ammonia, NH_3; ACF - anhydrous ammonia Cold-Flo, UAN --urea ammonium nitrate, $(NH_2)_2CO \cdot NH_4 NO_3$; UREA - $(NH_2)_2CO$; AS - ammonium sulfate. $(NH_4)_2SO_4$.

**Each value is the mean of 16 observations averaged across replications and N rates within sources. NUE was based on N removal in forage.

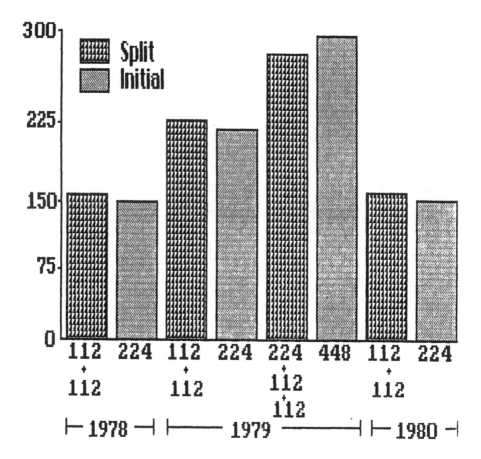

Figure 11.3:   Split N Effects on N Uptake in Bermuda Grass in kg ha[-1] at the
Eastern Research Station at Haskell, Oklahoma (Westerman and
McMurphy, 1986)

Split application of N has not increased grain yield of winter wheat (data
not reported) but has some advantages in allowing producers to hedge on
environmental factors.

Placement

The method of fertilizer placement may offer some advantages in
increasing yield and N uptake in the crop. Two main types of placement with
respect to N applications in Oklahoma are broadcast on the surface and
subsurface placement (~15 cm) behind thin shanked knives spaced at
approximately 45 cm or placed (~10 cm) under the V-blade during a tillage
operation.   Anhydrous ammonia is the major N source sold in Oklahoma;
therefore, most of the N applied to winter wheat is placed below the surface of

the soil. There were no differences in grain yield of winter wheat due to the method of N placement in the eight observation years reported in Table 11.10. This is typical for most observations; however, in some years with varying rainfall, temperatures, etc., slight advantages due to placement of N by one method or the other may be observed.

Table 11.10: Effect of Broadcast vs Deep Placement of N on Yield of Winter Wheat (Westerman and Edlund, 1985)

Method**	N rate kg ha^{-1}	Grain Yield, kg ha^{-1}*			
		1980	1981	1982	1983
			Altus		
	0	1250	2330	2320	2300
Broadcast	112	2140	3200	3510	3020
Deep Placement	112	1980	3340	3120	2910
			Stillwater		
	0	1930	1740	2380	1240
Broadcast	112	3020	2710	2320	1640
Deep Placement	112	2860	2750	2120	1380

*No statistically significant differences in yield between broadcast and deep placement applications of N using nonorthogonal single degree of freedom contrasts at P = 0.05 level of probability.

**N sources for broadcast and deep placement applications were NH_4NO_3 and NH_3, respectively.

Soil Erosion

The loss of topsoil by sheet, rill, and wind erosion is a contributing factor to the deterioration of water quality, and the sedimentation of lakes, reservoirs, and recreational waters. Soil nutrients are also lost when erosion takes place and NO_3^- is the ion of major concern.

Sediment and nutrient discharges in runoff vary greatly, depending on such factors as topography, soil-plant management practices, infiltration rates, and storm patterns (Legg and Meisinger, 1982). Menzel et al., (1978) pointed out that under Oklahoma conditions, long-term records are needed to compare nutrient and sediment losses due to large yearly variations in storm patterns. The maximum annual discharges were 13 kg ha^{-1} of total N and 4 kg ha^{-1} of NO_3^--N, with average values about 50 percent of the maximum. The annual deposition of N in rainfall averaged 5 kg ha^{-1} and thus exceeded the maximum discharge of NO_3^--N runoff.

Although some fertilizer N applied to soil may be lost with sediments and runoff, it is also a deterrent to soil erosion. Proper N fertilization allows for rapid crop growth and canopy development, resulting in increased yield which produces more plant residue thereby reducing runoff.

SUMMARY

Nitrogen usage in Oklahoma declined to 250,000 metric tons last year from the high of 274,000 metric tons observed in 1978-1979. Most of the N utilized was anhydrous ammonia applied to winter wheat and was deep placed in soil at the time of application. Losses due to erosion or runoff were minimal.

Nitrogen in soil is in constant turnover between organic and inorganic N forms. Plants take up both NH_4^+ and NO_3^- but, since NH_4^+ is oxidized rapidly to NO_3^- in soils, NO_3^- is taken up in larger quantities. Nitrogen in the form of NO_3^- is mobile and has the potential to leach. Extensive research utilizing nitrification inhibitors failed to increase yields or N uptake in crops. Therefore, it was concluded that leaching of fertilizer NO_3^- is not a major problem under climatic and rainfall conditions in most soils in Oklahoma. In addition, yields are often reduced because of lack of rainfall and significant quantities of water do not pass through the soil profile.

Split N applications and the method of fertilizer placement have not been shown to increase yield or nitrogen use efficiency in most agricultural production systems in Oklahoma. However, there may be some economic advantages to split applications of N to allow hedging on climatic factors which may affect yield.

The most effective management tool for minimizing the potential for NO_3^- contamination of ground water and streams is the use of the soil NO_3^--N test as a guide for N fertilizer recommendations.

The primary objective of most soil fertility researchers is to supply the crop with adequate nutrients to produce maximum economic yields, yet not in excess to minimize the potential for NO_3^- contamination of ground water and streams. Since N usage is increasing annually nationwide, research efforts to maximize N uptake in crops need to be intensified.

SELECTED REFERENCES

Bremner, J.M., and Hauck, R.D., "Perspectives in Soil and Fertilizer Nitrogen Research," Transactions of the 10th International Congress of Soil Scientists (Moscow), Vol. 9, 1974, pp 13-27.

Hauck, R.D., "Technological Approaches to Improving the Efficiency of Nitrogen Fertilizer Use by Crop Plants," Nitrogen in Crop Production, R.D. Hauck et al., (ed), American Society of Agronomy, Crop Science Society, and Soil Science Society of America, Madison, Wisconsin, 1984, pp 551-560.

Legg, J.O., and Meisinger, J.J., "Soil Nitrogen Budgets," Agronomy Journal, Vol. 22, 1982, pp 503-566.

Menzel, R.G. et al., "Variability of Annual Nutrient and Sediment Discharges in Runoff from Oklahoma Cropland and Rangeland," Journal of Environmental Quality, Vol. 7, 1978, pp 401-406.

Oklahoma State Department of Agriculture, "Tonnage Distribution of Fertilizer in Oklahoma Counties by Grades and Material," Annual Report (July 1, 1984 to July 1, 1985), Oklahoma City, Oklahoma.

Stangel, P.J., "World Nitrogen Situation--Trends, Outlook and Requirements," Nitrogen in Crop Production, R.D. Hauck et al., (ed), American Society of Agronomy, Crop Science Society of America, and Soil Science Society of America, Madison, Wisconsin, 1984, pp 23-54.

Tally, S.M., "A Study of Some Problems Regarding Soil Sampling, Soil Testing, and Fertilizer Recommendations in Oklahoma," M.S. Thesis, 1981, Oklahoma State University, Stillwater, Oklahoma.

Tucker, B.B., and Murdock, L.W., "Nitrogen Use in the South Central States," Nitrogen in Crop Production, R.D. Hauck et al., (ed), American Society of Agronomy, Crop Science Society of Agronomy, and Soil Science Society of America, Madison, Wisconsin, 1984, pp 735-749.

Westerman, R.L., and Edlund, M.G., "Deep Placement Effects of Nitrogen and Phosphorus on Grain Yield, Nutrient Uptake, and Forage Quality of Winter Wheat," Agronomy Journal, Vol. 77, 1985, pp 803-809.

Westerman, R.L., Edlund, M.G., and Minter, D.L., "Nitrapyrin and Etradiazole Effects on Nitrification and Grain Sorghum Production," Agronomy Journal, Vol. 73, 1981, pp 697-702.

Westerman, R.L., and Kurtz, L.T., "Isotopic and Non-isotopic Estimation of Fertilizer Nitrogen Uptake by Sudangrass in Field Experiments, "Soil Science Society of America Proceedings, Vol. 38, 1974, pp 107-109.

Westerman, R.L., Kurtz, L.T., and Hauck, R.D., "Recovery of ^{15}N-labeled Fertilizers in Field Experiments," Soil Science Society of America Proceedings, Vol. 36, 1972, pp 83-86.

Westerman, R.L., and McMurphy, W.E., "Nitrogen Boosts Forage/Animal Gains," Journal of Fertilizer Solutions, Vol. 30, No. 3, 1986, pp 53-54.

Westerman, R.L. et al., "Nitrogen Fertilizer Efficiency in Bermudagrass Production," Soil Science Society of America Journal, Vol. 47, 1983, pp 810-816.

CHAPTER 12

NITRATES AND PESTICIDES IN GROUND WATER:
AN ANALYSIS OF A COMPUTER-BASED LITERATURE SEARCH

by

L.W. Canter

Agricultural applications of fertilizers and pesticides are being increasingly recognized as significant sources of ground water pollution. Within the last decade there have been numerous studies and publications related to nitrate and pesticide contamination of ground water, and this chapter summarizes an analysis of a pertinent computer-based literature search. Sections are included on the search mechanics, nitrates in ground water, and pesticides in ground water.

COMPUTER-BASED LITERATURE SEARCH

A computer-based search of the literature was made by using the DIALOG System operated by Lockheed Corporation in Palo Alto, California. Ten data bases in DIALOG were used, and brief descriptions of each are in Table 12.1. Key descriptor words used in the search were pesticides, nutrients, herbicides, insecticides, fertilizers, and ground water.

A total of 1,317 abstracts were procured from the data bases. Each of the procured abstracts were read, and duplications were eliminated, along with abstracts deemed to be only minimally relevant to this chapter. Based on this review, 63 abstracts were selected for summarization in this chapter. Nitrates were addressed in 34 abstracts, and pesticides in 29 abstracts.

NITRATES IN GROUND WATER

A total of 34 references related to nitrates in ground water from the agricultural usage of fertilizers were selected, and they can be considered in terms of case studies, transport and fate, and management practices and mathematical models.

Case Studies

Table 12.2 contains brief comments on 20 references describing case studies of nitrate contamination of ground water. Seven cases are in Nebraska, and one each is in Arizona, California, Connecticut, Delaware, Iowa, Minnesota, Mississippi, New York, North Carolina, Ohio, Texas, Washington, and Wisconsin.

Nitrate concentrations in various Nebraska aquifer systems are described by Boyce et al. (1976), Exner and Spalding (1979), Gormly and Spalding (1979), Olsen (1974), Piskin (1973), Spalding (1975), and Spalding et al. (1978). Some

Table 12.1:   Descriptions of the Data Bases Utilized in the Search

1. <u>Agricola</u>.   Years   covered   1970-1978.   Contains   all   the information   of   the   National   Agricultural   Library   and comprehensive   coverage   of   worldwide   journals   and   monographic literature on agricultural subjects.

2. <u>Agricola</u>.   Years   covered   1979-1984.   See   above.   The   two Agricola data bases contain 2 million records.

3. <u>Compendex</u>.   Years   covered   1970-1984.   Contains   the   machine-readable   version   of   the   Engineering   Index   which   includes engineering and technological literature from 3,500 journals and selected government reports and books.   Size of data base--1.4 million records.

4. <u>Conference Papers Index</u>.   Years covered 1973-1984.   Contains records of more than 100,000 scientific and technical papers presented   at   over   1,000   major   regional,   national,   and international   meetings   each   year.   Size   of   data   base--1.0 million records.

5. <u>CRIS/USDA</u>.   Years   covered   1982-1984.   Contains   active   and recently completed agricultural research sponsored by the USDA or state agriculture institutions.   Size of data base--31,000 records.

6. <u>Dissertation   Abstracts</u>.   Years   covered   1861-1984.   Contains dissertations from U.S. institutions as well as Canada and some foreign schools.   Most of the abstracts are for degrees granted after 1980.   Size of data base--852,000 records.

7. <u>NTIS</u>.   Years covered 1964-1984.   Contains government sponsored research,   development   and   engineering   plus   analyses   prepared   by federal agencies, their contractors and grantees.   Unclassified, publically available reports are available.   Size of the data base--1.1 million records.

8. <u>Pollution   Abstracts</u>.   Years   covered   1970-1984.   Contains environmentally related literature on pollution, its sources and control.   Size of the data base--107,000 records.

9. <u>SSIE   Current   Research</u>.   Years   covered   1978-1982.   Contains reports   of   both   government   and   privately   funded   scientific research projects either in progress or recently completed in all fields of basic and applied research in the life, physical, social, and engineering sciences.   Size of data base--439,000 records.

10. <u>Water Resources Abstracts</u>.   Years covered 1968-1984.   Contains materials   collected   from   over   50   water   research   centers   and institutes in the United States and focuses on water planning, the water cycle, and water quality.   Size of data base--173,000 records.

Table 12.2:   Summary of References on Case Studies of Fertilizer-derived
              Nitrates in Ground Water

Author (Year)	Comments
Baier and Rykbost (1976)	Nitrates in Long Island, New York ground water due to potato fertilization.
Bingham, Davis, and Shade (1971)	Nitrate leaching losses from a citrus area in southern California.
Boyce et al. (1976)	Nitrates in ground water from irrigation areas in southwestern and central Nebraska.
Chichester (1976)	Influence of type of fertilizer and crop management practices on ground water nitrates in east-central Ohio.
De Roo (1980)	Nitrates in ground water at experimental study areas in Connecticut.
Exner and Spalding (1979)	Nitrates in Nebraska ground water from fertilization and irrigation of sandy soils.
Gast and Goodrich (1973)	Agricultural contribution of nitrates in Minnesota ground waters.
Gilliam, Daniels, and Lutz (1974)	Nitrates and ammonia nitrogen in shallow North Carolina coastal plain ground water.
Gormly and Spalding (1979)	Sources and concentrations of nitrates in Nebraska ground water.
McDonald and Splinter (1982)	Nitrates in Iowa ground water from anhydrous ammonia applications.
Olsen (1974)	Influence of fertilizer practices and irrigation on nitrates in ground water in Nebraska.
Piskin (1973)	Influence of unsaturated zone material on nitrate transport to Nebraska ground water.
Reeves and Miller (1978)	Nitrates in the Ogallala Aquifer in a 27-county area of west Texas.
Ritter and Chirnside (1984)	Land use influences on nitrates in ground water in southern Delaware.
Silver and Fielden (1980)	Potential sources of ground water nitrates in Arizona.
Spalding (1975)	Nitrates in ground water in Hall County, Nebraska.

Table 12.2:   (continued)

Author (Year)	Comments
Spalding et al. (1978)	Correlation of irrigated coarse-textured soils and nitrates in Nebraska ground water.
Spalding et al. (1982)	Sources of ground water nitrates in Washington.
Tryon (1976)	Influence of agricultural practices and karst areas on nitrates in Missouri ground water.
Wehking (1973)	Nitrates variations with well depths in Wisconsin.

nitrates may result from large quantities of geologic nitrate within the deep loess mantle in southwestern and central portions of the state (Boyce et al., 1976). Average nitrate concentrations in the ground water of Nebraska increased 25 percent in the decade spanned by the late 1960's and early 1970's (Olsen, 1974). The highest increases were found in several areas with intensive irrigation of sandy soils. During 1976-77, nitrate-nitrogen concentrations exceeded 10 mg/l in 183 of the 256 ground water samples collected from parts of Buffalo, Hall, and Merrick Counties in Nebraska (Gormly and Spalding, 1979). Comparison of the isotopic values with those of potential nitrate sources suggested that the primary source of contamination in most wells was fertilizer and that a small percentage of the wells contained significant concentrations of nitrate-nitrogen derived from animal wastes.

Field investigations in Paradise Valley, Arizona, conducted during 1974 and 1977, delineated areas of ground water with up to 132 mg/l nitrate (Silver and Fielden, 1980). Two alternative interpretations were developed as to possible sources of the excess nitrate. The first was a conventional interpretation identifying the use of nitrogenous fertilizers as the primary source and disposal of treated wastewater effluent as a secondary source. An alternative interpretation identified the source as a sand and gravel unit that is interpreted as a braided-stream deposit, located about 152 m (500 ft) below the land surface. The source of the nitrate may have been ammonium chloride leached from turfs in the adjacent Superstition Mountains, subsequently oxidized to nitrate, and deposited in abandoned channels of the braided-stream complex.

Bingham, Davis, and Shade (1971) reported on a study of irrigation management of a 960-acre mature orange grove in the inland region of southern California. The nitrogen requirements of the trees were determined, and nitrate effluent waters ranged up to 87 mg/l nitrate and averaged 50 to 60 mg/l nitrate; this nitrate loss is about 45 percent of the nitrogen applied each year as fertilizer. The possibility of drainage waters from cultivated and fertilized fields percolating into ground water basins was discussed. It is of interest to note that Exner and Spalding (1979) suggested that 50 percent of the applied nitrogen fertilizer infiltrates to the ground water in Holt County, Nebraska.

Farmland contributions to nitrate pollution of ground water in the Connecticut River Valley were studied by De Roo (1980). Ground water under the Valley Laboratory Farm averaged 3 mg/l nitrate nitrogen over three years, with temporary concentrations exceeding 10 mg/l after a heavy rainfall, as observed downstream from areas treated with generous amounts of fertilizer. On the Merrell Farm in Suffield, a rural environment, there were year-round high levels of nitrate nitrogen, averaging 20 mg/l.

Increases in nitrate concentrations in surface waters and shallow wells were found in long-term water quality studies in Iowa (McDonald and Splinter, 1982). High nitrate (more than 45 mg/l) values were common during periods of relatively low river flow in the fall and winter. Nitrate infiltration of shallow aquifers was also evident from shallow well samples. The increased levels of nitrate were not unexpected because of the steady increase in the use of anhydrous ammonia in Iowa. Ground water from deeper wells had much lower nitrate levels than ground water from shallow wells. Nitrate contamination has not yet reached the deeper aquifers.

Information from 675 water wells in and around Phelps County, Missouri, shows that discrete areas of differing ground water quality can be identified and mapped (Tryon, 1976). The best quality ground water, as judged by its low nitrate content and coliform bacteria density, is found in areas of relatively little agricultural (pasture or livestock) land use. The adverse effect of agricultural land use on ground water quality is more severe in the intensely developed karst areas than in the less intensely developed ones. Nitrate concentrations in ground water varies seasonally, and in response to rainfall, and decreases with increasing well depth.

The eastern portion of Long Island, New York, supports a productive agricultural industry whose main crop is potatoes (Baier and Rykbost, 1976). Ground water surveys have shown that the aquifer system of this area is contaminated with nitrate nitrogen. The average potato grower applies 200 to 250 lb N/a (224 to 278 kg N/ha) at planting time; and, depending upon a number of factors, the N recovered in harvested tubers varies from 75 to 150 lb N/a (83 to 167 kg N/ha). Losses to the ground water could vary from 50 to 175 lb N/a (55.5 to 194 kg N/ha). It should be noted that annual N losses of 50 lbs N per acre (55.5 kg N/ha) are sufficient to cause a concentration in the aquifer's surface layer of 10 mg/l nitrate-N.

The nitrate-nitrogen and ammonia-nitrogen concentrations in shallow (less than 3 m) ground water in the North Carolina coastal plain under a range of soil types, drainage conditions, and types of crop grown were monitored by Gilliam, Daniels, and Lutz (1974). The nitrate-nitrogen levels were always low (1 mg/l or less) in ground water under wooded areas, but were somewhat higher under cultivated fields, usually 1 to 5 mg/l, although several values of 10 to 20 mg/l were recorded. Concentrations in all wells were always higher during winter months with no relationship between cultivated crop and nitrate-nitrogen in the ground water. The nitrate-nitrogen concentration was almost always higher in the middle of the field than on the edge even though the direction of water flow was toward the edge. Denitrification may be responsible. There was little difference in ammonia-nitrogen concentrations in water under cultivated fields and under unfertilized woods or pasture, with concentrations normally in the range of 0.1 to 1 mg/l, with the higher levels being found under poorly drained soils. Based upon the characteristics of the surface sediments, it is concluded that very little of the nitrate-nitrogen present in the shallow ground water moves into deep aquifers in any area of the North Carolina Coastal Plain. However, the amount of nitrate-nitrogen that moves through the surficial sediments to the streams probably varies with location and characteristics of the confining beds.

The distribution of nitrate, chloride and dissolved solids from ground water from the Ogallala Aquifer in a 27-county area of west Texas illustrates widespread areas of poor water quality (Reeves and Miller, 1978). Elements studied increased in concentration from northwest to southeast across the Southern High Plains. Northeast of a line from about Clovis, New Mexico, to Lubbock, Texas, nitrate tends to be less than 45 mg/l; chloride is less than 20 mg/l, and dissolved solids are less than 400 mg/l. However, southwest of this line, nitrate concentrations may exceed 60 mg/l and, in some areas, they exceed 170 mg/l. Chloride commonly exceeds 500 mg/l and may be more than 2000 mg/l, and dissolved solids usually exceed 1000 mg/l and may be more than 8000 mg/l. Regional distribution may be the result of long-term migration of Ogallala ground water, but the present water quality and distribution, as well as time, distance, stratigraphy, and permeability, suggest contamination of Ogallala ground water by vertical rather than lateral migration. Most of the high nitrate values (more than 45 mg/l) occur in areas having sandy soils which have been intensively cultivated, thus leaching of nitrogen-based fertilizers is suspect. However, the high chloride and dissolved solids, which exist in essentially the same geographic area, probably represent vertical to local lateral seepage of saline water from large alkali lake basins and local vertical migration from saline cretaceous aquifers.

Transport and Fate

A number of factors influence the movement of nitrogen from applied fertilizers to ground water systems. For example, heavy rainfalls occurring shortly after fertilizer application can greatly increase losses of both nitrogen and phosphorus through a subsurface drainage system (Bottcher, Monke, and Huggins, 1981). Additional influencing factors include soil and crop types, and fertilizer application practices.

Thomas (1972) reported on a study wherein soils from several areas in Kentucky were placed in columns and leached with calcium nitrate. Subsoils high in iron oxide were found to significantly retard the leaching of nitrate. In other soils, the nitrate moved through as fast or slightly faster than the water. Field application of nitrogen to corn was most efficient in the spring or summer near the time that corn takes it up. The one exception to this was a red soil, where fall application of nitrogen resulted in little loss due to the retarding effect. Soils on which a sod or cover crop is killed and in which corn is planted lose little water by evaporation. Because of this, much more water movement occurs and nitrate moves out of the soil during the summer. In contrast, the soil without a killed sod mulch suffered no loss of nitrate during the growing season.

Substantial increases in ground water nitrate-nitrogen are of particular concern on intensively cropped irrigated sandy soils. A study was conducted by Hubbard, Asmussen, and Allison (1984) to compare shallow ground water nitrate-nitrogen concentrations under intensive multiple cropping systems with those from under nearby nonagricultural sites. The intensive multiple cropping systems involved center pivot irrigation and sprinkler-applied fertilizer. The study was conducted on a sandy Coastal Plain soil by weekly measurement and sampling of observation wells located under a center pivot area and at two forest sites. Nitrate-nitrogen concentrations ranged from less than 0.1 mg/l to about 133 mg/l, with a mean of 20 mg/l. In contrast, samples from the forest sites had nitrate-nitrogen concentrations ranging from less than 0.1 mg/l to just over 1 mg/l. Mean nitrate-nitrogen concentrations under the center pivot area were found to vary seasonally according to cropping and hydrologic patterns such that the mean values for March-May, June-August, September-November, and December-February were 7, 21, 27, and 21 mg/l, respectively. The lower value for March-May indicated that winter rains leached most of the root zone nitrate-nitrogen beneath the wells by March, and there was a 2 to 3 month lag between spring-applied nitrogen and its appearance in shallow ground water. The study raises concern about shallow ground water quality under intense multiple cropping systems on sandy soils.

Management Practices and Mathematical Models

Table 12.3 contains brief comments on 11 references dealing with management practices and mathematical models which can be used to minimize and/or predict agricultural contributions of nitrates to ground water. Mathematical models are addressed in three references (Barcelona and Naymik, 1984; Kanwar, Johnson, and Baker, 1983; and Letey et al., 1978), with the remaining eight references listed in Table 12.3 identifying management practices.

Management practices for ground water nitrate minimization include: (1) water application management via irrigation systems selection and irrigation scheduling; (2) fertilizer application management through control of the timing and amounts of fertilizer usage; (3) the use of nitrification inhibitors in applied fertilizers; and (4) combinations of the first three practices.

Sprinkler irrigation, furrow irrigation, subirrigation, automated subirrigation, and criteria for applying irrigation water were investigated as to their potential to decrease possible pollution from nitrate and other solutes in a loamy fine sand soil overlying a shallow aquifer in Knox County, Texas (Wendt et al., 1976). Less nitrate-nitrogen was available for leaching in subirrigation systems than furrow and sprinkler systems. Less irrigation water was applied with automated subirrigation systems than with the other irrigation systems. However, the crop water requirement was not significantly changed--the soil water was more efficiently used. Fertilizer remained in the root zone if the water applied was based on potential evapotranspiration and leaf area regardless of the irrigation system or the criteria used to apply the irrigation water.

Excessive irrigation of the sandy soils to which center pivot sprinklers are adaptable can result in leaching of significant amounts of nitrogen fertilizer as nitrate. Careful water management is an effective means of controlling nitrate-nitrogen. The U.S. Department of Agriculture's irrigation scheduling program was used successfully in a study by Duke, Smika, and Heerman (1978) to determine the time and amount of irrigation necessary to maintain high crop yields yet minimize leaching losses. Although significant nitrate-nitrogen losses were measured from fields in the study area, neither the fraction of that nitrate present in return flows nor the mechanism of possible denitrification occurred in the vicinity of the water table as the data suggested, the potential

Table 12.3:   Summary of References on Management and Modeling of
Fertilizer-derived Nitrates in Ground Water

Author (Year)	Comments
Barcelona and Naymik (1984)	Numerical solute transport model for nitrates in a sand and gravel aquifer in western Illinois.
Duke, Smika, and Heerman (1978)	Use of water management to control ground water pollution by nitrates.
Gerwing, Caldwell, and Goodroad (1979)	Effects of amounts and timing of fertilizer applications on nitrates in ground water.
Kanwar, Johnson, and Baker (1983)	Simulation of nitrate losses in tile drainage via use of a nitrate transport simulation model.
Letey et al. (1978)	Linear regression of nitrate leaching as a function of water and fertilizer application rates.
Owens (1981)	Use of a nitrification inhibitor to reduce nitrates in ground water.
Saffigna, Kenney, and Tanner (1977)	Effects of sprinkler irrigation and fertilizer application rates of nitrates in ground water.
Smika et al. (1977)	Effects of fertilizer management practices on nitrate losses from soils.
Smika and Watts (1978)	Management of water and fertilizer application to control nitrate movement beneath crop root zones.
Timmons (1984)	Effects of water application level and nitrification inhibitors on nitrate leaching from a sandy loam soil.
Wendt et al. (1976)	Effects of irrigation methods on ground water pollution by nitrates.

for ground water pollution by leached nitrate is considerably reduced. Even so, careful water management was considered to be an important factor in the irrigation program as it affects efficiency of fertilizer utilization, cost of energy for pumping, and ultimately, the yield of crops produced.

Smika et al., (1977) reported on a study to determine the magnitude and difference in nitrate-nitrogen losses from two widely different but current farmer-used fertilizer management practices with scheduled irrigation systems on sandy soils. The experiments were conducted during three consecutive corn growing seasons on three irrigation systems in Colorado. Nitrate-nitrogen percolation losses were measured in water samples collected at 150 cm below the soil surface in vacuum extractors. The actual amount of nitrate-nitrogen that will move through the soil is proportional to the concentration of nitrate-nitrogen in the soil and the amount of water that moves through the soil. Smika et al. (1977) found that the movement of nitrogen fertilizer below the major root zone of corn grown on loamy fine sand soils can be kept very small with proper water and fertilizer management, thereby minimizing the nitrate-nitrogen pollution potential of ground water.

Gerwing, Caldwell, and Goodroad (1979) conducted a field experiment on a Sverdrup sandy loam in central Minnesota to evaluate the effects of amounts and timing of nitrogen (N) fertilizer applications on N uptake by irrigated corn (Zea mays L.) and the movement of N into the aquifer 4.5 m below the soil surface. Nitrogen was applied as urea at rates of 179 and 269 kg N/ha in one application at planting or in split applications through the season. Soil solution analysis showed NO_3 moving below the rooting zone in all plots, with much higher concentrations below the one-time fertilizer applications. Split applications of N had only minimal effect on the concentration of nitrate-nitrogen in the aquifer, but one-time applications increased the concentration by 7 and 10 mg/l.

Field and lysimeter experiments were conducted in 1972 and 1973 on the Plainfield loamy sand in central Wisconsin to determine the effect of reducing the amount of sprinkler-irrigation and fertilizer N on potato tuber yields and on the field water and N balance (Saffigna, Kenney, and Tanner, 1977). Two main treatments were used. The conventional (CON) treatment received irrigation water and N fertilizer according to normal recommendations. The improved (IMP) treatment received less irrigation water than the CON treatment in both years. A satisfactory mass balance of water, N, and chlorides for potatoes grown in the lysimeters was obtained. Drainage accounted for considerable amounts of water, N, and chlorides output, with nearly all of the remainder going to evapotranspiration and plant uptake. The carefully managed IMP treatment decreased nitrate-nitrogen leaching from 200 to 120 kg N/ha, and lowered the overall average nitrate-nitrogen concentration of the leachate from 23 to 16 mg/l.

Nitrapyrin, which is the active ingredient of N-Serve, has specific toxicity to Nitrosomonas spp., which is the main microorganism responsible for the nitrification of ammonium to more leachable nitrogen forms in soil. Owens (1981) investigated the possible benefits of a nitrification inhibitor in terms of retardation or reduction in the amounts of nitrate leaching through the soil profile into the ground water. Nitrapyrin treated urea or an untreated urea was incorporated into the top 10 cm of prewetted sandy loam soil cores. The soil columns were leached periodically with distilled water. After 91 days, only 1 percent of the applied nitrogen had leached from the nitrapyrin-treated cores, while 9.7 percent had leached from the untreated cores. After 144 days, 53.0 percent of the applied nitrogen had leached from the nitrapyrin-treated cores. Examination of the cores after 144 days showed that the mineral nitrogen concentration was greater at all depths in the soil treated with nitrapyrin than in the untreated soil, and that the total mineral nitrogen remaining in the untreated soil was only 70 percent of that remaining in the nitrapyrin treated soil column. These short-term results indicate that nitrification inhibitors such as nitrapyrin may have an important role to play in controlling nonpoint source pollution problems by reducing both the rate of entry and the total quantity of nitrate-nitrogen leached into subsurface waters.

Timmons (1984) evaluated the effects of water application level and nitrification inhibitors on nitrate-nitrogen leaching from a sandy loam soil using soil columns and nonweighing field lysimeters. In soil solumns (297 mm diam. by 1.2 m deep) fertilized with 224 kg N/ha, addition of nitrapyrin reduced nitrate-nitrogen leaching losses by 51 and 30 kg/ha, respectively, for the 12.7- and 38.1-mm water application levels. Both nitrapyrin and terrazole inhibitors were equally effective in reducing total nitrate-nitrogen leaching losses from soil columns after 100 days for the 38.1-mm water application level. The change in flow-weighted nitrate-nitrogen concentration due to use of nitrification inhibitors was about -0.4 to -1.0 mmol/l. Annual nitrate-nitrogen leaching losses measured at the 1.2-m depth in field lysimeters cropped with corn (Zea mays L.) over a 3-yr period averaged about 12 kg/ha less (7%) for nitrapyrin-coated urea. Therefore, nitrification inhibitors may help to minimize nonpoint source pollution on sandy soil where supplemental irrigation is used by reducing and/or delaying nitrate-nitrogen leaching, particularly during the growing season when N is utilized by plants.

Mathematical models can be used for predicting the transport and fate of nitrates in the subsurface and for evaluating the potential effectiveness of management practices. For example, Barcelona and Naymik (1984) applied a numerical solute transport model in a study of massive inorganic nitrogen fertilizer contamination of a sand and gravel aquifer in western Illinois. Ground water monitoring in the early period of the project during near steady-state conditions disclosed that dissolved ammonium and nitrate levels exceeded 2,000 and 13,000 mg/l, respectively. The model predicted that approximately 420 days would be necessary after source removal to permit recovery of the aquifer water to near background levels within the study site. Subsequent monitoring results generally supported the model prediction and demonstrated its usefulness for predicting transport and transformations of inorganic forms of nitrogen.

PESTICIDES IN GROUND WATER

As noted earlier, a total of 29 references related to pesticides in ground water were selected for inclusion in this chapter. Several bibliographic searches on pesticides within the water environment are available, with some references relating to ground water (Brown, 1979; Brown, 1980; and National Technical Information Service, 1984). The 26 nonbibliographic references can be considered in terms of case studies, transport and fate, and mathematical models.

Case Studies

Table 12.4 contains brief comments on 11 references describing case studies of pesticide contamination of ground water. Five cases are in Wisconsin, two are in New York, and one each is in Florida, Nebraska, New Jersey, and Pennsylvania.

Table 12.4:   Summary of References on Case Studies of Pesticides in Ground
Water

Author (Year)	Comments
Greenberg et al. (1982)	Land use influences on pesticides in New Jersey ground water.
Hall and Hartwig (1978)	Atrazine mobility in Pennsylvania soils under conventional tillage.
Hansen (1983)	Aldicarb in Wisconsin ground water supplies.
Harkin et al. (1984)	Potential for natural degradation of aldicarb in Wisconsin ground water.
Hindall (1978)	DDT in ground waters of the sand plains of central Wisconsin.
Lemley and Janauer (1982)	Identification and treatment of aldicarb in eastern Long Island, New York ground water.
Mansell et al. (1977)	Pesticide movement to ground water from citrus groves in Florida.
McWilliams (1984)	Institutional measures for control of pesticides in Wisconsin ground waters.
Rothschild, Manser, and Anderson (1982)	Occurrence and movement of aldicarb in ground water in the central sand plain of Wisconsin.
Spalding, Junk, and Richard (1980)	Atrazine and other pesticides in Nebraska ground water.
Zaki, Moran, and Harris (1982)	Countermeasures for aldicarb in ground water used for water supply in Suffolk County, New York.

The systemic pesticide aldicarb (Temik) has been used in Wisconsin to protect potatoes from the Colorado potato beetle and nematodes. Rothschild, Manser, and Anderson (1982) have described a preliminary assessment of the occurrence and movement of aldicarb in ground water in the central sand plain of Wisconsin. Aldicarb concentrations in ground water beneath three main study fields and two subsidiary fields were monitored during the period December, 1980, to August, 1981. A total of 67 well points, some nested, and one multilevel sampler were installed for this study. Twenty-five private wells and seven irrigation wells were sampled. The highest concentrations of aldicarb were detected in shallow monitoring wells (those located immediately below the water table); no aldicarb was detected in any of the deep monitoring wells (those located roughly 20 m below the water table), although aldicarb was found in some of the irrigation wells finished at approximately the same depth. Aldicarb seems to be concentrated in roughly a 2 m layer near the water table. The presence of aldicarb in shallow ground water is due, in part, to the highly water-soluble nature of this pesticide. The presence of aldicarb in a few of the deep irrigation wells was probably a result of local vertical flow components near the pumping well. The detection of aldicarb in a few shallow private wells downgradient from some fields demonstrates the mobility of this contaminant in ground water.

The susceptibility of this region of Wisconsin to ground water pollution began thousands of years ago when glaciers retreated northward, depositing sandy, loamy soils in thin layers on top of igneous and metamorphic bedrock (McWilliam, 1984). Harkin et al., (1984) measured aldicarb residues in ground water under nine 160-acre fields in this central sand plain region. Potatoes were grown under irrigation in sandy soils over high water tables in each of these fields. Aldicarb was incorporated into soil once a year either with potatoes at planting at a rate of 3 lb/a or at shoot emergence at a rate of 2 lb/a. Regardless of application rate, timing, or frequency (single or repeated applications in consecutive or alternate years), aldicarb residues were found at some but not all locations tested within every field.

Minimization of aldicarb in the ground waters of Wisconsin, or other states for that matter, can be aided by both natural phenomena and institutional measures. Natural biodegradation can lead to reductions in aldicarb concentrations in ground water. For example, Harkin et al., (1984) noted that samples of ground water and aquifer sediments collected aseptically contained surprisingly high numbers of a large diversity of facultative anaerobic bacteria. These organisms can totally degrade aldicarb sulfoxide and aldicarb sulfone, the major metabolites of aldicarb which occur in ground water, at a rate which suggests a half-life of about 1.3 years for aldicarb residues in Wisconsin ground water.

Institutional measures for aldicarb control could include federal, state, and/or local legislation limiting the time and rate of application (McWilliams, 1984). Incentives to states that voluntarily implement ground water management, or programs that would aid innovative farmers in implementing practices that prevent ground water pollution could also be developed. Consideration could also be given to policies stipulating land use or zoning restrictions for activities which threaten ground water quality. Counties could also initiate an informational program to inform landowners and the general public of activities which could degrade the quality of their ground water (Hansen, 1983).

Aldicarb contamination of ground water has also been detected in samples from Suffolk County, New York (Zaki, Moran, and Harris, 1982). Although the pesticide reached human drinking water supplies, no cases of carbamate poisoning were reported in area hospitals. An extensive monitoring program conducted by the county in cooperation with federal and state agencies and the Union Carbide Corporation showed that 13.5 percent of the 8,404 wells sampled exceeded the state recommended limits for aldicarb of $7\mu g/l$. Residents whose wells exceeded this limit were advised not to use the water for drinking or cooking and to obtain an alternate source of potable water. The Union Carbide Corporation, manufacturer of the pesticide, provided these residents with activated carbon filtration units to remove most of the aldicarb and its metabolites. A novel approach to treating carbamate contaminated drinking water has been investigated and found to be successful on the laboratory scale (Lemley and Janauer, 1982). For example, acid catalyzed hydrolysis and nucleophilic cleavage of organophosphate pesticides are effectively achieved in-situ on ion exchange beds charged with protons or hydroxyl ions. Similar experiments with aldicarb and its environmental oxidation products, both in batch equilibrium studies and mini columns, have demonstrated virtually complete detoxification of part per million to part per billion level solutions. Base and acid hydrolysis degradation rates for aldicarb/metabolites in solution under a variety of conditions have also been determined.

Mansel et al., (1977) reported on the concentrations and discharge amounts of nitrate-nitrogen, phosphates, 2,4-D herbicide, terbacil herbacide, and chlorobenzilate acaricide in surface and subsurface drainage waters from a citrus grove located in an acid, sandy flatwood soil of southern Florida. The influence of fertilizer and pesticide upon water quality was examined for citrus growing in three soil management treatments: ST (shallow-tilled plowed to 15 cm); DT (deep-tilled and soil mixed within the top 105 cm); and DTL (deep-tilled to 105 cm and 56 Mt/ha of dolomitic limestone mixed with the soil). Average annual losses of nitrate-nitrogen in both surface and subsurface drainage from ST, DT, and DTL plots were equivalent to 22.1, 3.1, and 5.4 percent of the total N applied as fertilizer. Average annual losses of phosphate-phosphorus in both surface and subsurface drainage from ST, DT, and DTL plots were equivalent to 16.9, 3.6, and 3.5 percent of the total P applied as fertilizer. Deep tillage was thus observed to greatly decrease leaching losses of N and P nutrients. Loss of nutrients in surface runoff was very small for all three plots. Although the magnitudes were less, deep tillage also decreased leaching losses of terbacil and 2,4-D. Discharges of these herbicides in subsurface drainage were usually in the order: ST greater than DTL greater than DT. Discharge of 2,4-D was greater from drains with open outlets than from drains with submerged outlets. Discharge of terbacil did not differ for open or submerged drains. The chlorobenzilate pesticide was not detected in drainage water from any of the three soil treatments.

Samples of water from 14 irrigation wells in the Central Platte region of Nebraska were analyzed for residues of pesticides and nitrate-nitrogen (Spalding, Junk, and Richard, 1980). Atrazine, the most commonly used herbicide in this area, was found in all samples. Levels ranged from 0.06 to 3.12 μg/l. Nitrate-nitrogen concentrations, which ranged from 17.1 to 34.3 mg/l, correlated with atrazine levels. Alachlor levels above 0.01 μg/l appeared in two samples. It is believed that both atrazine and alachlor migrated vertically to the ground water. Although all reported levels were below limits for drinking water standards; the atrazine-nitrate combination is capable of forming nitrosamines.

Transport and Fate

Table 12.5 contains brief comments on 10 references dealing with the transport and fate of pesticides in the subsurface environment. Four references are related to field studies (Garrett, Maxey, and Katz, 1976; Hebb and Wheeler, 1978; LeGrand, 1970; and Schneider, Wiese, and Jones, 1977) and six describe laboratory experiments (Awad, Kilgore, and Winterlin, 1984; Boucher and Lee, 1972; Davidson et al., 1980; Dregne, Gomez, and Harris, 1969; Gile, Collins, and Gillett, 1980; and Virtanen et al., 1982).

Table 12.5:   Summary of References on the Transport and Fate of Pesticides
in Ground Water

Author (Year)	Comments
Awad, Kilgore, and Winterlin (1984)	Laboratory studies of the movement of aldicarb in eight different soil types.
Boucher and Lee (1972)	Laboratory studies of the adsorption of lindane and dieldrin on natural aquifer sand.
Davidson et al. (1980)	Laboratory studies of the adsorption, movement, and biological degradation of selected pesticides in four major soil orders.
Dregne, Gomez, and Harris (1969)	Laboratory studies of the movement of 2,4-D in three soils.
Garrett, Maxey, and Katz (1976)	Influence of subsurface materials on pesticide migration to ground water.
Gile, Collins, and Gillett (1980)	Transport and metabolism of selected herbicides in a terrestrial laboratory microcosm.
Hebb and Wheeler (1978)	Field study of bromacil transport to ground water in a sandy zone.
Le Grand (1970)	Natural environmental factors tending to mitigate pesticide transport to ground water.
Schneider, Wiese, and Jones (1977)	Field study of the transport of three herbicides into a sand aquifer.
Virtanen et al. (1982)	Study of DDT transport using a compartmentalized model ecosystem.

Factors related to the type and characteristics of the pesticide and the subsurface environment tend to influence pesticide migration to ground water. For example, pesticides with significant adsorptive properties in the soil structure and/or short-lived persistance may be of minimal concern (Garrett, Maxey, and Katz, 1976). LeGrand (1970) identified the following subsurface characteristics which would limit concern about pesticides in ground water: (1) a deep water table which allows for sorption of pollutants on earth materials slows the subsurface movement of pollutants and facilitates oxidation or other beneficial "die-away" effects; (2) sufficient clay in the path of pollutants so that retention or sorption of pollutants is favorable; (3) a gradient of the water table beneath an application area away from nearby wells; and (4) a large distance between wells and the application area. Pesticide transport to ground water can be facilitated by the improper application of pesticides or the existence of sandy soils or thin soils overlaying fissured rocks (Garrett, Maxey, and Katz, 1976).

Hebb and Wheeler (1978) evaluated the probable magnitude of leaching of pesticides into ground water under extreme conditions: a sandy soil low in organic matter, a persistent and mobile herbicide applied at a high rate, plentiful rainfall, and a water table within 6 m. Bromacil was applied at the rate of 22 kg/ha to a lakeland sand-bearing scrub vegetation of small oaks and poor grasses. Ground water (at depths ranging from 4.5 to 6 m) was sampled for bromacil residue at weekly intervals for two years. Residue was first found in the ground water three months after application and was highest (1.25 mg/l) one month later. Thereafter, the amount declined to less than 0.1 mg/l in about a year and less than 0.001 mg/l in two years. Peaks in residues generally followed periods of increased rainfall by about two weeks. Residues (0.24 mg/l) were still detected in the surface soil two years after application.

The movement of aldicarb was studied in different soil types under laboratory conditions and the effects of large volumes of water on this movement was also investigated (Awad, Kilgore, and Winterlin, 1984). Thirty-five grams of eight different but well defined soils were packed in individual glass columns and tapped lightly. Twenty-five mg of aldicarb active ingredient in formulated form were added to the top of the column and mixed thoroughly with the top 3 cm of soil. Three aliquots of 50 ml each of distilled water were then added to the top of the soil columns. Soil types included sand, loam, clay, sandy-loam, clay-loam, silt-loam, muck, and peat. Clay soil retained the highest amount of aldicarb residues followed by loam soil. The apparent retention of aldicarb residues by loam and clay soils may be a function of water flow as well as some type of physical binding to the soil particles. The lowest amounts of aldicarb residues were found in sandy, sandy-loam and peat soils.

Davidson et al., (1980) reported on a laboratory study to evaluate the adsorption, mobility, and degradation of large concentrations of atrazine, methyl parathion, terbacil, trifluralin, and 2,4-D in soils representing four major soil orders in the United States. Solution concentrations ranged from zero to the aqueous solubility limit for each pesticide. The mobility of each pesticide increased as its concentration in the soil solution phase increased. These results were in agreement with the adsorption isotherm data. Pesticide degradation rates and soil microbial populations generally declined as the pesticide concentration in soil increased; however, some soils were able to degrade a pesticide at all concentrations studied, while others remained essentially sterile throughout the incubation period (60 to 80 days). As shown by measurements of 14 CO_2 evolution, total CO_2 evolution was not always a good indication of pesticide degradation. Several pesticide metabolites were formed and identified. Bound residues of trifluralin and atrazine at the end of the incubation period appeared to be related to the types of metabolites formed.

Mathematical Models

Several types of mathematical models are available for usage in predicting the transport and fate of pesticides in the subsurface environment. Javandel, Doughty, and Tsang (1984) have developed a handbook which introduces the reader to various mathematical methods for estimating solute transport in ground water systems. It contains tables, figures, and simple computer programs that can be directly used for field studies. Three levels of mathematical methods are covered: (1) analytical, (2) semianalytical, and (3) numerical. The first two levels require relatively small amounts of data. At the third level, numerical approaches are discussed and a number of currently available numerical models are listed, indicating code capabilities and code developers to be contacted for further information. An example of the use of one such model is presented. A discussion on method selection and data requirements is also included in the handbook.

Enfield et al., (1982) describe three models for estimating the transport of organic chemicals through soils to ground water. The models consider mobility and first order degradation. The first model calculates linear sorption/desorption of the pollutant and first order degradation without considering dispersion. The second is similar but also considers dispersion. The third considers nonliner sorption following a Freundlich equation and first-order degradation, but does not consider dispersion. The models are then compared to field data for the pesticides aldicarb and DDT. Rao and Davidson (1982) have noted that octanol-water partition coefficients are good predictors of pesticide adsorption parameters.

Mansell et al., (1976) developed a mathematical model to simulate the transport and physical-chemical reactions for potassium, phosphorus, and 2,4-D in two representative Florida soils: Wauchula sand and Troup sand. They found that reactions such as adsorption-desorption, chemical precipitation, and immobilization (fixed) greatly influenced the movement and thus potential leaching of these solutes through the soil.

Finally, Dean, Jowise, and Donigian (1984) developed a methodology to assess potential pesticide leaching from the crop root zones in major (corn, soybean, wheat and cotton) crop growing areas of the United States. Use of the Leaching Evaluation of Agricultural Chemicals (LEACH) methodology provides an indication of the presence or absence of leaching past the rooting depth and, if such leaching is indicated, its severity. LEACH was developed through the use of long-term simulation (i.e., 25 years) of annual pesticide leaching using the Pesticide Root Zone Model. The user must evaluate key parameters for a pesticide-site-crop-management scenario to locate pesticide leaching cumulative frequency distributions. Each scenario has a unique distribution associated with it. The distribution functions indicate the chance that the annual quantity of pesticide leached past the crop rooting depth will exceed a given value.

CONCLUSIONS

Several concluding points can be made on this review of a computer-based literature search on nitrates and pesticides in ground water. These points are:

(1)   Nitrate contamination of ground water from agricultural applications of fertilizers is a wide-spread problem in the United States. While only 14 states with a wide-spread geographic distribution were identified as having such nitrate problems, most states have experienced some concerns, perhaps on a localized basis.

(2)   Nitrate concentrations in ground water may exceed, on a regular basis in some locations, the recommended drinking water standard of 10 mg/l nitrates as N.

(3)   Many factors influence nitrate concentrations in ground water, including the timing and rate of fertilizer application, subsurface soil characteristics, depth to ground water, crop nutrient needs, the timing and rate of irrigation, and natural rainfall.

(4)   Nitrate minimization in ground water can be achieved through water application management via irrigation system selection and irrigation scheduling, fertilizer application management through control of the timing and amounts of fertilizer usage, and the use of nitrification inhibitors. These approaches can be used either singly or in combination.

(5)   Agricultural applications of pesticides are causing pesticide contamination of ground water. While only six states were identified in this search, other information suggests that over half of the 50 states may be experiencing some problems with pesticide contamination. As more pesticide monitoring is done, the number of states with identified problems is expected to increase.

(6)   Many fctors influence pesticide concentrations in ground water, including the mobility and persistence characteristics of the pesticide used, the timing and rate of pesticide application, subsurface soil characteristics, depth to ground water, the timing and rate of irrigation, and natural rainfall.

(7)   Pesticide minimization in ground water can be achieved through water application management, pesticide selection, and pesticide application management. These approaches can be used either singly or in combination.

(8)   Mathematical models have been developed for predicting nitrate and pesticide concentrations in ground water. These models can also be useful in evaluating the potential effectiveness of management practices for reducing nitrates and pesticides in ground water.

SELECTED REFERENCES

Awad, T.M., Kilgore, W.W., and Winterlin, W., "Movement of Aldicarb in Different Soil Types", Bulletin of Environmental Contamination and Toxicology, Vol. 32, No. 4, April 1984, pp 377-382.

Baier, J.H., and Rykbost, K.A., "The Contribution of Fertilizer to the Ground Water of Long Island", Ground Water, Vol. 14, No. 6, November-December, 1976, pp 439-447.

Barcelona, M.J., and Naymik, T.G., "Dynamics of a Fertilizer Contaminant Plume in Groundwater", Environmental Science and Technology, Vol. 18, No. 4, April, 1984, pp 257-261.

Bingham, F.T., Davis, S., and Shade, E., "Water Relations, Salt Balance, and

Nitrate Leaching Losses of a 960-Acre Citrus Watershed", Soil Science, Vol. 112, No. 6, 1971, pp 410-417.

Bottcher, A.B., Monke, E.J., and Huggins, L.F., "Nutrient and Sediment Loadings from a Subsurface Drainage System", Transactions of the American Society of Agricultural Engineers, Vol. 24, No. 5, September-October, 1981, pp 1221-1226.

Boucher, F.R., and Lee, G.F., "Adsorption of Lindane and Dieldrin Pesticides on Unconsolidated Aquifer Sands", Environmental Science and Technology, Vol. 6, No. 6, June, 1972, pp 538-543.

Boyce, J.S., et al., "Geologic Nitrogen in Pleistocene Loess in Nebraska", Journal of Environmental Quality, Vol. 5, No. 1, January-March, 1976, pp 93-96.

Brown, R.J., "Ecology of Insecticide Water Pollution, Volume 2, 1974-1977 (A Bibliography with Abstracts)", NTIS/PS-79/0108/5, March, 1979, National Technical Information Service, U.S. Department of Commerce, Springfield, Virginia.

Brown, R.J., "Ecology of Insecticide Water Pollution, Volume 3, 1978-January, 1980 (Citations from the NTIS Data Base)", PB80-804727, February, 1980, National Technical Information Service, U.S. Department of Commerce, Springfield, Virginia.

Chichester, F.W., "The Impact of Fertilizer Use and Crop Management of Nitrogen Content of Subsurface Water Draining from Upland Agricultural Watersheds", Journal of Environmental Quality, Vol. 5, No. 4, October-December, 1976, pp. 413-416.

Davidson, J.M., et al., "Adsorption, Movement, and Biological Degradation of Large Concentrations of Selected Pesticides in Soils", EPA-600/2-80-124, August, 1980, Municipal Environmental Research Laboratory, U.S. Environmental Protection Agency, Cincinnati, Ohio.

Dean, J.D., Jowise, P.P., and Donigian, A.S., Jr., "Leaching Evaluation of Agricultural Chemicals (LEACH) Handbook", EPA-600/3-84-068, June, 1984, Environmental Research Laboratory, U.S. Environmental Protection Agency, Athens, Georgia.

DeRoo, H.C., "Nitrate Fluctuations in Ground Water as Influenced by Use of Fertilizer", Bulletin 779, June, 1980, Connecticut Agricultural Experiment Station, New Haven, Connecticut.

Dregne, H.E., Gomez, S., and Harris, W., "Movement of 2,4-D in Soils", Western Regional Research Project Progress Report, November, 1969, New Mexico Agricultural Experiment Station, University Park, New Mexico.

Duke, H.R., Smika, D.E., and Heermann, D.F., "Ground-Water Contamination by Fertilizer Nitrogen", Journal of the Irrigation and Drainage Division, American Society of Civil Engineers, Vol. 104, No. IR3, September, 1978, pp 283-291.

Enfield, C.G., et al., "Approximating Pollutant Transport to Ground Water", Ground Water, Vol. 20, No. 6, November-December, 1982, pp 711-722.

Exner, M.E., and Spalding, R.F., "Evolution of Contaminated Groundwater in Holt County, Nebraska", Water Resources Research, Vol. 15, No. 1, February, 1979, pp 139-147.

Garrett, D., Maxey, F.P., and Katz, H., "The Impact of Intensive Application of Pesticides and Fertilizers on Underground Water Recharge Areas Which May Contribute to Drinking Water Problems", EPA 560/3-75-006, January, 1976, U.S. Environmental Protection Agency, Washington, D.C.

Gast, R.G., and Goodrich, P.R., "Establishing the Impact of Agricultural Practices on Groundwater Quality", Proceedings of Conference on Toward a Statewide Ground Water Quality Information System, February, 1973, University of Minnesota, Minneapolis, Minnesota, pp 79-91.

Gerwing, J.R., Caldwell, A.C., and Goodroad, L.L., "Fertilizer Nitrogen Distribution Under Irrigation Between Soil, Plant, and Aquifer", Journal of Environmental Quality, Vol. 8, No. 3, July-September, 1979, pp 281-284.

Gile, J.D., Collins, J.C., and Gillett, J.W., "Fate of Selected Herbicides in a Terrestrial Laboratory Microcosm", Environmental Science and Technology, Vol. 14, No. 9, September, 1980, pp 1124-1128.

Gilliam, J.W., Daniels, R.B., and Lutz, J.F., "Nitrogen Content of Shallow Ground Water in the North Carolina Coastal Plain", Journal of Environmental Quality, Vol. 3, No. 2, April-June, 1974, pp 147-151.

Gormly, J.R., and Spalding, R.F., "Sources and Concentrations of Nitrate-Nitrogen in Ground Water of the Central Platte Region, Nebraska", Ground Water, Vol. 17, No. 3, May-June, 1979, pp 291-301.

Greenberg, M., et al., "Empirical Test of the Association Between Gross Contamination of Wells with Toxic Substances and Surrounding Land Use", Environmental Science and Technology, Vol. 16, No. 1, January, 1982, pp 14-19.

Hall, J.K., and Hartwig, N.L., "Atrazine Mobility in Two Soils Under Conventional Tillage", Journal of Environmental Quality, Vol. 7, No. 1, January-March, 1978, pp 63-68.

Hansen, S.C., "Groundwater: An Inventory of Wells and Contamination Potential Within the Silurian Aquifer of Calumet County", July, 1983, Fox Valley Water Quality Planning Agency, Menasha, Wisconsin.

Harkin, J.M., et al., "Pesticides in Groundwater Beneath the Central Sand Plain of Wisconsin", OWRT-A-094-WIS(1), 1984, Office of Water Research and Technology, U.S. Department of the Interior, Washington, D.C.

Hebb, E.A., and Wheeler, W.B., "Bromacil in Lakeland Soil Ground Water", Journal of Environmental Quality, Vol. 7, No. 4, October-December, 1978, pp 598-601.

Hindall, S.M., "Effects of Irrigation on Water Quality in the Sand Plain of Central Wisconsin", Information Circular No. 36, February, 1978, Wisconsin Geological and Natural History Survey, Madison, Wisconsin.

Hubbard, R.K., Asmussen, L.E., and Allison, H.D., "Shallow Groundwater

Quality Beneath an Intensive Multiple-Cropping System Using Center Pivot Irrigation", Journal of Environmental Quality, Vol. 13, No. 1, 1984, pp 156-161.

Javandel, I., Doughty, C., and Tsang, C.F., "Groundwater Transport: Handbook of Mathematical Models", Water Resources Monograph 10, 1984, American Geophysical Union.

Kanwar, R.S., Johnson, H.P., and Baker, J.L., "Comparison of Simulated and Measured Nitrate Losses in Tile Effluent", Transactions of the American Society of Agricultural Engineers, Vol. 26, No. 5, September-October, 1983, pp 1451-1457.

LeGrand, H.E., "Movement of Agricultural Pollutants with Groundwater", Agricultural Practices and Water Quality, 1970, Iowa State University Press, Ames, Iowa, pp 303-313.

Lemley, A.T., and Janauer, G.E., "An Investigation of the Chemistry of Degradation of Pesticides and other Serious Organic Contaminants in Drinking Water for Application to Domestic and Community Water Supplies", OWRT-A-083-NY(1), 1982, Office of Water Research and Technology, U.S. Department of the Interior, Washington, D.C.

Letey, J., et al., "Effect of Water Management on Nitrate Leaching", in National Conference on Management of Nitrogen in Irrigated Agriculture, 1978, University of California at Riverside, Riverside, California, pp 231-249.

Mansell, R.S., et al., "Movement of Fertilizer and Herbicide Through Irrigated Sands", Publication No. 38, September, 1976, Florida Water Resources Research Center, University of Florida, Gainesville, Florida.

Mansell, R.S., et al., "Fertilizer and Pesticide Movement from Citrus Groves in Florida Flatwood Soils", EPA-600/2-77-177, August, 1977, Environmental Research Laboratory, U.S. Environmental Protection Agency, Athens, Georgia.

McDonald, D.B., and Splinter, R.C., "Long-Term Trends in Nitrate Concentration in Iowa Water Supplies", Journal of the American Water Works Association, Vol. 74, No. 8, August, 1982, pp 437-440.

McWilliams, L., "Bumper Crop Yields Growing Problems", Environment, Vol. 26, No. 4, May, 1984, pp 25-33.

National Technical Information Service, "Ecology of Pesticide Water Pollution--1978-May, 1984 (Citations from the NTIS Data Base)", June, 1984, U.S. Department of Commerce, Springfield, Virginia.

Olson, R.A., "Influence of Fertilizer Practices on Water and the Quality of the Environment", OWRT-B-022-NEB(3), June, 1974, Office of Water Research and Technology, U.S. Department of the Interior, Washington, D.C.

Owens, L.B., "Effects of Nitrapyrin on Nitrate Movement in Soil Columns", Journal of Environmental Quality, Vol. 10, No. 3, July-September, 1981, pp 308-310.

Piskin, R., "Evaluation of Nitrate Content of Ground Water in Hall County, Nebraska", Ground Water, Vol. 11, No. 6, November-December, 1973, pp 4-13.

Rao, P.S.C., and Davidson, J.M., "Retention and Transformation of Selected Pesticides and Phosphorus in Soil-Water Systems" A Critical Review", EPA-600/3-82-060, May, 1982, Environmental Research Laboratory, U.S. Environmental Protection Agency, Athens, Georgia.

Reeves, C.C., Jr., and Miller, W.D., "Nitrate, Chloride and Dissolved Solids, Ogallala Aquifer, West Texas", Ground Water, Vol. 16, No. 3, May-June, 1978, pp 167-173.

Ritter, W.F., and Chirnside, A.E.M., "Impact of Land Use on Ground Water Quality in Southern Delaware", Ground Water, Vol. 22, No. 1, January-February, 1984, pp 38-47.

Rothschild, E.R., Manser, R.J., and Anderson, M.P., "Investigation of Aldicarb in Ground Water in Selected Areas of the Central Sand Plain of Wisconsin", Ground Water, Vol. 20, No. 4, July-August, 1982, pp 437-445.

Saffigna, P.G., Keeney, D.R., and Tanner, C.B., "Nitrogen, Chloride, and Water Balance with Irrigated Russet Burbank Potatoes in a Sandy Soil", Agronomy Journal, Vol. 69, No. 2, March-April, 1977, pp 251-257.

Schneider, A.D., Wiese, A.F., and Jones, O.R., "Movement of Three Herbicides in a Fine Sand Aquifer", Agronomy Journal, Vol. 69, No. 3, May-June, 1977, pp 432-436.

Silver, B.A., and Fielden, J.R., "Distribution and Probable Source of Nitrate in Ground Water of Paradise Valley, Arizona", Ground Water, Vol. 18, No. 3, May-June, 1980, pp 244-251.

Smika, D.E. et al., "Nitrate-N Percolation Through Irrigated Sandy Soil as Affected by Water Management", Agronomy Journal, Vol. 69, No. 4, July-August, 1977, pp 623-626.

Smika, D.E., and Watts, D.G., "Residual Nitrate-N in Fine Sand as Influenced by N Fertilizer and Water Management Practices", Soil Science Society of American Journal, Vol. 42, No. 6, November-December, 1978, pp 923-926.

Spalding, R.F., "Effects of Land Use and River Seepage on Groundwater Quality in Hall County, Nebraska", Nebraska Water Survey Paper No. 38, January, 1975, University of Nebraska, Lincoln, Nebraska.

Spalding, R.F., et al., "Nonpoint Nitrate Contamination of Ground Water in Merrick County, Nebraska", Ground Water, Vol. 16, No. 2, March-April, 1978, pp 86-95.

Spalding, R.F., et al., "Investigation of Sources of Groundwater Nitrate Contamination in the Burbank-Wallula Area of Washington, U.S.A.", Journal of Hydrology, Vol. 5, No. 3/4, September, 1982, pp 307-324.

Spalding, R.F., Junk, G.A., and Richard, J.J., "Pesticides in Ground Water Beneath Irrigated Farmland in Nebraska, August, 1978", Pesticides Monitoring Journal, Vol. 14, No. 2, September, 1980, pp 70-73.

Thomas, G.W., "The Relation Between Soil Characteristics, Water Movement and Nitrate Contamination of Groundwater", Research Report No. 52,

September, 1972, Kentucky Water Resources Institute, University of Kentucky, Lexington, Kentucky.

Timmons, D.R., "Nitrate Leaching as Influenced by Water Application Level and Nitrification Inhibitors", Journal of Environmental Quality, Vol. 13, No. 2, April-June, 1984, pp 305-309.

Tryon, C.P., "Ground Water Quality Variation in Phelps County, Missouri", Ground Water, Vol. 14, No. 4, July-August, 1976, pp 214-223.

Virtanen, M.T., et al., "Model Ecosystem for Environmental Transport of Xenobiotics", Archives of Environmental Contamination and Toxicology, Vol. 11, No. 4, 1982, pp 419-424.

Wehking, M.W., et al., "Nitrate Content of Well Water in West-Central Wisconsin", Wisconsin Academy of Sciences, Arts and Letters, Vol. 61, 1973, pp 259-265.

Wendt, C.W., et al., "Effects of Irrigation Methods on Groundwater Pollution by Nitrates and Other Solutes", EPA-600/2-76-291, December, 1976, U.S. Environmental Protection Agency, Washington, D.C.

Zaki, M.H., Moran, D. and Harris, D., "Pesticides in Groundwater: The Aldicarb Story in Suffolk County, NY", American Journal of Public Health, Vol. 72, No. 12, December, 1982, pp 1391-1395.

CHAPTER 13

BEHAVIOR AND SUBSURFACE TRANSPORT OF
AGROCHEMICALS IN CONSERVATION SYSTEMS

by

Thanh H. Dao

Conservation production systems have proven to be the most promising alternatives for reducing soil erosion by wind and water. Two major pollutants of lakes and streams are sediments and the man-made chemicals that are carried by them. Plant residues left on the surface of agricultural land under conservation tillage can effectively decrease the velocity and volume of runoff water; sediment loss from cultivated fields is reduced resulting in improved quality of surface waters. One purpose of soil tillage is to control weeds. The development and use of selective herbicides in place of cultivation for weed control have enhanced the feasibility and acceptance of reduced and no-till production systems. The latter system simulates the ideal permanent soil cover condition found on native rangeland. Environmental concerns have been expressed in conjunction with the increased usage of pesticides and commercial fertilizers under no-tillage. An improved understanding of the behavior and fate of agrochemicals in the changed microenvironment, associated with the lack of tillage and the presence of a surface residue mulch, is needed to understand the modes and mechanisms of contamination of the subsurface environment.

Ground water contamination is a more complex problem than pollution of surface waters from agricultural practices because it is more difficult to detect and rectify. This potential problem needs to be confirmed because there are other more obvious and prominent sources of ground water contamination such as hazardous wastes landfills, surface wastewater impoundments, industrial storage tanks, and septic tanks. However, the extensive use of chemicals in our agricultural production systems is sufficient cause to address the issue because agricultural uses of ground water account for about 70 percent of the national usage.

Much research has been done on the fate of fertilizers and pesticides in major agroecosystems. With widening acceptance of conservation practices to conserve the soil resource and maintain its productivity, an improved understanding of the impact of the changes in soil surface conditions on agrochemicals must be gained in order to better manage them, and to minimize risks to ground water such as the Ogallala and Madison Aquifers of the Great Plains.

The changes in soil surface ecology in conservation tillage systems as compared to plowed ones have been reported to include:

(1)   an increase in soil microbial biomass (Doran, 1980a) and soil enzyme activity (Dick, 1984) due to increased energy substrates and water content at the surface;

(2)   a more reduced environment due to reduced aeration of the soil surface under the mulch, paralleled by an increase in facultative anaerobes and denitrifiers (Doran, 1980b; and Broder et al., 1984);

(3)   higher surface soil water content caused by reduced evaporation (Johnson, Lowery, and Daniel, 1984; and Lonkerd and Dao, 1985);

(4)   attenuated temperature fluctuations due to the surface mulch acting as a cover and insulator; and

(5)   increased soil surface acidity resulting from current nitrogen fertilization practices and the lack of soil mixing without tillage (Blevins, Thomas, and Cornelius, 1977).

This study was undertaken to assess the impact of the surface ecology on the fate of agricultural chemicals in conservation tillage systems. The effects of three soil management practices on field persistence of ammonium nitrate fertilizer N and the herbicide BAY SMY 1500 were measured. Their subsurface movement was evaluated to formulate improved management practices to safeguard ground water resources underlying agricultural areas.

MATERIALS AND METHODS

Soils from field plots under three residue management systems of winter wheat (<u>Triticum</u> <u>aestivum</u>) production were periodically sampled throughout 1983-86 to make a comparative assessment of the fate of agrochemicals. The field plots were established on a Bethany-Renfrow-Norge association (Pachic-, Udertic-, and Udic Paleustolls) and at a second site on a Port loam (Cumulic Haplustolls) at El Reno, Oklahoma. El Reno is located in central Oklahoma in the 75 cm annual rainfall area. The mean annual temperature is 15 °C. The surface texture of the Bethany-Renfrow-Norge association was silt loam; the soils contained 0.7-1.0 percent organic carbon and have a pH of 4.7 (in 0.01 M $CaCl_2$). The Port loam is a well-drained soil, containing 0.6 percent organic carbon and has a pH of 4.4 (in 0.01 M $CaCl_2$).

Both sites have been under continuous wheat cultivation for less than a decade. Three tillage systems: moldboard plow-disk, chisel-disk (subtill), and no-tillage were established in July, 1983, on the Bethany-Renfrow-Norge soils and in July,1985, on the Port loam. The fields were maintained weed-free during the summer fallow by secondary tillage (sweeping-disking) and by herbicides (glyphosate-cyanazine) in appropriate plots. At planting, fields were seeded with TAM 101, a semidwarf and medium maturity hard red winter wheat in rows spaced 20-cm apart at rates of 100 kg/ha with a no-till drill. Ninety kg N as ammonium nitrate was surface-broadcast along with 40 kg P_2O_5/ha in the fall.

A comparative study of the fate of N fertilizer was made in the spring of 1985 following an application of 55 kg N/ha in the no-tillage and the moldboard-plow treatments. Soil cores were taken to 0.6 m depths at two-week intervals, and sectioned into 5-, 10- and 20-cm sections. Field soil samples were immediately brought back into the laboratory and extracted with 1 N KCl. Ammonium and nitrate-N content were measured by steam distillation and reduction with Devarda's alloy (Bremner, 1965). A chemical availability index was also used to quantify the effect of tillage on soil nitrogen turnover. Soil samples collected annually during the 1983-86 period were analyzed by the autoclave method (Stanford and Smith, 1976) after preextraction with 0.01 M $CaCl_2$. A 15.0 g sample (oven-dry basis) was twice extracted with 0.01 M $CaCl_2$, then autoclaved with fresh 0.01 M $CaCl_2$ for 16 hrs, 120°C. The solution phase was decanted into distillation flasks, along with the additional washings of the soil residue, and steam distilled. The NH_3-N released was trapped in a 2 percent boric acid solution and determined by titration to an end-point of pH 4.0 (Bremner, 1965).

The herbicide BAY SMY 1500 was applied at a rate of 1.0 kg/ha to control Bromus sps. in the tillage plots. Soil cores were removed to a depth of 10 cm with a hand-held soil sampler prior to application and periodically thereafter during the fall and winter of 1985. Soil samples were frozen for later herbicide residue analysis. BAY SMY 1500 was extracted from 50 g samples (oven-dry basis) by refluxing the sample in an acetonitrile:water (9:1, v/v) mixture. An aliquot of the extract was evaporated to dryness; the residue was redissolved and made to volume for high-performance liquid chromatographic analysis. A mixture of methanol:H_2O (7:3, v/v) was used as the mobile phase. Chromatographic separation was achieved with a C-18, 4 μm particle size column (Nova-pak Cartridge, Water Associates, Millipore Corp.).

Regression analysis and a first-order kinetic model were used to describe the disappearance of the herbicide under field conditions:

$$C_t = C_o e^{kt} \qquad (1)$$

where C_t = observed herbicide concentrations at time t

C_o = initial herbicide concentration

t = days after application and

k = reaction rate constant

Measurements of adsorption of BAY SMY 1500 on the 4 soils were also made using the batch equilibration technique (Dao et al., 1982). Retardation factors (R) were calculated to quantify the influence of adsorption and to estimate the relative subsurface mobility of BAY SMY 1500 in these soils (Rao et al., 1979). R is calculated as follows:

$$R = 1 + P/_\theta \ K \qquad (2)$$

where K = adsorption partition coefficient

P = soil density, g/m^3

θ = volumetric water content, cm^3/cm^3

RESULTS AND DISCUSSION

Comparative N Status Among Tillage Systems

The distribution of mineral N as a function of time and soil depth is presented in Table 13.1. The data of the untreated check plots indicated that higher concentrations of residual nitrogen from the fall application were present in the soil, to a depth of 0.6 m, under no-tillage than under moldboard plowing at the time of the spring topdress N treatment. These results suggest that no-tillage has decreased the possibility of N losses with surface water runoff and slower leaching, resulting in a higher availability of N in the root zone.

Table 13.1:   Mineral N Distribution as a Function of Time and Soil Depth Under Two Tillage Systems.

Nitrogen Application Rate	Tillage	Soil Depth	Mineral N Content				
			Days After N Application				
			0	6	13	28	51
(kg/ha)		(cm)	(μg N/g soil)				
0	Moldboard	0- 5	9.3	5.9	5.3	14.7	5.8
		5-10	5.8	6.3	3.9	4.2	5.9
		10-20	4.5	5.0	3.4	1.3	3.8
		20-40	8.6	11.9	8.3	1.9	6.9
		40-60	13.4	21.7	7.7	4.4	11.3
	No-Till	0- 5	28.4	18.6	25.3	5.7	9.1
		5-10	11.3	10.6	12.8	1.7	3.8
		10-20	9.2	15.6	7.7	2.4	2.9
		20-40	15.8	13.0	8.3	3.8	2.5
		40-60	21.1	17.8	11.2	4.7	7.0
55	Moldboard	0- 5	114.8	121.3	61.5	72.8	13.8
		5-10	9.5	26.8	17.3	21.5	12.9
		10-20	8.9	12.1	8.3	3.4	6.2
		20-40	8.6	12.0	15.2	1.7	3.6
		40-60	11.1	21.0	15.5	4.6	7.5
	No-Till	0- 5	167.1	40.6	59.6	56.4	12.5
		5-10	27.8	23.3	17.8	14.8	4.9
		10-20	8.7	16.7	10.3	12.1	4.4
		20-40	11.0	11.2	10.1	11.9	4.3
		40-60	10.5	11.5	15.2	12.6	4.3
Cumulative precipitation incurred since N application (cm)			0	5.2	8.6	10.8	21.6

During the first two weeks after N application in the spring, the N distribution was similar in the soils under the two tillage systems except for the surface 5 cm (Table 13.1). Patterns of convective transport of mineral-N were not affected by tillage and the physical properties of the soil under saturated conditions. However, there was a more rapid decline in N concentration from the untilled surface 5 cm as rain occurred which possibly could have indicated runoff loss or enhanced leaching from the no-till plots beyond the sampling depth 6 days after N application. That would have been the case, except that the N distribution as a function of depth on the 28th day after treatment was still equivalent to that of N on the 13th day even though an additional 2.3 cm of precipitation occurred. During that same period the concentration of nitrogen in the plow treatment decreased by 50 percent (Table 13.2).

Table 13.2: Mineral N Content in Soil, to a Depth of 0.6 m, as a Function of Time Following Fertilizer Application

Nitrogen Application Rate (kg/ha)	Tillage	Mineral N Content (kg/ha) Days After N Application				
		0	6	13	28	51
0	Moldboard	86.0	124.5	60.9	32.2	70.8
	No-Till	154.1	138.5	93.9	34.9	42.1
55	Moldboard	138.0	199.1	150.2	72.0	59.1
	No-Till	177.4	129.8	136.2	132.3	42.9
Cumulative precipitation (cm)		0	5.25	8.55	10.84	21.56

Expressed on an area basis, to a depth of 0.6 m, the results more clearly indicated that fertilizer N was retained in the straw mulch and surface soil layers to minimize leaching losses out of the wheat root zone. While first minimizing N losses initially following application in early spring, the mulch layer can act as a controlled-release source of N for plant growth requirements, when coupled with the increased temperatures of later spring.

The stratification of soil properties with depth is again observed with the N availability index data (Table 13.3). The maintenance of surface residues resulted in an increased immobilization of nitrogen as organic forms in the surface 10 cm of soil paralleled the increased microbial biomass in the chisel-till and no-till treatments. The 10-20 cm soil depth in the chisel-till system also showed a similar increase in immobilization of N due to the shallow partial incorporation of straw. It is also noteworthy that the size of this potentially available pool of N is gradually increasing with the length of time the field plots are in conservation or no-tillage (Table 13.4). This conservation of soluble N in the soil organic N pool should eventually stabilize and be potentially available for crop requirements. Mineralization-immobilization turnover can be altered by edaphic as well as management factors and mineralization of formerly immobilized fertilizer N can occur even within a growing season (Carter and Rennie, 1984).

Table 13.3:   Nitrogen Availability Index in Soil as a Function of Depth, Under Three Tillage Systems, 1983-86

| Tillage | Soil Depth | N Availability Index | | |
		3 Nov 83	12 Dec 84	3 Mar 86
	(cm)	(µg N/g)		
Moldboard	0-10	16.5	15.1	17.5
	10-20	17.7	10.3	15.2
	20-30	7.6	9.7	8.3
Chisel	0-10	23.8	25.5	33.4
	10-20	15.4	16.9	21.2
	20-30	5.6	7.8	8.7
No-Till	0-10	25.1	24.7	25.3
	10-20	12.5	15.2	15.1
	20-30	4.4	3.5	5.0
Noncultivated Check	0-10	--	--	24.5
	10-20	--	--	13.8
	20-30	--	--	15.2

Table 13.4:   Nitrogen Availability Index of the 0-30 cm Soil Depth Under Three Tillage Systems, 1983-86

| Tillage | N Availability Index | | |
	3 Nov 83	12 Dec 84	3 Mar 86
	(µg N/g)		
Moldboard	13.8	11.6	13.5
Chisel	14.7	16.5	20.8
No-Till	13.8	14.2	14.9
Noncultivated Check	--	--	17.8

Herbicide Dissipation in Tillage Systems

The change in surface ecology of the tilled and untilled soils also modified the persistence of the herbicide BAY SMY 1500 (Table 13.5).   A higher microbial activity, resulting from surface mulching, enhanced the herbicide dissipation from the surface 10 cm of the untilled soil. DT_{50} or the time required for 50 percent reduction in soil concentrations, as affected by tillage were as follows: chisel-plowing > moldboard-plowing > no-tillage. Retention of herbicides by the mulch reduces their soil concentration, as straw has been observed to intercept surface-applied herbicides at the time of application (Banks and Robinson, 1982; Ghadiri, Shea, and Wicks, 1984; Ghadiri et al., 1984; and Crutchfield, Wicks, and Burnside, 1985). In this study however, retention of BAY SMY 1500 by straw weathered for 3 months of summer must not have been appreciable as recovery of the applied herbicide, in extracts of soil sampled immediately following herbicide application or 6 days later, exceeded 98 percent in no-till plots. In addition, the performance of this soil-activated herbicide was not hampered as >90 percent control of winter annuals was achieved in all tillage plots with an early post-emergence application of 1.0 kg/ha.

Table 13.5: Field Persistence of BAY SMY 1500 in Port Loam as Affected by Three Tillage Systems at an Initial Concentration of 1 kg/ha

Tillage	Disappearance Rate Constant	DT_{50}[a]	r_2[b]
	day^{-1}		
Moldboard	0.0434	16.0	0.70
Chisel	0.0365	19.0	0.63
No-Till	0.0599	11.6	0.55

[a]Time for 50 percent disappearance (days)

[b]Goodness of fit to first-order reaction model $A=A_o e^{kt}$.

Along with the degradation process, soil solution-phase concentrations of reactive solutes such as organic herbicides are also determined by the process of adsorption or retention by the soil surface. Adsorption has a significant impact on the subsurface movement of agrochemicals by reducing their solution concentration and thereby their redistribution in the environment. Estimates of transport potential of BAY SMY 1500 in two benchmark soils, one of each from the two tillage experimental sites, indicated that BAY SMY 1500 is a potentially mobile herbicide (Table 13.6). Its retardation is highest in the surface horizon in both soils and declined with lower soil depths. Retardation did not appear to be a function of the clay content of horizon soil (Table 13.6).

In conservation systems the volumetric water contents of surface soil was found to be higher than plowed soil (Johnson, Lowery, and Daniel, 1984) throughout most of the year on both of these experimental sites (Lonkerd and Dao, 1985). These moisture conditions are more conducive for herbicide mobility as an inverse relationship exists between R and θ as shown in equation (2). Field verification of this relationship is currently underway.

Table 13.6:   Retardation Factors for BAY SMY 1500 by Soil from Each
               Horizon of Two Oklahoma Benchmark Soils

Soil	Horizon Texture and Depth	Retardation Factor and 95% Confidence Limit
	(cm)	
Renfrow	silt, 0- 20	3.9 (2.5-6.7)
	clay-loam, 20- 40	1.8 (1.3-3.0)
	clay, 40-140	2.1 (1.4-3.9)
Port	loam, 0- 30	3.4 (2.1-5.6)
	clay-loam, 30- 90	2.5 (1.1-5.3)
	clay, 90-140	1.5 (1.2-2.6)

SUMMARY

In summary, a comparative study of the transformation of ammonium nitrate showed that soluble N was tied up in the residue layer which behaved as a controlled-release source. The result was a less rapid loss of fertilizer by leaching, out of the root zone, enhancing the effectiveness of fertilizer and indigenous soil nitrogen for crop production. Disappearance of BAY SMY 1500 from the surface 10 cm of soil was observed to be more rapid with the no-tillage treatment as compared to the subtill and plowed treatments. Higher volumetric water contents in surface soil contributed to an optimal growth environment for microbial decomposers explaining the rapid disappearance rate of the herbicide. The adsorption of BAY SMY 1500 onto the four soils under study was relatively low and estimated retardation factors for soil from various soil depths indicated the herbicide would be mobile in these soils. However, the shorter persistence tends to negate its potential for subsurface movement with no-tillage. Therefore, the ecological differences at the soil surface between clean-till and conservation till systems have affected the behavior and subsurface transport of these agrochemicals to apparently reduce the potential risks to ground water quality.

The surface microenvironmental differences also point to the need for new management practices for agrochemical inputs. New formulations, delivery and placement techniques, timing of applications to coincide with plant nutritional requirements, and effective weed management must be developed to optimize agrochemical efficacy and reduce the potential for environmental contamination. The safeguarding of ground water resouces from the potential contamination by agricultural chemicals depends a great deal upon the management practices in place at the soil surface, just as much as on the natural hydrologic processes acting upon the contaminant, once it has migrated into the subsurface environment. Conservation tillage practices are effective in controlling soil erosion while improving the quality of surface waters; they may also prove to be the best production practices to protect ground water resources as well.

ACKNOWLEDGMENT

The technical assistance of R.D. Meyer and L.S. Pellack in these studies is sincerely acknowledged.

SELECTED REFERENCES

Banks, P.A., and Robinson, E.L. "The Influence of Straw Mulch on the Soil Reception and Persistence of Metribuzin," Weed Science, Vol. 30, 1982, pp 164-168.

Blevins, R.L., Thomas, G.W., and Cornelius, P.L., "Influence of No-tillage and Nitrogen Fertilization on Certain Soil Properties after 5 Years of Continuous Corn," Agronomy Journal, Vol. 69, 1977, pp 383-386.

Bremner, J.M., "Inorganic Forms of Nitrogen, In: Methods of Soil Analysis, Part 2, C.A. Black et al., (eds), American Society of Agronomy, Madison, Wisconsin, 1965, pp 1179-1237.

Broder, M.W. et al., "Fallow Tillage Influence on Spring Populations of Soil Nitrifiers, Denitrifiers, and Available N," Soil Science Society of America Journal, Vol. 48, 1984, pp 1060-1067.

Carter, M.R., and Rennie, D.A., "Nitrogen Transformations Under Zero and Shallow Tillage, Soil Science Society of America Journal, Vol. 48, 1984, pp 1077-1081.

Crutchfield, D.A., Wicks, G.A., and Burnside, O.C., "Effect of Winter Wheat (Triticum aestivum) Straw Mulch Level on Weed Control," Weed Science, Vol. 34, 1985, pp 110-114.

Dao, T.H. et al., "Effect and Statistical Evaluation of Soil Sterilization on Aniline and Diuron Adsorption Isotherms," Soil Science Society of America Journal, Vol. 46, 1982, pp 963-969.

Dick, W.A., "Influence of Long-term Tillage and Crop Rotation Combinations on Soil Enzyme Activities," Soil Science Society of America Journal, Vol. 48, 1984, pp 569-574.

Doran, J.W., "Microbial Changes Associated with Residue Management with Reduced Tillage," Soil Science Society of America Journal, Vol. 44, 1980a, pp 518-524.

Doran, J.W., "Soil Microbial and Biochemical Changes Associated with Reduced Tillage," Soil Science Society of America Journal, Vol. 44, 1980b, pp 765-771.

Ghadiri, H., Shea, P.J., and Wicks, G.A., "Interception and Retention of Atrazine by Wheat (Triticum aestivum) Stubble," Weed Science, Vol. 32, 1984, pp 24-27.

Ghadiri, H. et al., "Atrazine Dissipation in Conventional-till and No-till Sorghum," Journal of Environmental Quality, Vol. 13, 1984, pp 549-552.

Johnson, M.D., Lowery, B., and Daniel, T.C., "Soil Moisture Regimes of Three Conservation Tillage Systems," Transactions of the American Society of Agricultural Engineers, 1984, pp 1385-1395.

Lonkerd, W.E., and Dao, T.H., "Effects of Tillage on Soil Water Under Continuous Wheat," Agronomy Abstracts, 1985.

Rao, P.S. et al., "Evaluation of Conceptual Models for Describing Nonequilibrium Adsorption-desorption of Pesticides During Steady-flow in Soils," Soil Science Society of America Journal, Vol. 43, 1979, pp 22-28.

Stanford, G., and Smith, S.J., "Estimating Potentially Mineralizable Soil Nitrogen from a Chemical Index of Soil Nitrogen Availability," Soil Science, 1976, Vol. 122, pp 71-76.

CHAPTER 14

IMPACTS OF AGRICULTURAL CHEMICALS ON
GROUND WATER QUALITY IN IOWA

by

Robert D. Libra
George R. Hallberg
Bernard E. Hoyer

Since 1980, the Iowa Geological Survey, in conjunction with numerous state, federal, and local agencies and university researchers, has been investigating the impact of agricultural chemicals, specifically nitrogen fertilizers and pesticides, on ground water. These investigations (Hallberg and Hoyer, 1982; Hallberg et al., 1983, 1984; and Libra et al., 1984) have documented the magnitude of ground water contamination resulting from ag-chemicals, yielded insights into the mechanisms that deliver ag-chemicals to groundwater, and identified hydrogeologic settings that are susceptible to agricultural contamination. The susceptible hydrogeologic settings that have been most fully investigated include karst areas where aquifers lie close to the land surface and alluvial aquifers.

GROUND WATER QUALITY IN A KARST-SHALLOW BEDROCK AQUIFER: THE BIG SPRING BASIN STUDY

Much of the work has focused on the Big Spring Ground Water Basin area of northeast Iowa (Figure 14.1). The Big Spring Ground Water Basin includes 267 km^2 in Clayton County, Iowa, that discharges to the Turkey River. It was chosen for study because of prior studies in the area, a local concern with water-quality problems, and because state-owned structures built at a trout hatchery afford the rare opportunity to gage ground water discharge at Big Spring (a large, carbonate ground water spring).

An extensive database has been developed defining the hydrogeology, soils, and land utilization in the basin. A few items of particular interest are described here (Hallberg et al., 1983, 1984). The ground water basin has been defined by study of the potentiometric surface in the Galena Aquifer, by dye-trace studies, and through assessment of gaining and losing stream reaches. These data, combined with spring and stream gaging show that Big Spring accounts for 85-90 percent of the ground water discharged from the basin.

The Galena Aquifer, which supplies the water discharging at Big Spring, is composed primarily of dolomite and limestone, collectively termed carbonate rocks. Carbonate-rock aquifers possess two properties that often result in anomalous hydrologic characteristics, relative to clastic-rock aquifers. These properties are: (1) generally low primary permeability; and (2) solubility in water which is undersaturated with respect to carbonate minerals. The low primary permeability results in ground water recharge and flow being

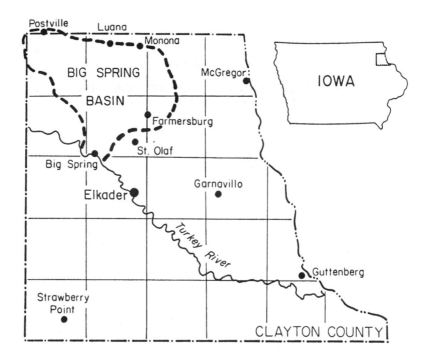

Figure 14.1: Location of Big Spring Basin, Clayton County, Iowa (modified from Hallberg et al., 1983)

concentrated within fractures and along bedding planes, while the solubility of carbonate rocks allows for solutional enlargement of these features.

The wide range of permeabilities that are possible in carbonate aquifers results in varying types of ground water flow and recharge/discharge mechanisms. White (1977) used the terms conduit flow to describe ground water movement through large open cavernous zones or conduits, and the term diffuse flow to characterize flow through relatively unmodified fractures and bedding planes. Carbonate aquifers characterized by conduit flow are recharged largely by the partial or complete capture of surface runoff by sinkholes. Flow is exceedingly fast, relative to most ground water systems, and may be turbulent. Discharge is generally concentrated in a small number of related large springs or gaining-stream reaches. The response of such a system to precipitation is extremely fast and often analogous to the response of a surface water system.

Diffuse-flow carbonate aquifers receive recharge through infiltration along relatively unenlarged fractures and the low-permeability rock matrix. Flow through the system is generally more analogous to flow in clastic aquifers. Discharge is through numerous small springs, seeps, and gaining-stream reaches. The response of a diffuse-flow system to precipitation is slow and similar to the response of a clastic aquifer.

The ground water system that discharges at Big Spring has both conduit- and diffuse-flow components. The location of the main conduit-flow parts of the system has been identified. Figure 14.2 indicates areas where surface drainage is swallowed by sinkholes; these areas account for about ten percent of the basin. These sinkholes and their associated surface drainage basins form the direct recharge points for the conduit-flow system. Figure 14.3 shows the potentiometric/water table surface of the Galena Aquifer. Note the pronounced north-south trending troughs in the surface leading from the sinkhole recharge areas to Big Spring. The potentiometric lows delineate the main conduit zones. Away from these main conduits, the ground water system receives recharge via infiltration through a thin (generally less than 5 m) soil mantle, and ground water flow is along relatively unenlarged fractures and bedding planes. This part of the system, which accounts for the bulk of the Galena Aquifer, is dominated by diffuse-flow. The main conduits act as drains for the diffuse-flow system, as shown schematically on Figure 14.4.

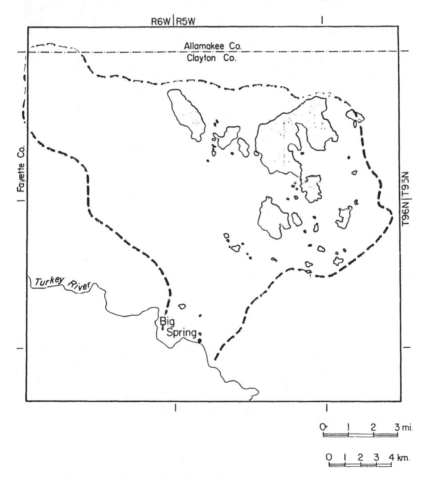

Figure 14.2: Location of Surface Drainage Basins in the Big Spring Basin That Are Swallowed by Sinkholes. Dashed Line Is Basin Boundary /Ground Water Divide (modified from Hallberg et al., 1983)

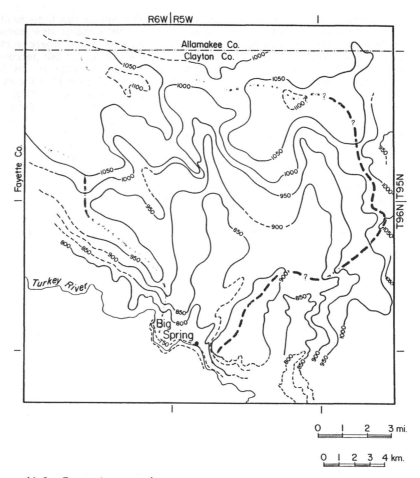

Figure 14.3: Potentiometric/Water Table Surface of the Galena Aquifer in the
Big Spring Basin. Elevation in Feet Above msl (modified from
Hallberg et al., 1983)

Landuse mapping has shown that the Big Spring Basin is wholly
agricultural (with no significant industries, municipalities, ag-chemical
facilities, etc.); essentially 91 percent of the area is used for corn, pasture, or
hay, in rotations for grain and to support the common small dairy and livestock
operations. Except during the PIK program in 1983, about 50-60 percent of the
area has been planted in corn since 1979. With the assistance of the U.S.D.A.
Soil Conservation Service and Iowa State University Cooperative Extension
Service, ag-chemical use has also been monitored.

Since 1981, the discharge from the Big Spring Ground Water Basin has
been gaged. Water quality sampling has also been conducted from a network of
surface water sites, tile lines, wells, the Big Spring, and other small springs.
Sampling intensity at Big Spring has varied for different constituents, in

relation to research goals and hydrologic conditions; for nitrate (NO_3) and pesticides, the sampling interval has ranged from weekly to hourly (or less) since the 1982 water-year. All water analyses have been performed using standard methods by the University Hygienic Laboratory, which has an EPA approved quality assurance/quality control plan.

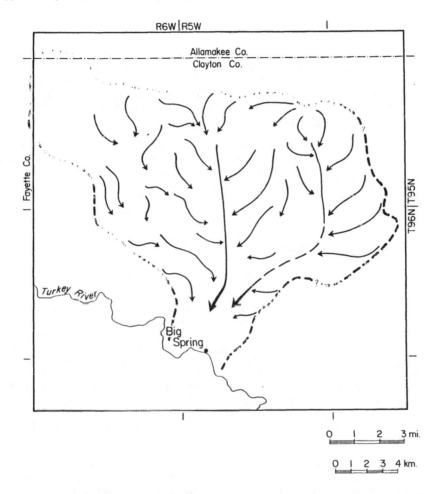

Figure 14.4: Schematic Ground Water Flow in the Big Spring Basin; Long, Continuous Arrows Identify the Main Conduit Zones (modified from Hallberg et al., 1983)

RESULTS OF MONITORING

Several aspects of the monitoring results from the Big Spring Ground Water Basin will be discussed. These include: (1) the similar seasonal trends of nitrate concentrations in all parts of the basin's hydrologic system, from recently infiltrated, shallow ground water, to the large streams that act as ground water discharge zones; (2) the relative importance of the sinkhole-recharged conduit-flow ground water system versus the infiltration-recharged diffuse-flow system in delivering water and ag-contaminants to ground water; and (3) the magnitude of ag-chemical losses into the hydologic system. Long-term changes in ground water quality at Big Spring will also be discussed.

Seasonal Trends in Nitrate Concentrations

The delivery of nitrate (and other surface-derived contaminants) to ground water is largely controlled by the timing and volume of ground water recharge. The effects of ground water recharge to the Galena Aquifer within the basin are reflected by discharge rates and responses at Big Spring. Therefore, an examination of the Big Spring discharge hydrograph (from the first year of monitoring) begins this discussion.

The relationships between precipitation, recharge, and discharge at Big Spring are complex, but show definite seasonal trends. During the winter months at the beginning of monitoring (November-February, 1981-1982), discharge from Big Spring was relatively low, about 1.0-1.1 cubic meters/second (cms) (35-40 cfs) and followed a slowly decreasing trend (Figure 14.5). Temperatures during this period generally remained well below freezing, and precipitation fell as snow. As a result, little recharge occurred, and flow from the spring represented water draining from storage within the diffuse-flow parts of the aquifer. The gradual decrease in discharge is caused by declining heads, and, therefore, declining hydraulic gradients, within the aquifer, as ground water storage is depleted. This is analagous to baseflow-recession conditions in a surface stream.

Spring snowmelt, sometimes accompanied by rainfall, occurred during March and April. Rapid snowmelt generated significant runoff and yielded large volumes of direct recharge to the conduit-flow system. Discharge at Big Spring responds quickly to this direct conduit recharge, and the resulting hydrograph (Figure 14.5) during these months is punctuated by numerous high-flow peaks, with a maximum discharge of about 7.4 cms (260 cfs). During May and June, rain storms produced similar results, though of lesser magnitude. Wet conditions prevailed throughout the entire March-June period, and total basin discharge remained generally high, averaging over 1.7 cms (60 cfs). This four month interval accounted for nearly 50 percent of the total discharge for the water year.

While the high, peak-discharge events that occurred during March through June resulted from runoff recharge to sinkholes and the conduit system, the persistently elevated flows between and following peak events are, to a large degree, the result of significant infiltration recharge to the diffuse-flow system. This infiltration recharge increases the amount of ground water in storage within the Galena Aquifer, raises water table/potentiometric elevations and imposes steeper hydraulic gradients upon the system. This results in increased discharge from the diffuse-flow parts of the aquifer.

The hydrograph for the late summer-fall (July-October) period contrasts markedly with that for the preceeding months (Figure 14.5). Although numerous rainfalls greater than 19 mm (0.75 inches) occurred, no significant runoff was generated and thus, there was little direct recharge to the conduit-flow system, and, in turn, no significant discharge events occurred at Big

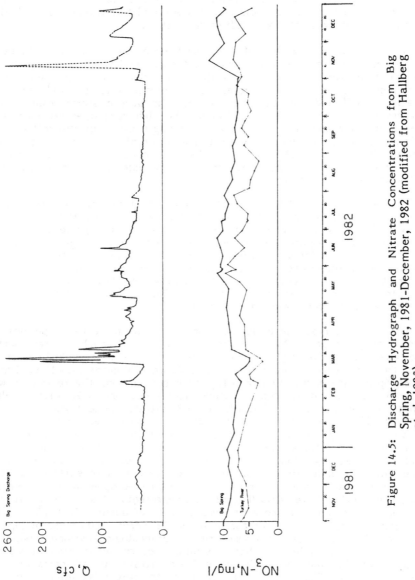

Figure 14.5: Discharge Hydrograph and Nitrate Concentrations from Big Spring, November, 1981-December, 1982 (modified from Hallberg et al, 1983)

Spring. Additionally, discharge-flow rates steadily decreased across this period, indicating base-flow conditions with little infiltration recharge occurring. As in the winter months, discharge from Big Spring during July-October was primarily ground water released from storage within the aquifer. The lack of recharge to the aquifer during these months is caused by high-water uptake by crops and other plants and hot summer temperatures. These factors result in very high rates of evapotranspiration and relatively low soil moisture levels, leaving little or no precipitation available for runoff or infiltration.

November and December of 1982 experienced very different climatic conditions compared to the same months of the preceeding water year. Daily maximum temperatures were generally well above freezing, and most precipitation fell as rain. Several fairly intense rain storms occurred generating runoff and high-flow events, with the largest storm resulting in discharges greater than 7.1 cms (250 cfs). The high discharges are partially related to the size and intensity of the preceeding storms, but also reflect the effects of rainfall on harvested fields during a period of low evapotranspiration potential. These factors allow for a significant amount of the precipitation to run off, causing direct recharge to the conduit-flow system, and resulting in the high discharges at Big Spring.

Seasonal trends in nitrate concentrations in ground water discharging from the Big Spring Ground Water Basin show a strong relationship to certain aspects of discharge and, therefore, as stated above, recharge to the ground water system. During periods of no recharge, where the discharge is recessing (such as November-January, 1981-82, and July-October, 1982), nitrate concentrations also decrease (Figure 14.5). When recharge occurs and discharge increases (such as February-June and November-December, 1982), nitrate is leached from the soil to saturated parts of the Galena Aquifer; this ultimately increased nitrate concentrations at Big Spring.

This seasonal response of nitrate concentrations to recharge periods is evident in all parts of the basin's hydrologic system. Although different parts of the system have differing "lag-times" in their response to individual recharge events, the overall trends are consistent, as shown in Figures 14.6-14.9. Data from Big Spring are included on each figure for comparison. Figure 14.6 shows median and quartile nitrate concentrations from a network of 20 private wells completed in the Galena Aquifer; Figure 14.7, nitrate concentrations from a deep main tile drainage outlet and the stream it discharges to; Figure 14.8, from two sites along the main surface stream within the basin; and Figure 14.9, from the Turkey River, a major perennial stream that acts as an important ground water discharge zone in the area. Analyses of the tile-line effluent (Figure 14.7) are indicative of nitrate concentrations in recently recharged shallow ground water collected from a fertilized, row-cropped area of several hectares; the Turkey River data (Figure 14.9) represents nitrate concentrations from ground water and surface water over an area of about 2500 km^2. While the absolute concentrations decrease as the scale increases (because of inputs from unfertilized areas and less susceptible hydrogeologic environments), the similarity of trends indicates that a similar process, infiltration of water through nitrate-bearing soil horizons, delivers nitrates to all parts of the hydrologic system. The data also demonstrates the responsive interconnected nature of the soil water-ground water-surface water system in northeast Iowa (and geologically similar areas).

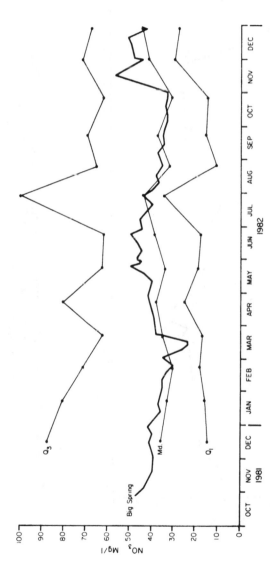

Figure 14.6: Median and Quartile Nitrate (as NO₃) Concentrations from Galena Aquifer Wells. Data from Big Spring Included for Comparison (Hallberg et al., 1983).

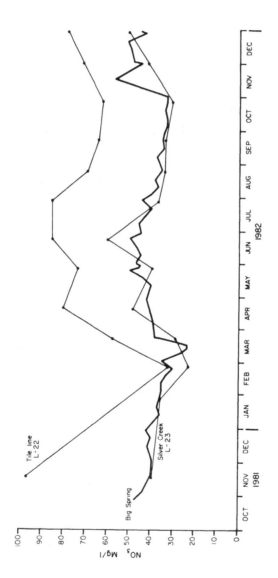

Figure 14.7: Nitrate (as NO_3) Concentrations from a Tile-line and the Stream It Discharges into, with the Big Spring Basin. Data from Big Spring Included for Comparison (Hallberg et al., 1983).

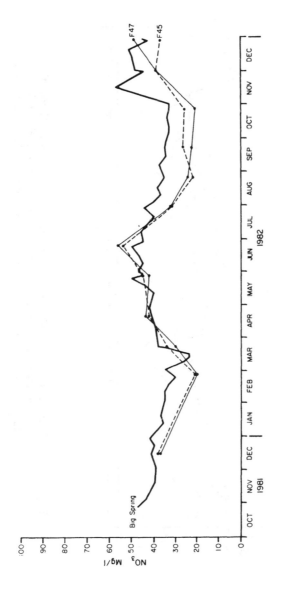

Figure 14.8: Nitrate (as NO₃) Concentrations from Two Sampling Points on the Main Surface Drainage Within the Big Spring Basin. Data from Big Spring Included for Comparison (Hallberg et al., 1983)

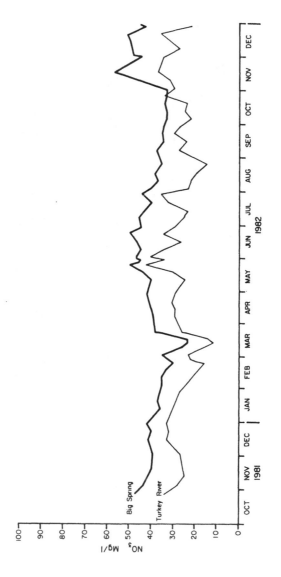

Figure 14.9: Nitrate (as NO_3) Concentrations from the Turkey River, the Surface Discharge Area for the Big Spring Ground Water Basin. Data from Big Spring Included for Comparison (Hallberg et al., 1983)

Recharge Mechanisms and Contaminant Delivery

Recharge to the Galena Aquifer in the Big Spring Ground Water Basin takes place in two ways: by surface runoff that is swallowed by sinkholes and by conventional infiltration. These dissimilar recharge mechanisms deliver different concentrations of ag-chemicals to the Galena Aquifer. To define the relationship between contaminant concentrations and recharge mechanisms, several major, runoff- and infiltration-producing recharge (precipitation or snowmelt) events have been monitored intensively. Big Spring itself has been the focus of the majority of the event-related well and tile-line sampling. Surface streams (including those draining to sinkholes) were also sampled, though less frequently.

Figure 14.10 shows the Big Spring discharge hydrograph for a large event during the summer of 1983. Approximately 230 mm of rain fell intermittently during a three-day period, which produced significant infiltration recharge and generated a sequence of three runoff events that, in turn, caused three abrupt rises of discharge at Big Spring (Figure 14.10). The third event produced the highest discharge which reached 7.1 cms (250 cfs) on 1 July. Prior to these rains, ground water was undergoing slow, base-flow recession, and the discharge was 1.4 cms (50 cfs).

Figures 14.11-14.14 show water-quality data superimposed on the hydrograph. These data illustrate the effects of sinkhole-captured surface runoff water on ground water quality. During the abrupt discharge rises, the ground water chemistry changes in response to the mixing of different sources of recharge water: runoff vs. infiltration. Runoff provides constituents typical of surface water, such as suspended sediment and chemicals of low mobility; infiltration provides constituents more typical of ground water, including various mobile dissolved ions such as NO_3. A few examples illustrate this.

Suspended sediment (Figure 14.11) increases from negligible values to concentrations over 4,000 mg/l (a load of over 87,000 kg/hr) as the runoff component discharges at the spring. In marked contrast is the change in specific conductance (SpC), which integrates the effects of dissolved ions. During base-flow (before and after the discharge events), when the ground water quality must be controlled by infiltration processes, the SpC is between 700-730 μmhos/cm^2. During the major discharge peak, the SpC drops sharply from 670 to 450 μmhos/cm (Figure 14.12), reflecting the dilution of infiltration-derived ground water by the surface runoff water, which has very low concentrations of dissolved constituents.

Similar changes can be seen in the concentrations of pesticides and NO_3, whose maximum concentrations are related to the runoff and infiltration components, respectively (Figures 14.13 and 14.14). Atrazine is the most widely used herbicide in the basin, and it is the dominant pesticide found in the ground water. Prior to the discharge events (June 24-26), atrazine concentrations were about 0.2 μg/l. During the discharge events, three peak periods of atrazine and total pesticide concentrations occur, corresponding to the three discharge events and the three periods of captured runoff recharge. The first two pesticide-concentration peaks are out-of-phase with the discharge peaks because of time lags between discharge rises and the arrival of water-quality changes at these lower discharges (Hallberg et al., 1984); the third and highest pesticide peak roughly coincides with the peak of suspended sediment. During the influx of runoff recharge, several other pesticides also occur in the

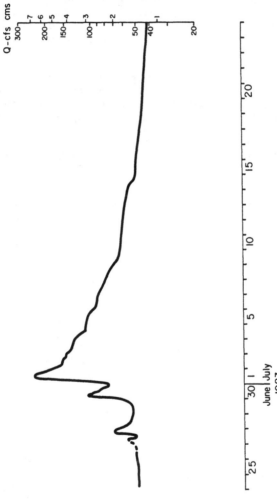
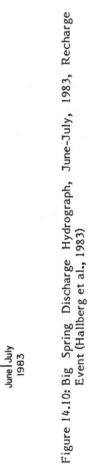

Figure 14.10: Big Spring Discharge Hydrograph, June-July, 1983, Recharge Event (Hallberg et al., 1983)

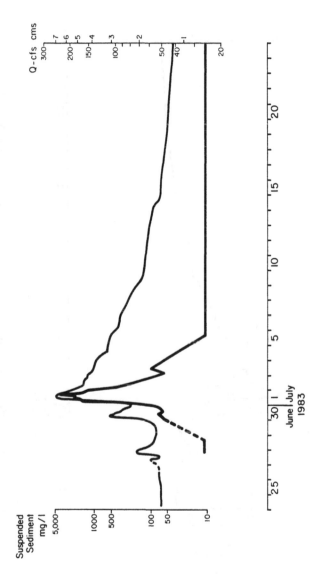

Figure 14.11: Sediment Concentrations in Big Spring Discharge Water, June–July, 1983, Recharge Event. Discharge Hydrograph also Shown (Hallberg et al., 1983)

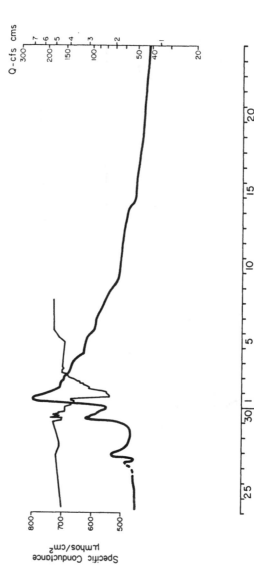

Figure 14.12: Specific Conductance of Big Spring Discharge Water, June-July, 1983, Recharge Event. Discharge Hydrograph also Shown (Hallberg et al., 1983)

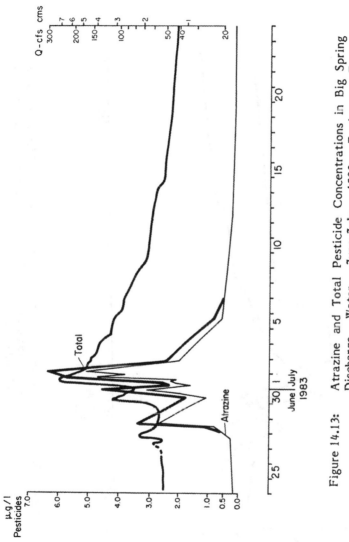

Figure 14.13: Atrazine and Total Pesticide Concentrations in Big Spring Discharge Water, June-July, 1983, Recharge Event. Hydrograph also Shown (Hallberg et al., 1983)

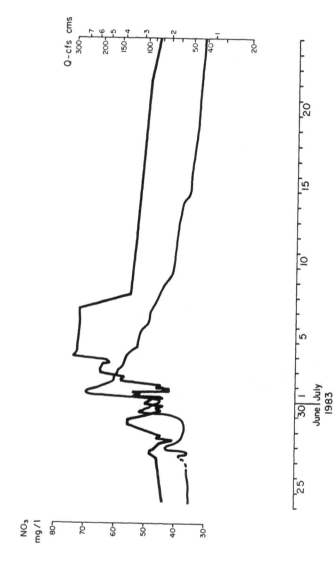

Figure 14.14:   Nitrate (as NO$_3$) Concentrations in Big Spring Discharge Water
June-July, 1983, Recharge Event. Discharge Hydrograph also
Shown (Hallberg et al., 1983)

ground water (in the water phase). The various pesticides and their maximum concentrations (μ g/l) were: atrazine, 5.1; cyanazine, 1.2; alachlor, 0.6; metolachlor, 0.6; and fonofos, 0.1. In the Big Spring Ground Water Basin ground water atrazine is the only pesticide which persists year-round, though alachlor and cyanazine have appeared intermittently during ground water baseflow periods.

Nitrate concentrations show a complex record (Figure 14.14) that is almost directly out-of-phase with the pesticides. Nitrate concentrations are lowest when sinkhole-captured runoff dominates the discharge. As the "flood wave" of this runoff recharge passes and discharge begins to recess (from July 3 on), infiltration recharge dominates, NO_3 concentrations rise dramatically to values of 16.5 mg/l NO_3-N (75 mg/l NO_3) for a short time, and then decrease to typical to baseflow concentrations of 10-11 mg/l NO_3-N.

Other water-quality parameters show the same effects and vary in relation to their recharge-delivery mechanism. Parameters generally associated with surface water runoff (e.g., K, PO_4, Fe, organic-N, and ammonium-N) are found in their highest concentrations, or only occur in ground water, coincident with the high sediment and peak pesticide concentrations occurring in the sinkhole captured runoff recharge. Also, bacterial occurrence in wells peak during these periods. Other water-quality parameters, typically associated with slower, infiltration recharge and diffuse-flow (e.g., Cl, and SO_4) coincide with elevated SpC and NO_3.

Detailed monitoring during such rainfall-runoff events at Big Spring shows that the high surface water concentrations of parameters such as suspended sediment and pesticides enter the ground water and move as a "slug" through the system. They discharge from the ground water in essentially the same concentration that they entered, at least during the larger hydrologic events.

The detailed analysis of such events not only demonstrates what happens to ground water quality during a 'discharge' event in the unique karst setting, but also affords a calibration for quantitative estimates of the relative delivery of recharge water and contaminants by runoff vs. infiltration. A ready analogy can be made between the Big Spring ground water hydrograph and a humid climate, surface-stream hydrograph. The large peaks in the ground water discharge are related to surface water runoff and rapid flow through conduits, while in a stream this storm or flood flow is more gradual, falling, or receding. The prolonged, stable portions of the hydrograph are related to typical ground water components, infiltration recharge, and bank-storage effects in a stream.

Based upon this analogy, various hydrograph-separation techniques, typically used in surface water hydrology (e.g., Singh and Stull, 1971), can be adapted to these ground water hydrographs. Just as these methods are used for quantitative assessment of 'flood flow' and 'base flow' in a stream, they provide a quantitative estimate of the 'runoff-conduit flow' component of recharge and the 'infiltration-diffuse flow' component (Hallberg et al., 1983, 1984). Several analytical and chemical methods (e.g., Freeze and Cherry, 1979, pp. 219-229) have been used to provide a check on the consistency and range of derived values. The various methods produce different results which provide insights into hydrologic behavior. Overall, the methods have produced quantitative results which are quite consistent ($\pm$ 10 percent) and compare quite well with the actual chemical monitoring data. This allows some refined conclusions.

Synthesis of three water-years data shows that the infiltration component contributes about 90 percent of the water, about 95 percent of the NO$_3$-N, and 50-85 percent of the pesticides delivered to the ground water system. The runoff component delivers about 10 percent of the water, only about 5 percent of the nitrate, and from 15-50 percent of the pesticides. While the runoff component delivers contaminants to the ground water which are of concern for public health on the local level, the infiltration component is responsible for regional aquifer contamination. In summary, these studies show that: the infiltration component delivers to ground water the largest mass and the highest concentrations of nitrates (and other mobile ions), and the largest mass of mobile pesticides, but in generally low concentrations; the runoff component delivers to ground water high concentrations and large loads of pesticides and other relatively insoluble or highly adsorbed chemicals, peak turbidity and sediment loads, and other peak loads of organics and pathogenic organisms, but for short periods of time.

In the past, many water-quality problems in the Iowa karst areas had been attributed to sinkholes and captured runoff. Monitoring at Big Spring has shown that simple infiltration is the major mechanism of contaminant delivery, even in these karst areas. Infiltration is the recharge mechanism common to all aquifers, which gives these data much broader implications and clearly affects the nature of management changes needed to mitigate these problems.

Magnitude of Agricultural Chemical Losses

Gauging of ground water and surface water discharge, coupled with ag-contaminant concentrations in discharge water, has allowed for the calculation of ag-chemical losses from row-cropped fields to the Big Spring basins hydrologic systems. Table 14.1 summarizes ground water, surface water, and nitrate-N discharge from the basin, along with precipitation for three complete years of monitoring. Total losses have varied from 820,000-1,300,00 kg of nitrate-N. The observed variability is related to the total water flux, with losses during a "wet" year (i.e., 1983) being higher than during a "normal" year (i.e., 1982 or 1984). Table 14.2 utilizes results from landuse and chemical-use inventories to put the nitrate-N losses in perspective. These losses represent 50 to 80 kg-N/ha from the long-term corn acreage in the basin and are equivalent to 33-55 percent of the fertilizer-N applied annually. This is not to imply that the NO$_3$-N lost in water discharging from the basin is derived directly from nitrogen fertilizer. However, the losses that occur are a direct response to the large amount of fertilizer applied (Hallberg, 1986).

These are minimum figures for the amount of N lost, because only the NO$_3$-N losses are computed. Other forms of N are also discharged in the water and losses by denitrification have not been estimated. Comparison with other regional and local studies suggest that these losses are probably typical for Iowa and much of the midwest under current management practices (e.g., Baker and Johnson, 1977, 1981; Gast, Nelson, and Randall, 1978; Kanwar, Johnson, and Baker, 1983; Hallberg, 1986b; and Hallberg et al., 1984). Beyond the environmental impact, the magnitude of the N-losses are of economic concern as well. When such substantial amounts of N are not being utilized for crop production, there is obvious room for improved efficiency and economic gain.

In contrast to the nitrogen losses, the total pesticide losses are quite small. For example, the total amount of atrazine discharged in ground water at

Big Spring, during the three water-years, ranged from 7-18 kg.   The loss of atrazine (and total pesticides) in ground water equaled less than 0.1 percent of the amount applied.   Pesticide concentrations in surface water were not monitored in sufficient detail (because of costs) to warrant calculating mass losses with surface water.   However, pesticide concentrations in surface waters commonly were an order of magnitude higher than in ground water, and total pesticide losses are estimated at about one percent of the amount applied.

As previously discussed, annual nitrate-N losses from the basin are a function of water flux.   Figure 14.15 summarizes ground water discharge and flow-weighted mean nitrate and atrazine concentrations at Big Spring for four years of monitoring.   Note that the flow-weighted mean nitrate concentrations in ground water are also tied to water-flux.   This is in sharp contrast to the trend shown by annual flow-weighted atrazine concentrations (Figure 14.15), which have increased during each year of monitoring, independent of water flux and climatic conditions.   While the mean annual atrazine concentration for water-year 1985 was still below 1.0 μg/l, it has increased more than 50 percent per year for the period of monitoring.

As shown by Figure 14.15 and Table 14.1 and 14.2, water-year 1983 had the highest losses and flow-weighted mean nitrate-N concentrations for the period of record at Big Spring.   The year 1983 was also the year of the Payment in Kind (PIK) program.   PIK reduced corn acreage in the basin and, therefore, the amount of fertilization-N applied, by about 43 percent (Hallberg et al., 1984), while nitrate losses and concentrations increased.   This indicates that after many years of nitrogen fertilization, in excess of crop uptake of nitrogen, a large amount of nitrate-N is stored in the soil-water system.   A significant decrease in N-application, for only one year, has had no immediate impact on water quality over an area the size of the basin (267 km^2).

Long-term Impacts of Agricultural Contaminants

Evaluation of the long-term impacts of ag-contaminants is difficult, largely because of the paucity of historic data for nitrates and particularly pesticides.   Big Spring was sampled for nitrate in September, 1951, and again in 1968, where monthly samples were collected from January-September.   Nitrate concentrations from these samples ranged from 1.5-3.0 mg/l (7-14 mg/l as NO_3) and averaged 2.7 mg/l (12 mg/l as NO_3).   Normal climatic years during the present monitoring at Big Spring yield flow-weighted mean nitrate-N concentrations of about 9.0 mg/l (40 mg/l as NO_3), indicating a three-fold increase in ground water nitrate concentrations.   Figure 14.16 shows the changes in nitrate concentrations at Big Spring, along with estimates of annual fertilizer and manure-nitrogen applications in the basin.   Increases in cattle and hog populations suggest a 30 percent increase in the manure-N produced in the basin over the past two decades.   During the same period, the combination of increased rates of nitrogen fertilizer application and increased corn acreage have resulted in a 250 percent increase in the total amount of fertilizer-N applied in the basin.   The increase in ground water nitrates directly parallels the increase in N-fertilization.   Existing ground water quality data from many parts of Iowa show this response during the last two decades.   Experimental farm studies and other agronomic research also provide corroborative results, as summarized by Hallberg (1986a).

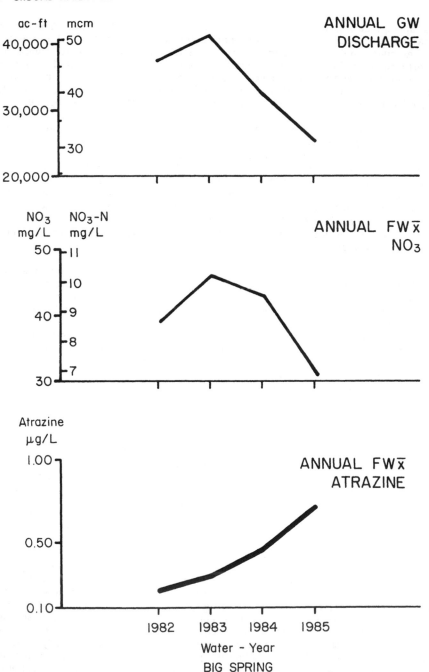

Figure 14.15: Annual Ground Water Discharge, Flow-weighted Mean Annual Nitrate, and Atrazine Concentrations, 1982-1985 Water Years (Hallberg, 1985)

Table 14.1:   Summary of Three Water-years' Discharge, Precipitation, and
NO₃-N Discharge Data from the Big Spring Basin

		Water-Year	
	1982*	1983*	1984
1. Water Yield	$x10^6$ cubic meters		
Total ground water discharge	58.5	63.4	52.5
Streamflow discharge	42.9	85.9	45.1
Total discharge	101.3	149.2	97.6
2. Precipitation and Discharge	mm		
Precipitation	864	1,130	833
Water Yield			
A. Less change in gw storage	332	559	353
	%		
B. As % precipitation	33	49	42
3. NO₃-N Discharged	$x10^3$ kg		
In ground water	478	614	466
In surface water	343	687	350
Total	821	1,300	817
	mg/l		
Flow-wtd mean NO₃-N concentration in ground water	8.7	10.2	9.6

*From Hallberg et al., 1984.

GROUND WATER QUALITY IN SHALLOW BEDROCK VERSUS DEEP BEDROCK REGIONS: FLOYD AND MITCHELL COUNTIES, IOWA

During the initial phases of Iowa Geological Survey investigations into ag-contaminants and ground water, existing geologic data and water quality analyses were compared, and criteria for evaluating the susceptibility of areas to surficial contamination were established (Hallberg and Hoyer, 1982). On a regional basis, some simplistic, yet effective, geologic areas were defined: (1) deep bedrock areas--where 15 m (50 ft) or more of low permeability glacial till or shale overlies an aquifer; (2) shallow bedrock--areas where less than 15 m (50

Table 14.2:   Summary of NO3-N Loss from the Big Spring Basin in Relation to
Landuse for Three Water Years (shown as kg-N/ha equivalent)

	1982[a]	1983[a]	1984
kg-N/ha of total basin	31	49	31
kg-N/ha of long-term row crop (land in corn rotation)	52	83	52
% of applied chemical-N	33%	53%[b]	33%

[a]Data from Hallberg et al., 1983, 1984.
[b]PIK year; % based on 1982 applied chemical -N.

ft) of low permeability materials overlie an aquifer; and (3) karst areas--where
significant clusters of sinkholes exist. This initial study showed that significant
concentrations of nitrate were present in ground water in the shallow bedrock
and karst areas to depths of 45-60 m (150-200 ft), while in deep bedrock areas
such contamination is largely absent (Hallberg and Hoyer, 1982). Investigations
carried out in Floyd and Mitchell Counties have further demonstrated the
applicability of these geologic regions (Libra et al., 1984).

The location of Floyd and Mitchell Counties is shown in Figure 14.17.
Bedrock geology consists of Devonian strata, largely carbonate units with lesser
amounts of interbedded shales and shaly carbonates. Shale and shaly carbonate
strata act as confining beds, separating the carbonate units that are the major
aquifers in the area.   Stratigraphic (Witzke and Bunker, 1984, 1985) and
hydrogeologic (Libra et al., 1984; and Libra and Hallberg, 1985) investigations
have indicated that the Devonian strata are best viewed as a layered three
aquifer system. Most rural residents rely on the two stratigraphically highest
carbonate aquifers for their drinking water supplies.

Mapping of depth to bedrock and sinkhole locations allowed for the
delineation of geologic regions in the area (Figure 14.18).   Note that an
additional region, termed "incipient karst", has been added. The incipient karst
area is characterized by very shallow (< 5m) depths to bedrock, and the presence
of numerous very shallow, closed, soil-filled depressions. These depressions
may represent an early stage of karst development, where a limited amount of
settling of the thin unconsolidated deposits, into fractures and other solutional
openings, has occurred. These shallow "incipient" sinkholes have not undergone
collapse, are not "open" to the surface, and act as no impediment to row-crop
agriculture. They do, however, limit surface runoff and increase infiltration.

Figure 14.18 also shows the results of nitrate analyses from about 50
private wells (collected in December, 1982) in the area, superimposed on the
geologic regions. Note that with few exceptions, nitrate concentrations in the
deep bedrock areas are less than 5 mg/l (as NO_3; < 1 mg/l NO_3-N).

Table 14.3 summarizes means and ranges of nitrate (as NO_3-N) and
bacteria concentrations, and the percentage of samples that showed detectable

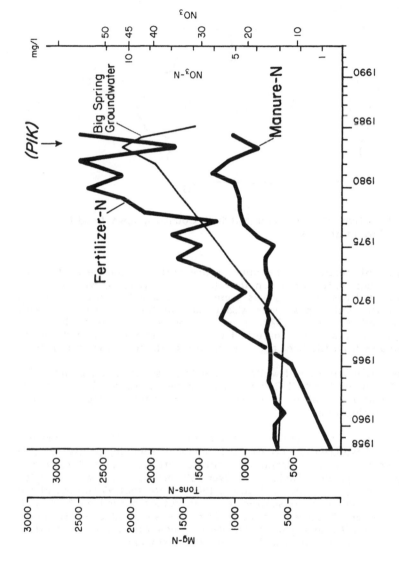

Figure 14.16: Graph Showing Changes in Nitrate (as NO₃) Concentrations at Big Spring with Estimates of Fertilizer and Manure–N Applications in the Big Spring Basin (Hallberg et al., 1983)

pesticide residues, from about 20 wells in the area sampled monthly during the period December, 1982-December, 1983.   Again, the lowest nitrate concentrations occur in deep bedrock regions, the highest in the karst, particularly in the high infiltration-recharge incipient karst regions.   Shallow bedrock areas have intermediate values.   Note that over 50 percent of the samples from wells in the "susceptible" geologic regions showed detectable levels of pesticides; 70-80 percent of these wells had detectable pesticides at least once during the year.

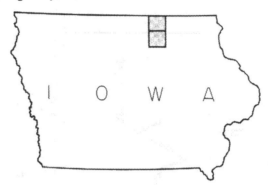

Figure 14.17: Location of Floyd and Mitchell Counties, Iowa (modified from Libra et al., 1984)

The type of pesticides, maximum concentrations, and percentage of detection in ground water are given in Table 14.4.   Atrazine was the most commonly detected pesticide and was present in all samples collected from a number of wells.   Other pesticides also persist in the environment.   Well samples collected in April, 1983, prior to spring chemical applications, contained cyanazine, alachlor, metolachlor, and metribuzin, in addition to atrazine. Samples from tile lines and streams also contained these pesticides.

The monthly monitoring of wells, the tile lines, small streams, and major rivers in the Floyd and Mitchell County area documented seasonal variations of nitrate concentrations, in all parts of the hydrologic system analogous to those seen in the Big Spring area.

Additional work is currently in progress in Floyd and Mitchell Counties. Ongoing projects have greatly increased the number of sampling points, relative to the work described above. Preliminary nitrate data from one such project in Mitchell County are given in Table 14.5. Over 300 domestic wells were sampled for nitrate during one weekend in October, 1985. Note that 90 percent of the wells from deep bedrock areas showed less than detectable (< 1 mg/l, NO_3-N) nitrates, while over 20 percent of those from the susceptible environments exceeded the 10 mg/l drinking water standard. Median nitrate concentrations in these susceptible areas were equal to one-half the standard.

GROUND WATER QUALITY IN ALLUVIAL AQUIFERS

In much of western Iowa, bedrock aquifers are deeply buried by clay-rich

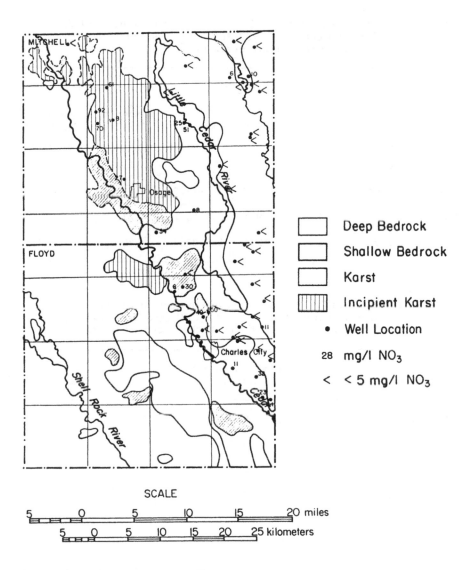

Deep Bedrock

Shallow Bedrock

Karst

Incipient Karst

● Well Location

28 mg/l NO₃

< < 5 mg/l NO₃

SCALE

Figure 14.18: Geologic Regions and Nitrate (as NO₃) Concentrations in 50 Private Wells, Sampled in December, 1982, Floyd and Mitchell Counties, Iowa (Libra et al., 1984)

Table 14.3: Summary Data from 20 Wells, Sampled December, 1982,
Through December, 1983, for Nitrate and Pesticides, by
Geologic Region (Libra et al., 1984)

	NO$_3$-N (mg/l)			Pesticides	
Geologic Region	Number of Samples	Mean	Range	Number of Samples	Number of Detections
Deep Bedrock	44	<1	0 - 1	13	1
Shallow Bedrock	66	4.9	<1 -12.9	29	17
Karst	44	9.1	3.5-26.4	23	14
Incipient Karst	40	15.1	5.5-29.3	20	7

Table 14.4: Types of Pesticides and Maximum Concentrations Detected in
Ground Water. Samples (N=85) Collected December 1982 Through
December, 1983, Floyd and Mitchell Counties (Libra et al., 1984)

Pesticides	Maximum Concentrations (μg/l)	Detections %
Atrazine	0.85	27
Alachlor	16.0	8
Cyanazine	0.5	8
Metolachlor	0.1	1
Metribuzin	4.4	16

glacial till, and/or have naturally high total dissolved solids concentrations.
Therefore, bedrock aquifers in western Iowa are not utilized as drinking water
sources to the degree they are in other parts of the state. Alluvial sand and
gravel deposits, located along major stream valleys in western Iowa, offer large
quantities of water with good natural quality coupled with inexpensive drilling
and well construction costs. These alluvial aquifers are widely used by
municipalities, rural water districts, and individual rural residents. As the
alluvial aquifers occur beneath relatively low relief floodplains, they are the
sites of intensive row-crop production for corn and soybeans. In some areas,
the sandy soils are irrigated, generally with sprinkler systems.

The coarse-grained nature and relatively thin (< 3m) unsaturated zones of

Table 14.5: Nitrate (as NO₃-N) in Ground Water from Mitchell County,
October, 1985, by Geologic Region

Geologic Region	Number of Wells Sampled	% <5 mg/l	% >45 mg/l	Median
Deep Bedrock	78	91	5	<1 mg/l
Shallow Bedrock	119	40	21	4.4 mg/l
Karst	28	25	21	4.9 mg/l
Incipient Karst	55	31	27	5.1 mg/l

these alluvial deposits offer little protection, from surficial contaminants, to the ground water they contain. Existing data (summarized by Hallberg, 1986b) and in-progress studies (Thompson, Libra, and Hallberg, 1986, in prep.) concerning alluvial aquifers indicate: (1) seasonal trends in nitrate are very similar to those seen in the bedrock aquifers of eastern Iowa; (2) nitrate concentrations are generally similar to or higher than those in eastern Iowa bedrock aquifers; (3) nitrate concentrations in many alluvial aquifers have increased greatly over the past two decades, paralleling those changes noted at Big Spring; and (4) a pronounced vertical stratification of contaminants exists. Sampling by the Iowa Department of Water, Air, and Waste Management (IDWAWM) showed that 20 of 41 municipal wells utilizing alluvial systems (mainly in western Iowa) contained detectable pesticides, including atrazine, cyanazine, alachlor, metolachlor, metribuzin, and fonofos (Kelley, 1985).

SUMMARY

The susceptible hydrogeologic settings discussed in this chapter occupy a significant part of the State of Iowa, and similar areas are present across much of the midwest and other parts of the country as well. The findings of the work described are applicable to many areas where intensive row-crop agriculture, utilizing significant applications of nitrogen fertilizers and pesticides, is practiced in a susceptible hydrogeologic environment.

The presence of significant concentrations of nitrate and low, yet persistant and possibly increasing concentrations of pesticides in ground water drinking water supplies are of concern from an environmental and public health viewpoint. However, susceptible geologic settings are not needed for the delivery of ag-contaminants into the hydrologic system. Where sufficient thicknesses of low permeability materials protect underlying aquifers, ag-contaminants are still leaching to ground water, but this shallow ground water generally moves laterally, discharging to streams, rather than vertically to aquifers used as drinking water sources.

The magnitude of nitrogen losses from agricultural lands has economic as well as environmental implications. Data from Big Spring and many experiment station farms in the upper midwest (Hallberg et al., 1984; and Hallberg, 1986a) indicate that corn crops are not utilizing 50 percent or more of the nitrogen applied. Such inefficiencies in fertilizer utilization suggest an area for obvious improvement and economic gain. In Iowa alone, farmers spend about $400 million annually on nitrogen fertilizers.

Ground water quality problems related to agriculture can only be resolved through a more holistic approach to agricultural management and research. We must couple our standard concerns for soil conservation and surface water quality with the need to protect ground water. New combinations of many current practices may be needed, and undoubtedly better chemical and nutrient management must play a part. A coordinated effort is needed to define cost-effective management alternatives, new technology, and education programs that can be applied to these problems. Resolving these problems will require a concerted effort by all segments of agriculture to gain the experience and data necessary to effect a satisfactory balance between efficient agricultural production and the protection of our water supplies.

SELECTED REFERENCES

Baker, J.L., and Johnson, H.P., "Impact of Subsurface Drainage on Water Quality," Proceedings of the Third National Drainage Symposium, American Society of Agricultural Engineers, St. Joseph, Missouri, 1977.

Baker, J.L., and Johnson, H.P., "Nitrate-nitrogen in Tile Drainage as Affected by Fertilization," Journal of Environmental Quality, Vol. 10, 1981, pp 519-522.

Freeze, R.A., and Cherry, J.A., Groundwater, 1979, Prentice-Hall, Inc., Englewood Cliffs, New Jersey.

Gast, R.G., Nelson, W.W., and Randall, G.W., "Nitrate Accumulation in Soils and Loss in Tile Drainage Following Nitrogen Application to Continuous Corn," Journal of Environmental Quality, Vol. 7, 1978, pp. 258-262.

Hallberg, G.R., "Agricultural Chemicals and Groundwater Quality in Iowa: Status Report 1985," in Proceedings of Iowa's 38th Annual Fertilizer and Ag-chemical Dealers Conference, Iowa State University, Cooperative Extension Service, CE-2158q, 1986a.

Hallberg, G.R., "Nitrates in Groundwater in Iowa," in Proceedings of Iowa's 38th Annual Fertilizer and Ag-Chemical Dealers Conference, Iowa State University, Cooperative Extension Service, CE-2158, 1986b.

Hallberg, G.R., and Hoyer, B.E., "Sinkholes, Hydrogeology, and Ground-water Quality in Northeast Iowa," Open-file Report No. 82-3, 1982, Iowa Geological Survey, Iowa City, Iowa.

Hallberg, G.R. et al., "Hydrogeology, Water Quality, and Land Management in the Big Spring Basin, Clayton County, Iowa," Open-file Report No. 83-3, 1983, Iowa Geological Survey, Iowa City, Iowa.

Hallberg, G.R. et al., "Hydrogeologic and Water-quality Investigations in the Big Spring Basin, Clayton County, Iowa: 1983 Water-Year," Open-file Report No. 84-4, 1984, Iowa Geological Survey, Iowa City, Iowa.

Kanwar, R.S., Johnson, H.P., and Baker, J.L., "Comparison of Simulated and Measured Nitrate Losses in Tile Effluent," Transactions of the American Society of Agricultural Engineers, Vol. 26, 1983, pp 1451-1457.

Kelley, R.D., "Synthetic Organic Compound Sampling Survey of Public Water Supplies," Iowa Department of Water, Air and Waste Management, April, 1985, Des Moines, Iowa.

Libra, R.D. et al., "Ground-water Quality and Hydrogeology of Devonian-Carbonate Aquifers in Floyd and Mitchell Counties, Iowa," Open-file Report No. 84-2, 1984, Iowa Geological Survey, Iowa City, Iowa.

Libra, R.D., and Hallberg, G.R., "Hydrologic Observations from Multiple Core Holes and Piezometers in the Devonian-carbonate Aquifers in Floyd and Mitchell Counties, Iowa," Open-file Report No. 82-2, 1985, Iowa Geological Survey, Iowa City, Iowa, pp 1-19.

Singh, K.P., and Stull, J.B., "Derivation of Base Flow Recession Curves and Parameters," Water Resources Research, Vol. 7, No. 2, 1971, pp 293-303.

Thompson, C.T., Libra, R.D., and Hallberg, G.R., "Water Quality Related to Ag-Chemicals in Alluvial Aquifers in Iowa," in preparation, 1986.

White, W.B., "Conceptual Models for Carbonate Aquifers: Revisited," in Dilamarter, R.R., and Csallany, S.C., (eds), Hydrologic Problems in Karst Regions, 1977, Western Kentucky University, Bowling Green, Kentucky, pp 176-187.

Witzke, B.J., and Bunker, B.J., "Devonian Stratigraphy of North Central Iowa," Open-file Report No. 84-2, 1984, Iowa Geological Survey, Iowa City, Iowa, pp 107-149.

Witzke, B.J., and Bunker, B.J., "Stratigraphic Framework for the Devonian Aquifers in Floyd and Mitchell Counties Iowa," Open-file Report No. 85-2, 1985, Iowa Geological Survey, Iowa City, Iowa, pp 21-32.

CHAPTER 15

ASSESSING SOME POTENTIALS FOR CHANGING AGRONOMIC PRACTICES AND IMPROVING GROUND WATER QUALITY— IMPLICATIONS FROM A 1984 IOWA SURVEY

by

Steven Padgitt and James Kaap

A scant 25 years ago only a few alarmists seemed concerned about environmental quality. Now the public is showing signs of growing impatient on a variety of natural resource issues (Dunlap, 1985). One of these is ground water quality (Dillman and Hobbs, 1982). Upon entering the second half of the 1980s decade, scientists are establishing unmistakable links between ground water quality and certain agricultural practices, especially increased use of agrichemical fertilizers and pesticides (Council for Agricultural Science and Technology, 1985).

Farmers, particularly, are caught in the middle. They share the concerns about potential health hazards in the water they and the rest of the population drink, but they perceive limited alternatives to the chemical agriculture they have adopted, most of which has been upon the recommendations of university and industry scientists. Furthermore, farmers are staunchly independent entrepreneurs, and they become frustrated over potential regulation of their industry. Their frustrations have become even more exacerbated during the present era of financial difficulty throughout many farming regions.

The challenge has emerged to find management alternatives that neutralize environmental degradation but still maintain agricultural productivity and profitability. Administrators of land grant Agricultural Experiment Stations and Cooperative Extension Services have recently placed increased priority on both research and educational activities to solve the agriculture--ground water quality puzzle (Joint Council on Food and Agricultural Sciences, 1985).

In Iowa, like elsewhere, there has been growing and convincing documentation of the link between agricultural practices and contaminants in shallow ground water (Hallberg, 1986). Special attention has been given to Northeast Iowa where problems were identified and where studies were initiated earlier than in other parts of the state. Of special interest has been the Big Spring Basin (Figure 15.1) where several state and federal agencies and organizations* have cooperated in coordinating and sharing their research and

---

*The cooperators have assumed an identity of the Iowa Consortiums on Agriculture and Ground Water Quality: members of the consortium include the Iowa Department of Water, Air and Waste Management; Iowa State University Agriculture Experiment Station and Cooperative Extension Service; U.S. Environmental Protection Agency; Iowa Geological Survey; Iowa Department of Soil Conservation; U.S. Soil Conservation Service; Northeast Iowa

educational efforts. Because of its special geological features, this 66,000 acre basin has provided a unique laboratory to intensively study and better understand problems surrounding agriculture and ground water quality. These features are described in greater detail elsewhere (Hallberg et al., 1984), but essentially the spring from which the basin gets its name provides a "window" through which both surface runoff and shallow ground water can be monitored continuously and quite precisely. The basin is also of special interest because it is an area of intense agriculture--both row crops and livestock. Another characteristic that makes it ideal for studying agriculture and ground water relationships is the absence of cities, towns, or industries within the basin. There are, however, approximately as many nonfarm as farm households in the basin.

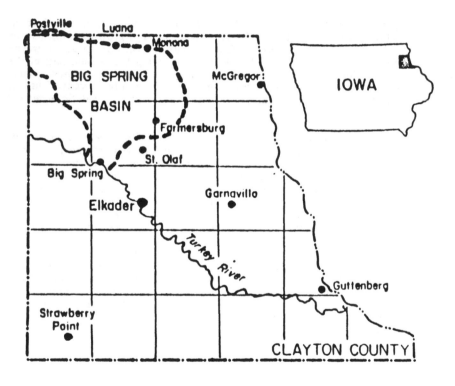

Figure 15.1: Location of Big Spring Basin

Among the problems documented in the basin have been particularly high levels of nitrate in infiltrated ground water discharge at the spring. For the 1982-83 water year, Iowa Geological Survey research estimated this to be 43.3

Conservancy District; Iowa Fertilizer and Chemical Association; and University of Iowa Institute of Agricultural Medicine.

pounds of nitrate per basin acre (Hallberg et al., 1984). In addition, traces of the pesticides Atrazine, Bladex, Lasso, Dual, and Dyfonate were also found. Because of the seriousness of this problem, along with a bit of skepticism that so much nitrate could result from farmers following recommended fertilization rates, the Iowa State University Cooperative Extension Service and the U.S. Soil Conservation Service conducted a detailed survey inventorying individual farming practices in the Big Spring Basin. The survey augmented earlier estimates of fertilizer and pesticide application rates as well as livestock numbers which had been extrapolated from secondary data sources such as retail sales of fertilizer and pesticides and the Census of Agriculture.

Beyond documenting the existence of the problem, local farmers and public agencies have become interested in seeking alternatives in order to start addressing the growing problem. A goal of all the agencies involved in the Big Spring research consortium has been to encourage voluntary adjustments in farming practices so that environmental degradation would be reduced. Farmers in the basin appear to understand those motives and share in the goal. They have been quite cooperative in the various research and educational efforts.

PROFILES OF FARMING OPERATIONS IN BIG SPRING BASIN

Personal interviews were conducted with 209 farmers during the summer, 1984. These 209 farmers operated 92 percent of the land (or 60,600 acres) in the basin (Kaap and Padgitt, 1985). Findings from the study confirm the intensiveness of agriculture in the area, and that livestock and cropping are more intensified than in Clayton County, the site of the basin (Table 15.1). Although the average Big Spring Basin farm was 4 percent larger than the average for the county, Big Spring farms have 50 percent more corn acres and 40 percent more alfalfa acres. Fifty-five percent of the farmland was in corn, compared to 38 percent for the county as a whole. Hay comprised 18 percent, pasture 12 percent, small grain 7 percent, and timber and other uses, 9 percent. Nearly all of the farming operations (95%) involved corn-alfalfa-livestock. Sixty-one percent of the farms had swine enterprises, 58 percent dairy, and 41 percent beef. The livestock inventory revealed approximately 1.8 hogs per basin acre, 0.4 cattle, and 0.9 chickens. Even though the basin comprised only 12 percent of the farmland in the county, it accounted for approximately 28 percent of the hogs, 22 percent of the dairy cows, 55 percent of the chickens, and 10 percent of the beef cattle.

Several sources of nitrogen are available as plant nutrients, and, cumulatively, they may eventually contribute to the high levels of nitrate found in the ground water. In this analysis only three sources were used--chemical fertilizer applications, nitrogen released from plowed legumes, and the nitrogen content of animal waste.

Chemical Fertilizer Applications of Nitrogen

Iowa farmers currently apply about 1,000,000 tons of nitrogen fertilizer annually (Hergett and Berry, 1985). This is 10 times greater than in 1960, but appreciably unchanged since 1976. It averages to between 135 and 145 pounds of nitrogen per corn acre statewide. Farmers in the Big Spring Basin, on average, were applying 150 pounds of nitrogen per acre on third and succeeding years of corn*. The range among individual farmers was 70 to 210 pounds per acre. On corn land following "decent" stands of alfalfa**, the average nitrogen application was 100 pounds per acre. On second year corn land following alfalfa, the average nitrogen application was 145 pounds per acre. (Nearly all of the nitrogen was applied as preplant. Five percent of the farmers reported some fall nitrogen applications. Only eight percent of the farmers reported any sidedress applications.) When these rates are aggregated across the number of acres in the different rotations, approximately 4,950,000 pounds of chemical fertilizer nitrogen was reported applied to corn acreage in the basin in 1984.

Table 15.1:   Big Spring Basin Farming Operations

Characteristics	Big Spring Basin	Clayton County
Land Use		
Percent of farm land in county	12 percent	100 percent
Average farm size	290 acres	277 acres
Average corn acres/farm	157 acres	105 acres
Average alfalfa acres/farm	54 acres	38 acres
Average oats acres/farm	21 acres	23 acres
Average pasture and timber acres/farm	58 acres	111 acres
Livestock		
Farms with livestock enterprise*	95 percent	
Swine	61 percent	
Dairy	58 percent	
Beef	41 percent	
Poultry	<1 percent	

*For inventory of numbers of livestock, see Table 15.2.

---

*Because of the rolling terrain, the shallow layer of topsoil, and other factors, farmers in the Big Spring Basin do not utilize soybeans in the land use rotations as is found in many other parts of the state. The survey identified only 30 acres of soybeans throughout the basin.

**"Decent" was self-defined by the farmer respondents and was used to exclude land where only minimal stands of alfalfa remained.

Nitrogen Released from Plowed Legumes

Within the basin in 1984 about 6,500 acres, or 20 percent, of the corn acres were in a first-year corn following alfalfa rotation. Some 8,200 acres, or 25 percent, was second year following alfalfa. Estimates from the Iowa State University Experiment Station and Cooperative Extension Service indicate that on a 20 percent stand of plowed alfalfa 100 pounds of nitrogen per acre are likely released (Voss and Killhorn, 1984). There also would be a small carryover to a second and succeeding years. If only the first year rotation of nitrogen availability is considered and second year carryover is omitted in making a projection, a conservative estimate of nitrogen available from plowed legumes in the basin in 1984 would have been 650,000 pounds. This converts to 20 pounds per corn acre or approximately 11 pounds per basin acre.

A more liberal set of assumptions could be used. For example, if one assumes a 50 percent stand with 140 pounds of nitrogen for a first year crop and 30 pounds carried over to the second year, then as much as 35 pounds of nitrogen per corn acre would be available.

Nitrogen from Livestock Manures

The volume of livestock in the basin means significant amounts of plant nutrients would be potentially available from livestock manures. Estimates of nitrogen, phosphorus and potassium were made utilizing findings from the survey and guidelines from the Midwest Planning Service's Livestock Facilities and Waste Handbook (1985). Some 3,940,000 pounds of nitrogen were estimated to be generated annually from collectable livestock manures in the basin (Table 15.2). Under proper management practices, half of the nitrogen could be captured and made available for plant nutrients (Midwest Planning Service, 1985). Using a 50 percent availability assumption, this converts to approximately 70 pounds of nitrogen per corn acre in the basin.

Total Available Nitrogen vs. Nitrogen Requirements for Corn

If the three major sources of nitrogen (inorganic applications, legume released and livestock manures) are summed for the corn acreage in the basin, then approximately 240 pounds of nitrogen would be available per corn acre.

Nitrogen Requirements for Crop Yields

The three to five-year average corn yield in the basin was reported by the farm operators to be 130 bushels per acre*. Current recommended rates of fertilization by Iowa State University agricultural scientists suggest that a 130 bushel corn yield requires no more than 160 pounds of nitrogen per acre (Voss and Killorn, 1984). Thus, in the Big Spring Basin, potentially 80 pounds of surplus nitrogen per corn acre were present. Incidentally, when this amount was prorated across the total basin, it was very close to the amount the Iowa

---

*The basin is comprised of mostly Downs and Fayette soils. Corn suitability ratings for these soils suggest realistic corn yield goals to be in a 120 to 160 bushels per acre range.

Geological Survey measured at the spring for the 1982-83 water year (42.9 vs. 43.3 pounds of nitrate per basin acre).

Table 15.2:   Livestock Numbers and Collectable Manure Nutrients in Big Spring Basin

Livestock	Average Daily Numbers	Nitrogen	P_2O_5	K_2O
Swine	49,244	1,477,500	1,024,000	1,031,000
Beef (feedlot animals)	6,956	710,000	521,000	612,000
Dairy cows	6,468	1,358,000	750,000	1,074,000
Dairy heifers	3,000	337,500	135,000	267,000
Layers	55,000	57,000	51,000	30,000
Total		3,940,000	2,481,000	3,014,000

Note: Animals on pasture are omitted from table.   This included 5,999 dairy heifers (for a 6-month period) and 2,816 beef animals (full year).

MANAGEMENT CONCERNS AND ALTERNATIVES

The challenge to farmers and extension service personnel, as well as others in the Iowa Consortium on Agriculture and Ground Water Quality has been to encourage changes in management practices that reduce the volume of nitrogen applied to the soils to a level more nearly coinciding with the nutrient needs of crops.

Unlike some other environmental concerns linked to farming practices, such as soil erosion, the nitrate issue may be one that eventually can be approached on the basis of short-run economic advantages rather than merely concern about the quality of the environment. At least this appears to be the case in the Big Spring Basin. Also, unlike soil erosion, there does not appear to be the same "proximity" effect whereby farmers become more concerned about an issue as it becomes nearer their own operations (Padgitt, 1986).

In an effort to assess the economic value of potentially available plant nutrients not presently being fully acknowledged by many basin farmers, some projections were made. These projections used the following assumptions and definitions:

Best manure management practices

Nutrient availability – N = 50%, P_2O_5 = 90%, K_2O = 90%

Uniform application
Immediate incorporation

Average manure management practices

Nutrient availability – N = 35%, P_2O_5 = 70%, K_2O = 70%
Uniform application
No incorporation

Nutrient Values

Nitrogen	= $0.15/lb
P_2O_5	= 0.205/lb
K_2O	= 0.11/lb

Using basinwide averages and using the above assumptions/definitions for best management manure handling practices, collectable manures could supply nitrogen, phosphorus, and potassium valued at approximately $1,051,000 basinwide (Table 15.3). This converts to $5,300 per farm in the basin. Under an "average" management practice for manure handling, nutrients worth as much as $800,000 would be available basinwide, or on average, $3,800 per farm. Alfalfa rotated to corn was estimated to supply nitrogen worth $97,500 basinwide or about $450 per farm.

When asked specifically about taking manure credits, 60 percent of the basin farmers indicated they did not adjust for nitrogen and 40 percent did not adjust for potassium and phosphorus. Additionally, 50 percent of the farmers underestimated the amount of nitrogen available to the first-year corn following alfalfa by as much as one half of the amount likely available as plant nutrients.

If one assumes for illustrative purposes that nutrients in the livestock manures and the nitrogen released in the first year following alfalfa were uniformly available throughout the basin, then these sources could provide approximately two thirds of the nitrogen and all the potassium and phosphorus needs for the corn acreage.

This preliminary analysis, while based on a detailed inventory of livestock numbers and management practices, contains a number of assumptions about animal weights and daily livestock inventories. The nitrogen estimated to be available in animal manures was considered by the authors to be conservative. A more refined cost-benefit analysis would include additional factors such as additional sources of nitrogen and cost incurred in adapting best management practices. The weight of the evidence, however, remains convincing. A surplus, and a rather large surplus, of nitrogen existed in comparison to nutrient needs of corn plants. Given existing farming practices, some, if not most, farmers could reduce current chemical fertilizer applications and not experience reduced yields. If improved manure management was adopted even further reductions could be made without appreciable reduction in corn yields. Currently, farmers have not adapted best management practices for livestock manures. Just over one-half (56%) have some kind of manure storage facility. Only 10 percent of the manures are incorporated into the soil at the time of application.

Table 15.3: Fertilizer Value of Collectable Manure[1] Under Best Management and Average Management Practices

	Best Management Practices[2]			Average Management Practices[3]		
	Nitrogen	P_2O_5	K_2O	Nitrogen	P_2O_5	K_2O
Total Nutrients (pounds)	3,940,000	2,480,000	3,310,000	3,940,000	2,480,000	3,310,000
Available Nutrients (pounds)	1,970,000	2,230,000	2,710,000	1,380,000	1,740,000	2,110,000
Market Value[4]	$ 295,000	$ 458,000	$ 298,000	$ 207,000	$ 356,000	$ 232,000
Total Value	$1,051,000			$795,000		

[1]Estimate omits animals in pasture settings

[2]Assumes 50 percent nitrogen, 90 percent P_2O_5, and 90 percent K_2O availability

[3]Assumes 35 percent nitrogen, 70 percent P_2O_5, and 70 percent K_2O availability

[4]Assumes the following market values: 15¢/lb for nitrogen, 20.5¢/lb for phosphate (P_2O_5), and 11¢/lb for potash (K_2O).

EDUCATION EFFORTS TO ENCOURAGE VOLUNTARY CHANGE IN MANAGEMENT PRACTICES

For the past two years the local county extension program has devoted increased attention to water quality. The local extension director has prepared a quarterly newsletter, "Water Watch", whereby farmers and other residents in the basin have been informed of research findings from the several studies underway in the basin. Additionally, the extension director and the extension area's soil, water, and waste management specialist have held numerous meetings where better manure management has been stressed. They have also sought cooperation from basin farmers in setting up demonstration plots. These demonstrations have documented the potential for reducing nitrogen applications while maintaining current yield averages (Table 15.4). More extensive demonstrations have been planned but are pending funding.

Table 15.4: Results from Demonstration Plots* in Big Spring Basin

Nitrogen Application		1985
Urea	Starter & Urea	Yield
lb N/acre		bu/acre
0	23	179
30	53	176 N.S.**
30 sidedress	53	181 N.S.
60	83	182 N.S.
90	113	180 N.S.
120	143	191 N.S.

*Plots were on fields following 50 percent alfalfa stand (3-4 plants square foot).

**Using least significant difference (LSD) test and 0.05 level of probability.

SUMMARY

Given the production increases farmers have experienced over the years with applications of nitrogen, immediate changes in fertilizer management by large numbers of farmers should not be expected. Some inroads might be starting, however. A 1986 survey just being completed has updated the 1984 inventory. Farmers' plans for continuous corn in 1986 were for 140 pounds of chemical nitrogen per acre, down 10 pounds from the 1984 survey. This estimate was based on a sampling of approximately 80 percent of the basin acres; consequently, the difference might be a function of sampling error.

When asked if they had made changes in their corn management practices since 1984, one third of the Big Spring farmers said they had reduced the rate of fertilizer applications. In addition to these indicators, the most recent survey found an attitudinal disposition toward acknowledging the nutrient value of manure. Of 161 basin farmers responding, 66 percent indicated the nutrient value of manure was more than the costs to retrieve it. (Twenty-one percent said the cost of capturing the manure nutrient value outweighed the return, and 13 percent were undecided.) Clearly, if 80 pounds of excess nitrogen is being applied per corn acre, then more than attitudinal change is needed, especially if voluntary measures are going to solve the problem. But the attitudinal change in nitrogen application is a start, and it is in the right direction.

SELECTED REFERENCES

Council for Agricultural Science and Technology, "Agriculture and Ground Water Quality," CAST Report No. 103, May, 1985, Ames, Iowa.

Dillman, A., and Hobbs, D.J., Rural Society in the U.S.: Issues for the 1980's, 1982, Westview Press, Boulder, Colorado.

Dunlap, R., "Public Opinion: Behind the Transformation," Environmental Protection Agency Journal, July-August, 1985, pp 15-17.

Hallberg, G.R., "Nitrates in Ground Water in Iowa," Proceedings from Nitrogen and Ground Water, Iowa Fertilizer and Chemical Association, Des Moines, Iowa, 1986.

Hallberg, G.R. et al., "Hydrogeologic and Water Quality Investigations in the Big Spring Basin, Clayton County, Iowa; 1983 Water Year," 1984, Iowa Geological Survey, Iowa City, Iowa.

Hergett, N.H., and Berry, J.T., "Fertilizer Summary Data," Bulletin Y189, 1985, National Fertilizer Development Center, Muscle Schoals, Alabama.

Joint Council on Food and Agricultural Sciences, "FY 1987 Priorities for Research, Extension, and Higher Education: A Report to the Secretary of Agriculture," 1985.

Kaap, J.D., and Padgitt, S., "Effects of Agronomic Practices on Ground Water Quality: Results from a 1984 Iowa Survey," Agronomy Abstracts, 1985, p. 27.

Midwest Planning Service, Livestock Facilities and Waste Handbook, Second Edition, PM 429, 1985, Iowa State University, Ames, Iowa.

Padgitt, S., "Agriculture and Ground Water Quality as a Social Issue," Proceedings from National Water Well Conference, Omaha, Nebraska, August, 1986 (forthcoming).

Voss, R., and Killorn, R., "Understanding Your Soil Test Report," MWPS 18, 1984, Cooperative Extension Service, Iowa State University, Ames, Iowa.

CHAPTER 16

ASSESSMENT OF EMPIRICAL METHODOLOGIES FOR PREDICTING GROUND WATER POLLUTION FROM AGRICULTURAL CHEMICALS

by

Debra S. Curry

In 1985 the U.S. Environmental Protection Agency placed ground water pollution by agricultural chemicals at the top of their priority list (Amsden, 1986). Through a joint effort of the U.S. Environmental Protection Agency Offices of Drinking Water and Pesticide Programs, a nationwide survey has been designed to characterize the occurrence of approximately 50 pesticides in ground water. The problem of possible ground water contamination by agricultural chemicals is as large as it is important. With over 475 million acres of cropland in the United States (U.S. Department of Agriculture, 1985), it is important to have a reliable method of determining where problems are most likely to occur.

The U.S. Environmental Protection Agency has funded the development of two methodologies that evaluate the probability of agricultural chemicals contaminating the ground water in any location of interest. These two methodologies, DRASTIC (Aller et al., 1984) and LEACH (Dean, Jowist, and Donigian, 1984), direct their focus on different aspects of the pollution mechanism. DRASTIC focuses on classic hydrogeologic parameters, whereas LEACH concentrates on the movement of pesticides through soil. Neither of these methodologies has been tested for accuracy against a field study.

In this paper these two methodologies were applied to the Big Spring Ground Water Basin in northeastern Iowa and their results were compared to a field study of this area using nonparametric statistics.

THE DRASTIC METHODOLOGY

DRASTIC was created by the National Water Well Association as a "methodology that will permit the ground water pollution potential of any hydrogeological setting to be systematically evaluated with existing information anywhere in the U.S." (Aller et al., 1985). The word DRASTIC is an acronym that stands for the 7 parameters that the National Water Well Association found to be the most important mappable factors that control ground water pollution potential:

    D - Depth to water
    R - (net) Recharge
    A - Aquifer material
    S - Soil media
    T - Topography (slope)
    I - Impact of vadose zone
    C - (hydraulic) Conductivity of the aquifer

Each of these parameters has been assigned a weight that describes its importance in the pollution process as compared to the other parameters. This methodology was initially a generic assessment of ground water pollution, but it was soon recognized that agricultural pollution is a special case. This led to a different weight assignment for "Agricultural DRASTIC" that reflects the comparatively greater importance of soil media and topography. The two weighting systems can be found in Table 16.1 (Aller et al., 1985).

At the site in question, each factor is rated on a scale of 1 to 10 indicating the relative pollution potential of the given factor at that site. Once all factors have been assigned a rate, each rate is multiplied by the assigned weight and resultant numbers are summed:

$$D_r D_w + R_r R_w + A_r A_w + S_r S_w + T_r T_w + I_r I_w + C_r C_w = \underline{\text{Pollution Potential}}$$

where r = rating for the site

       w = weight for the parameter

The pollution potential is a relative evaluation. The higher numbers indicate a greater pollution potential than the lower numbers. Because of this, it is imperative to be consistent in one's evaluation, always labeling the same scenario in the same way.

The DRASTIC parameters were delineated for an area of approximately 100 acres around each of the 19 wells that were monitored for nitrates and pesticides in the Big Spring Basin. All the wells were completed into the Galena Aquifer. Table 16.2 shows the evaluations of the 7 parameters at each of the well sites.

The depth to water was found by taking the difference between the elevation of the wells on U.S. G.S. topographic maps and the elevation of the water table/potentiometric surface on the map of the Galena Aquifer.

The net recharge of 6.8 inches/year given in the Big Spring Basin Report (Hallberg et al., 1983) was used for most of the wells. For the 4 wells that are located in sink hole basins, the net recharge was taken to be the total water yield (ground water and surface water) of 13.1 inches/year.

For the evaluation of the aquifer material, the 2 wells in the west were rated as bedded limestone because the clay/shale aquiclude overlying it has prevented the karst development there. In the karst region, the sites lying amidst the sink holes or close to the major conduits were rated 10, the highest rating that could be given, and the other sites were rated 9, the lowest karst rating to be given.

The soils in the areas, according to the SCS Soil Survey of Clayton County, were virtually all silt loams, with the exception of one site near Turkey Creek that was a sandy loam.

In DRASTIC, topography is defined by the slope of the land surface. The topography was found by approximating the weighted average of the slopes represented at each site, according to the Soil Survey of Clayton County.

Table 16.1: Assigned Weights for DRASTIC Features (Aller
et al, 1985)

	Weight	Agricultural Weight
Depth to Water Table	5	5
Net Recharge	4	4
Aquifer Material	3	3
Soil Media	2	5
Topography	1	3
Impact of Vadose Zone	5	4
Hydraulic Conductivity	3	2

The impact of the vadose zone was the parameter that varied the most. There is a thick clay-shale aquiclude in the west. In the central area, the formation immediately above the Galena Aquifer is composed of shaley carbonates. In the east the Galena Aquifer itself outcrops.

The hydraulic conductivity was estimated using typical ranges given in Freeze and Cherry (1979), as was suggested in the DRASTIC Handbook. The basin was again broken into the western nonkarst area, which was rated at the highest bedded limestone rating of 10 GPD/ft^2 (which coincides with the lowest karst rating), the karst area that is associated with the sink holes and conduits (rated as 2000+ GPD/ft^2), and the karst area that was not associated with conduits (rated as 1000 to 2000 GPD/ft^2).

Once the sites were evaluated, the ratings were multiplied by the weights and the pollution potential was acquired, as shown in Table 16.3. The pollution potentials are moderately high with the exception of 3 wells in the west. These potentials were mapped as shown in Figure 16.1 and compared to a nitrate map of the basin (Figure 16.2, Hallberg et al., 1983) and an atrazine map of the basin (Figure 16.3, Hallberg et al., 1983). Some relationship can be seen between the pollution potential, nitrates, and atrazine depicted on these three maps.

THE LEACH METHODOLOGY

The LEACH methodology was created to determine whether agricultural chemicals will move in significant quantities past the crop root zone. It is assumed that once past the crop root zone it will enter the phreatic aquifer (Dean, Jowist, and Donigian, 1984). LEACH is also an acronym standing for:

Table 16.2: Site Evaluations for the Assignment of Rating

Site No.	Depth to Water (ft)	(Net) Recharge (in)	Aquifer Material	Soil Media	Topography (% Slope)	(Vadose) Impact	(Hydra.) Conduct. (GPD/ft^2)
11	147	13.1	no sink hole	silt loam	½ 9-14 ½ 5-9	in sink hole basin, no sink hole within 100 ac	1000-2000
15	95	6.8	no sink hole	silt loam	5-9	bedded limestone	1000-2000
16	98	13.1	by sink hole	silt loam	5-9	in sink hole basin with sink hole within 100 ac	1000-2000
26	29	13.1	by conduit	silt loam	½ 5-9 ½ 9-14	in sink hole basin, no sink hole within 100 ac	2000+
30	105	6.8	by sink hole	silt loam	½ 2-5 ¼ 5-9 ¼ 9-14	bedded limestone and karst, close to sink holes	1000-2000

Table 16.2: (continued)

Site No.	Depth to Water (ft)	(Net) Recharge (in)	Aquifer Material	Soil Media	Topography (% Slope)	(Vadose) Impact	(Hydra.) Conduct. (GPD/ft^2)
37	125	6.8	by sink hole	silt loam	5-9	bedded limestone, close to sink holes	1000-2000
39	105	6.8	by conduit	silt loam	½ 5-9 ½ 9-14	bedded limestone	2000+
45	130	6.8	bedded limestone, no karst	silt loam	5-9	thick clayshale	10
47	85	6.8	bedded limestone No karst	silt loam	½ 0-2 1/6 5-9 1/6 9-14 1/6 14-18	2/3 bedded limestone 1/3 clayshale	10
49	240	6.8	by conduit	silt loam	½ 2-5 ½ 9-14	3/4 bedded limestone ¼ karst	2000+

Table 16.2: (continued)

Site No.	Depth to Water (ft)	(Net) Recharge (in)	Aquifer Material	Soil Media	Topography (% Slope)	(Vadose) Impact	(Hydra.) Conduct. (GPD/ft^2)
52	180	6.8	no sink hole	silt loam	½ 2-5 ~½ 5-9 some 9-14	bedded lime-stone	1000-2000
56	210	13.1	no sink hole	silt loam	½ 5-9 ½ 9-14	sink hole basin, no sink hole within 100 ac	1000-2000
57	55	6.8	by sink hole	silt loam	2-5	karst with sink hole within 100 ac	1000-2000
61	224	6.8	no sink hole	silt loam	½ 5-9 ¼ 9-14 ¼ 14-18	bedded lime-stone	1000-2000
72	98	6.8	no sink hole	silt loam	½ 5-9 ¼ 9-14 ½ 14-18	2/3 clayshale 1/3 bedded lime-stone	10

Table 16.2: (continued)

Site No.	Depth to Water (ft)	(Net) Recharge (in)	Aquifer Material	Soil Media	Topography (% Slope)	(Vadose) Impact	(Hydra.) Conduct. (GPD/ft^2)
75	223	6.8	no sink holes	silt loam	9-14	3/4 clay shale ¼ bedded limestone	1000-2000
81	30	6.8	no sink holes	sandy loam	½ 0	karst no sink holes	1000-2000
82	10	6.8	by conduit and sink holes	silt loam	½ 0	2/3 karst 1/3 bedded limestone	2000+
84	202	6.8	by conduit and sink holes	silt loam	9-14	3/4 bedded limestone ¼ karst	2000+

Table 16.3: Parameters and Pollution Potential for DRASTIC Methodology

Site No.	Depth (5)*	(Net) Recharge (4)	Aquifer Material (3)	Soil Media (5)	Topo (3)	(Vadose) Impact (4)	(Hydra.) Conduct. (2)	Pollution Potential
11	5	36	27	20	15	32	16	151
15	10	24	27	20	21	24	16	142
16	10	36	30	20	21	40	16	173
26	35	36	30	20	15	32	20	188
30	5	24	30	20	24	36	16	155
37	5	24	30	20	21	32	16	148
39	5	24	30	20	15	24	20	138
45	5	24	18	20	21	8	2	98
47	10	24	18	20	21	16	2	111
49	5	24	30	20	21	28	20	148
52	5	24	27	20	24	24	16	140
56	5	36	27	20	15	32	16	151
57	15	24	30	20	27	40	16	172
61	5	24	27	20	15	24	16	131
72	10	24	27	20	15	12	2	110
75	5	24	27	20	20	8	16	120
81	30	24	27	30	30	32	16	189
82	45	24	30	20	30	28	20	197
84	5	24	30	20	12	28	20	139

*Assigned weights for Agricultural DRASTIC

LEACH

Leaching
Evaluation of
Agricultural
Chemicals
Handbook

This methodology was developed on the basis of 19 representative sites; 6 sites in the wheat growing regions, 7 in the corn regions, 4 in the soy bean regions, and 2 in the cotton regions. The Big Spring Ground Water Basin is part of the representative site No. 10 in the basic LEACH methodology (Dean, Jowist, and Donigian, 1984).

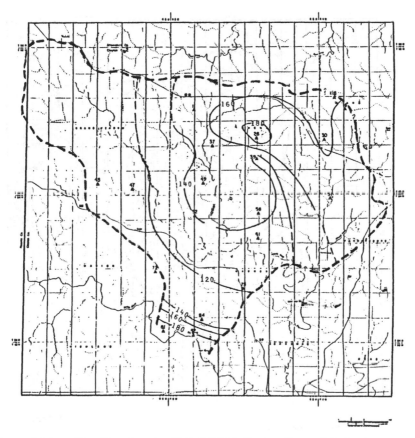

Figure 16.1:   Map Showing the DRASTIC Pollution Potentials in the
Big Spring Ground Water Basin

There are three parameters to be found (Dean, Jowist, and Donigian,
1984):

(1)   The retardation factor (R) for the pesticide/soil combination

$$R = 1 + \frac{(Kd)\ (ps)}{fc}$$

where: ps = soil bulk density

fc = average soil water content at field capacity

Kd = chemical partition coefficient

= K_{oc} (OC)/100

where: K_{oc} = organic carbon partition coefficient

OC = % soil organic carbon

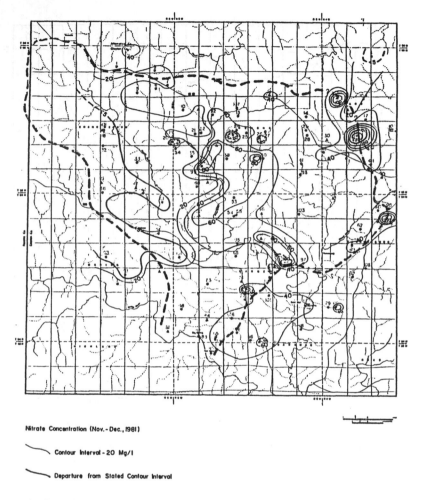

Nitrate Concentration (Nov.-Dec., 1981)

⌒ Contour Interval - 20 Mg/I

⌒ Departure from Stated Contour Interval

Figure 16.2:   Contour Map of Average Nitrate Concentrations from the Galena Aquifer Inventory Water Samples (Hallberg et al., 1983).

(2)   The decay rate, K_s, of the chemical for the appropriate soil pH, temperature, organic matter, and moisture conditions involved. Some decay rates are given in the LEACH Handbook.

(3)   The U.S. Soil Conservation Service curve number, CN, which incorporates the soil characteristics and land management practices. This is also given in the LEACH Handbook.

Once these values have been obtained, one consults a table, such as Table 16.4, which is based on the representative site number to determine which graphs to use for transport prediction (Dean, Jowist, and Donigian, 1984).

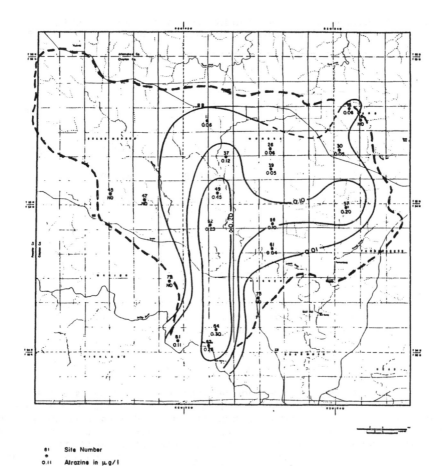

8'   Site Number
●
0.11   Atrazine in µg/l

Figure 16.3:   Atrazine Concentrations in the Galena Aquifer,
June 7-8, 1982 (Hallberg et al., 1983)

The decay rate of atrazine in soils comparable to those found in the Big Spring Basin is 0.0064. The values that were used for the retardation factor at each well site were acquired by finding the weighted average of the retardation factors of the soils in the 100 acre area around the well. Using this information, one finds that all the sites correspond to Graphs A10-11 and A10-15 from the LEACH methodology, as shown in Table 16.5.

If one looks at Figure 16.4 (Graph A10-11; Dean, Jowist, and Donigian, 1984), one sees that 10 percent of the atrazine will never leach below the crop root zone; however, at least one percent of the atrazine will always leach below the root zone. The curves are so close together that the CN's can be ignored. Looking at Figure 16.5 (Graph A10-15; Dean, Jowist, and Donigian, 1984), it shows that none of the atrazine is expected to leach. Significant leaching is defined as more than 0.05 percent (Dean, Jowist, and Donigian, 1984); therefore, these two graphs can be used as the basis for division between areas

Table 16.4: Graphs Corresponding to Retardation Factors (R) and Decay
Rates at Representative Site No. 10; Crop: Corn; CN: 71-78

R	Ks				
	.001	.005	.010	.050	.100
1	A10-1	A10-2	A10-3	A10-4	A10-5
3	A10-6	A10-7	A10-8	A10-9	
5	A10-10	A10-11	A10-12	A10-13	
20	A10-14	A10-15			
50	A10-16	A10-17			

in which pollution is and is not expected. A map of the basin, Figure 16.6, was constructed based on the differences between the two graphs. The stippled area shows where pollution is most likely to occur.

ANALYSIS OF RESULTS

To see if there is a correlation between the methodologies and the field data, they were compared using the Spearman Rank Correlation Coefficient. This is a nonparametric statistical method which makes no assumption about the linearity or the normality of the data. In this test, one ranks the data in each set, starting with 1 for the lower data point, 2 for the next lowest, and so on. Once the data are ranked, the correlation coefficient can be found according to the equation (Walpole and Myers, 1978):

$$r_s = 1 - \frac{6 \sum_{i=1}^{n} d_i 2}{n(n^2-1)}$$

where:

r_s = Spearman Rank Correlation Coefficient

n = sample size

d_i = arithmetic difference between the variables ranks

An r_s close to 1 indicates a good correlation; an r_s close to 0 indicates very little correlation, and an r_s close to -1 indicates a reverse correlation. This method was chosen for two reasons:

(1)   The DRASTIC Pollution Potentials are not linear; that is, an area

Table 16.5: Retardation Factors and Graph Number for the Well Sites in Big Spring Basin

Site #	Approx Amount	Soil Number	Retardation Factor (R)	R Group	Graph Number
11	1/3	162	12.54	20	A10-15
	1/3	163	7.34		
	1/6	487	26.18		
	1/6	589	36.62		
15		162	12.54	20	A10-15
16	1/2	162	12.54	5	A10-11
	1/2	163	7.34		
26	1/3	162	12.54	20	A10-15
	1/3	163	7.34		
	1/9	902	16.60		
	1/9	981	20.33		
	1/9	1212	26.74		
30	1/3	162	12.54	20	A10-15
	1/3	163	7.34		
	1/3	981	20.33		
37	1/3	162	7.34	20	A10-15
	1/3	163	12.54		
	1/6	981	20.33		
	1/6	1212	26.74		
39		162	12.54	20	A10-15
45	1/2	162	12.54	5	A10-11
	1/2	163	7.34		
47	1/4	163	7.34	20	A10-15
	1/4	323	28.71		
	1/4	978	18.95		
	1/4	612	11.88		

Table 16.5:   (continued)

Site #	Approx Amount	Soil Number	Retardation Factor (R)	R Group	Graph Number
49	1/3	163	7.34	20	A10-15
	1/3	162	12.54		
	1/3	487	26.18		
52	2/3	162	12.54	20	A10-15
	1/3	489	31.90		
56		162	12.54	20	A10-15
57	1/2	162	12.54	20	A10-15
	1/6	98	17.30		
	1/6	487	26.18		
	1/6	183	7.23		
61		163	7.34	5	A10-11
72	1/3	163	7.34	5	A10-11
	1/3	162	12.54		
	1/3	129	3.22		
75		162	12.54	20	A10-15
81	1/4	41	19.06	20	A10-15
	1/4	323	28.71		
	1/4	284	18.41		
	1/4	98	17.30		
82	1/4	158	4.78	5	A10-11
	1/4	163	7.34		
	1/4	162	12.54		
	1/8	478(rock out crop)	0.0		
	1/8	977	22.43		
84		163	7.34	5	A10-11

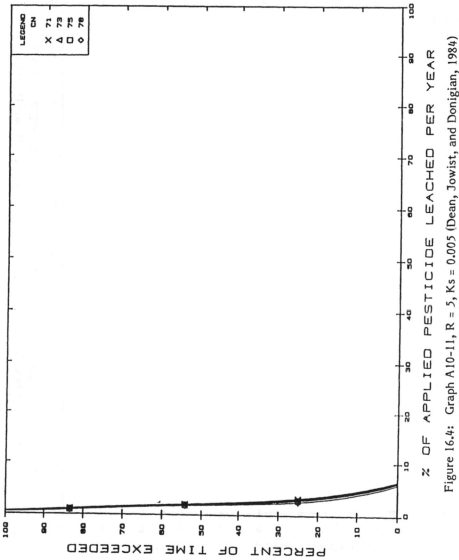

Figure 16.4: Graph A10-11, R = 5, Ks = 0.005 (Dean, Jowist, and Donigian, 1984)

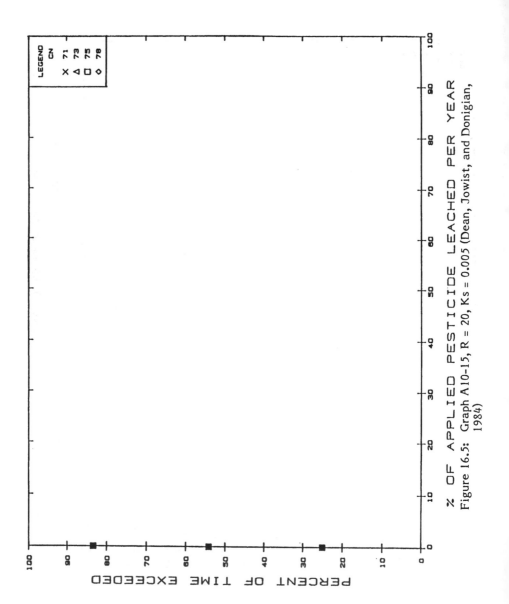

% OF APPLIED PESTICIDE LEACHED PER YEAR

Figure 16.5: Graph A10-15, R = 20, Ks = 0.005 (Dean, Jowist, and Donigian, 1984)

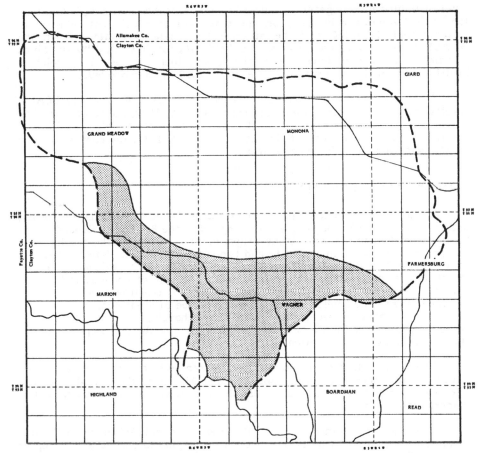

Figure 16.6: Areas Predicted by the LEACH Methodology to be Contaminated

with a potential of 200 is not twice as likely to be polluted as an area with a potential of 100.

(2)   Some of the field data may be somewhat biased by the method of the well completion or its depth. The ranking of the data helps to alleviate some of this bias.

When the Spearman Test was run for DRASTIC versus the nitrate field data, an r_S of -0.036 was found, indicating virtually no correlation.

When the test was run for DRASTIC versus atrazine field data from June 7-8, 1982, the r_S was found to be 0.418. For this number of data points (19), if one takes the level of significant correlation to be 0.05, then r_S must exceed the critical value of 0.476 for these data to be considered related. If one takes the level of significance to be 0.10, the critical value is 0.399. However, the level of significance is generally taken to be 0.05, in which case r_s is less than the critical value, and it should be said that there is insufficient evidence to indicate that DRASTIC and the atrazine field data correlate.   When the DRASTIC results were compared to the field data for July 28, 1982, in which atrazine was found in fewer wells, the r_s was found to be 0.182, which is insufficient to indicate correlation at even the 0.10 significance level.

Using the Spearman Rank Correlation Coefficient to test the LEACH results against the atrazine field data from June 7-8, 1982, the r_s was found to be -0.263.   This indicates a very small negative correlation. When the LEACH results were compared to the field data for July 28, 1982, the r_s was equal to 0.186, indicating virtually no correlation.

SUMMARY AND CONCLUSION

In this study the DRASTIC (Aller et al., 1985) and LEACH (Dean, Jowist, and Donigian, 1984) Methodologies were applied to the Big Spring Ground Water Basin in northeastern Iowa.   Using nonparametric statistics, the results were compared to one set of nitrate data and two sets of field data defining atrazine concentrations in wells throughout the basin.   There was no evidence that the methodologies correlated with the field data.   The lack of correlation between the field data and LEACH is expected.   In a karst situation, the interaction between the soil and applied chemicals would not be of great consequence.   The lack of correlation between the DRASTIC Methodology and the field study could indicate either failure of the weighting system or the inappropriateness of the methodology in this location.

The field study itself could also have detracted from the correlation.   The data was not acquired for the purpose of comparing it to these methodologies. The wells and data points are possibly too few, too far apart, and/or inappropriately placed.

There are some other elements that should be included in the evaluation of pollution potential.   One is the location in the aquifer, whether it is up or down gradient from most of the chemical application.   DRASTIC approached this point with the hydraulic conductivity parameter.   Another point that was incompletely addressed was infiltration.   Macropores, through which chemicals and water can flow virtually unimpeded, can be created by plant roots and animal burrowings.   Also, a more complete look at rainfall and/or irrigation amount and timing, in conjunction with the solubility of the chemical in question, would be of help.

SELECTED REFERENCES

Aller L., et al., "DRASTIC: A Standardized System for Evaluating Ground Water Pollution Potential Using Hydrogeologic Settings", May, 1985, Robert S. Kerr Environmental Research Laboratory, U.S. Environmental Protection Agency, Ada, Oklahoma.

Amsden, T., Personal communication, Director, EPA Region VII Office of Ground Water Protection, January, 1986, Kansas City, Missouri.

Dean, J.D., Jowist, P.P., and Donigian, A.S., Jr., "Leaching Evaluation of

Agricultural Chemicals (LEACH) Handbook", EPA-600/3-84-068, June, 1984, Environmental Research Laboratory, U.S. Environmental Protection Agency, Athens, Georgia.

Freeze, R.A., and Cherry, J.A., <u>Groundwater</u>, Prentice Hall, Englewood Cliffs, New Jersey, 1979.

Hallberg, G.R., et al., "Hydrogeology, Water Quality, and Land Management in the Big Spring Basin, Clayton County, Iowa", Open File Report 83-3, June, 1983, Iowa Geological Survey, Iowa City, Iowa.

U.S. Department of Agriculture, "1985 Agricultural Chartbook", Agricultural Handbook No. 652, December, 1985.

Walpole, R.E., and Myers, R.H., "Nonparametric Statistics", <u>Probability and Statistics for Engineers and Scientists</u>, 2nd Ed., Macmillan, New York, 1978, pp 492-495.

CHAPTER 17

INVESTIGATION OF NITRATE CONTAMINATION IN
SHALLOW GROUND WATERS NEAR WOODWARD, OKLAHOMA

by

B.K. Gopal

The possible pollution of ground water with nitrates from man-made activities is a problem of wide interest in Oklahoma, because of potential human and animal health hazards. Many farm families in Woodward, Oklahoma, use ground water as a source of drinking water. High nitrate levels in drinking water are of concern because of an infant disease, methemoglobinemia, commonly known as "Infant Cyanosis." In the past, concentrations of nitrate exceeding the 1962 U.S. Public Health Service Limit of 10 mg/L for drinking water, have been noted in Woodward ground water by various state investigators (American Water Works Association, 1981). Most excursions were in samples collected from shallow wells.

This investigation was conducted to determine the nature of the sources which contributed to the high nitrate levels in the ground water around Woodward, Oklahoma, and suggest potential corrective measures. While high nitrate levels have been monitored in this area for many years, the sources were too obscure to be determined without any intensive analysis of the levels, and areal and temporal extent of the nitrate pollution. An intensive sampling program was conducted between March and December of 1982.

The results from this investigation were used to identify problem areas, suggest better permitting activities for point sources, to refine and revise Best Management Practices (BMPs), and to develop ground water quality standards. The results are reported in the final report for 208 Task 861, "Ground Water Prototype Project--Nitrate Contamination in Ground Water," under a grant (#P-006315-01-0) awarded in 1980 by the U.S. Environmental Protection Agency (EPA) through the Pollution Control Coordinating Board to the Oklahoma Water Resources Board (OWRB).

LITERATURE REVIEW

Gopal (1982) discussed causes and occurrences of nitrates in ground water, and notable investigations and procedures for control. Presented in the paper is a selective review of the literature on man-caused ground water nitrate pollution.

A 1971-72 study in Hartsville and Florence, South Carolina, by Peele and Gillingham revealed excessive nitrate concentrations in ground water and tile drainage effluent. They concluded from their study that excessive applications of nitrogen fertilizers above the amounts needed for plant growth will increase the nitrate content of shallow ground water and the soil profile.

Several investigators have observed fairly high concentrations of nitrogenous compounds leaching into the ground water systems resulting from fertilizer applications. Feth (1966) reported high concentrations of nitrate in ground water due to nitrate fertilizer applications, and runoff from barnyards and silos.

DeWalle and Schaff's study (1980) detected gradually increasing ground water contamination in the densely populated central Pierce County, Nebraska, over a 30-year period, with most of the pollution originating from septic tank drainfields. Depth profiles showed high nitrate concentrations near the surface, decreasing to lower values with depth. The highest nitrate and coliform concentrations in ground water were noted in the winter when infiltrating rainfall dissolved and leached these contaminants downward.

Enberg (1967) conducted a survey for Hold County, Nebraska, and reported a wide range of nitrate concentrations collected from 0.1 to 409 ppm as NO_3. Ground water pumped close to feedlots was generally more enriched in nitrate than when pumped from distant wells.

DESCRIPTION AND SITE BACKGROUND

The study area for this investigation (Figure 17.1) is west and southwest of the City of Woodward, Oklahoma, on a predominantly rural area of 315 square miles. The study area was divided into four zones. Sixteen wells were identified for sampling in Zone I, seven in Zone II, 26 in Zone III (including 24 municipal wells), and seven in Zone IV. Row crop farming is common in this region. Rainfall averages 28 inches per year. The surface topography of the region is generally an undulating erosional plain. Soils of the area are highly permeable and rainfall percolation is rapid (vertical hydraulic conductivity of 20 m/day for Dune Sand), resulting in reduced surface runoff. The soils in the study region are the Pratt-Tivoli and Nobscott-Brownfield-Miles types: tan, loamy, and fine grained sandy soils. The Caney-St. Paul is a reddish soil with loam subsoils developed in loamy red beds. These soils are relatively thin and overlie the Ogallala and Rush Springs ground water formations.

Logs of two test holes which closely represent the lithology in Zone I, Township 23N, Range 21W, Sections 28-30 are shown in Figure 17.2.

GROUND WATER AVAILABILITY

In Woodward County, ground water occurs principally in the terrace deposits and alluvium in the valley of the North Canadian River and its major tributaries--Wolf, Indian, Persimmon, and Bent Creeks; and in the Ogallala Formation which covers the southwestern part of the county. Ground water is also present in small or moderate quantities in the Bedrock Formations (Red Beds).

Wells have been a primary source of water for domestic and stock use in Woodward County since the first settlement. Windmills have been and still are a major source of power for pumping wells, although many rural wells now are powered by internal-combustion engines or electric motors.

Most large capacity wells in the county are used for irrigation, but a few are used for public water supply or industrial purposes. Table 17.1 shows the

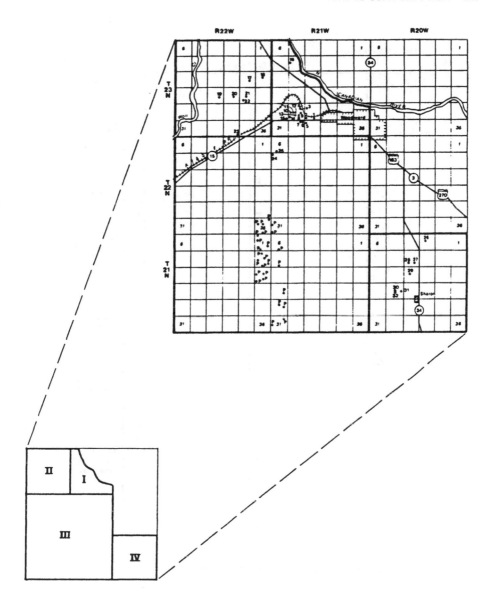

Figure 17.1: Well Sampling Network in Woodward County, Oklahoma

average depth of wells, depth to water, and water use in each zone.

METHODS AND MATERIALS

The objectives of the sample design were to obtain ground water samples

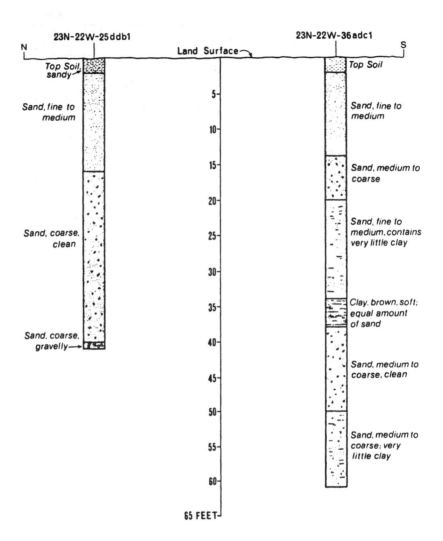

Figure 17.2: Logs of Two Test Holes in the Study Area

at intervals such that virtually no change in quality would pass unnoticed between sampling times. The sampling schedule usually is a compromise between the accuracy and detail desired in the water quality record and the availability of funds and personnel. The chemical quality of the ground water at any sampling point may vary in response to changes in the rate of movement, to pumpage, or to differences in the rate and chemical composition of recharge from direct precipitation.

Changes in ground water quality can occur on a short-term, seasonal, or long-term basis. This study is concerned with short-term and seasonal changes

Table 17.1:   Well Depth, Depth to Water and Water Use in Each Zone

ZONE	DEPTH TO WATER	WELL DEPTH	USE
I	20- 50 feet	40-110 feet	D,S
II	20- 50 feet	40-110 feet	D,S (M,I)*
III	50-100 feet	50-200 feet	D,S,I
IV	20- 50 feet	40-100 feet	D,S (M,I)*

*Yields are adequate where thickness of saturated material above the red bed is sufficient.

D - domestic
S - stock
M - municipal
I - irrigation

of nitrate in ground water. Therefore, samples were collected about once every three weeks.   Twelve sampling events over a nine-month period were accomplished.   This frequency permitted the observation of short-term and seasonal trends of nitrate levels in the wells sampled (Gopal, 1981a).   The sampling frequency design is in the paper titled "Development of an Experimental Design Program for the Problem" (Gopal, 1981b).

For hydrogeologic sampling, it is desirable to sample water directly from the aquifer. However, due to practical constraints, samples were taken from the well discharge. Wells were generally pumped for 15 minutes so that water quality in the well reasonably represented that of the aquifer.

Nitrate levels in well water were measured in the field. An Orion Model 407-A/F specific-ion meter in conjunction with an Orion 93-07 nitrate electrode and 90001 double junction reference electrode was used for nitrate measurements.

DATA ANALYSIS

The nitrate data collected from wells in Zone I in Woodward County, Oklahoma, covering the sampling period from March through December of 1982, is given in Table 17.2. The nitrate data for Zone I is presented in computer map form (SYMAP). SYMAP is a computer program for the production of maps which graphically depict quantitative and qualitative information in space (Dudnik, 1971). A discussion of contour maps for Zone I follows. Fifteen out of sixteen wells were used for analysis (well #16 was too far from the other wells and would be inappropriate for use in the interpolation). Nine of the sixteen wells are in the Rush Springs Sandstone, five are in the Dune Sand which covers certain outcrop areas of the Ogallala, one in the Ogallala, and one in the alluvium of the North Canadian River. The general movement of ground water

Table 17.2: Nitrate Data Collected from Wells in Zone I (mg/l)

WELL	MARCH	MAY*		JUNE	JULY	AUGUST*		SEPT.	OCT.	NOVEMBER*		DEC.	RANGE
1	24.0	40.0	45.0	66.5	24.1	27.2	24.8	38.5	27.3	33.2	36.6	31.1	24.0-66.5
2	6.3	7.3	5.9	7.6	6.4	6.4	7.6	8.1	6.5	8.0	9.0	8.0	5.9- 9.0
3	3.0	3.0	3.0	4.0	4.1	3.0	3.2	3.4	3.2	6.3	6.0	4.0	3.0- 6.3
4	19.5	20.0	23.2	28.7	27.4	27.7	35.9	30.5	29.8	35.4	34.7	28.3	19.5-35.9
5	19.1	15.9	16.3	13.5	11.7	15.7	15.0	13.7	15.9	20.4	24.2	11.5	11.5-24.2
6	1.0	1.6	0.4	0.4	1.4	1.5	1.0	0.9	1.2	1.2	1.2	1.2	0.4- 1.6
7	2.6	0.7	1.2	0.8	1.2	1.0	1.4	1.7	1.2	1.2	1.2	1.5	0.7- 2.6
8	13.4	15.3	13.1	16.1	19.0	21.5	21.0	18.7	18.9	19.8	17.6	11.7	11.7-21.5
9	5.3	6.8	5.1	6.9	6.9	8.7	8.7	7.7	15.1	13.6	11.8	15.0	5.1-15.1
10	12.2	10.3	12.4	21.4	53.5	91.0	80.0	62.0	75.8	55.7	52.1	39.0	10.3-91.0
11	1.7	1.5	0.9	0.7	2.1	1.0	0.8	4.0	2.1	6.3	6.4	7.1	0.7- 7.1
12	4.0	5.5	4.1	3.5	4.3	5.8	7.8	3.0	5.2	7.2	5.7	4.1	3.0- 7.8
13	15.0	18.2	29.0	29.4	23.5	36.4	33.5	21.9	42.7	43.7	44.8	31.0	15.0-44.8
14	5.0	6.7	5.7	7.9	8.8	7.8	9.0	6.7	13.1	11.9	15.8	11.4	5.0-15.8
15	30.0	35.0	34.0	36.7	25.2	34.7	20.3	38.0	11.1	20.5	24.8	20.9	11.1-38.0
16	-	-	-	3.1	3.2	3.4	3.8	3.9	4.0	6.0	4.6	6.0	3.1- 6.0

*Samples collected twice within the month designated.

in the study area is northward and northeastward. Nitrate levels as N ranged from a minimum of 0.4 mg/L in well #6 to a maximum of 91 mg/L in well #10.

The SYMAPs depicting ispoleths of nitrate concentrations for Zone I are in chronological order. Nitrate data were collected only in the left half of the grid which represented the unsewered area. For convenience of contouring, background levels (4 mg/L) were introduced in Sections 22, 27, and 34, Township 23N, Range 21W. June and July of 1982 were fairly dry. As a result, nitrate levels on July 15, 1982, were relatively low (Figure 17.3). The OWRB had been informed in the last week of July, 1982, that about 30,000 gallons of nitrogen based effluent water (stored in holding ponds) was sprayed as soil conditioner onto the surface of an unlined pond under construction. Waste in the lagoons consists of wastewater associated with the production of urea and ammonium nitrate fertilizers. A general layout of the plant is shown in Figure 17.4. Three vertical cross-sections, AA', BB', and CC', are shown in Figure 17.5.

Soils underlying the ponds are of the Pratt Loamy fine sand type. The surface layer of brown or grayish-brown loamy fine sand is underlain by a subsoil of loamy fine sand that takes water rapidly. Available water holding capacity and water intake rate of these soils are 0.6 to 1.1 inches (low) and 5-10 inches per hour (rapid) respectively (U.S. Soil Conservation Service, 1963). Rapid changes of water table fluctuations are observed during heavy precipitation events. Ground water flowing in the general northeasterly direction discharges into the North Canadian River.

Due to the landspreading of nitrogenous waste around the new pond under construction, an increased sampling frequency was initiated by the OWRB. The nitrate plume in Section 29 (left center of the SYMAPs) was spreading and the concentration was increasing. In well #10 the concentration increased to 91 mg/L nitrate on August 5, 1982 (Figure 17.6). The fertilizer industry's sample of August 27, 1982, showed a concentration of 112 mg/L, and then the plume in section 29 started slowly shrinking around wells #10, 11, 12, 13, and 14 (Figure 17.7). As seen from the SYMAP for December 6, 1982 (Figure 17.8), the slug injection which caused the nitrate plume around well #10 had essentially dispersed.

The dilution of nitrates is due to hydrodynamic dispersion. The mixing mechanisms (microscopic and macroscopic) produced spreading of the NO_3-N within the unconfined ground water so that the volume of ground water affected increased and the resulting concentration decreased with time, after the slug injection. This dilution is the most important mixing mechanism after a pollutant reaches the water table.

Well #1 is located downslope from a vegetable patch and barn in Section 28, Township 23N, Range 21W (Figure 17.1). Precipitation in the month of September was 1.35 inches, substantially higher than in August, when only 0.34 inches fell in the area. Rainwater from September storms infiltrated rapidly through the highly permeable sand in Zone I. Thus most of the nitrogen produced by these sources was carried by the infiltrating rainwater to the water table and contaminated well #1. This resulted in a dramatic increase in nitrate at the well on September 22, 1982 (Figure 17.9). There is a direct correlation between precipitation data and nitrate levels in this well (Figure 17.10). Due to dry conditions in October, and lowering of the water table, the nitrate concentration in the well decreased, as shown on the SYMAP for October 6, 1982 (Figure 17.11). The mobility of nitrogenous waste in shallow permeable

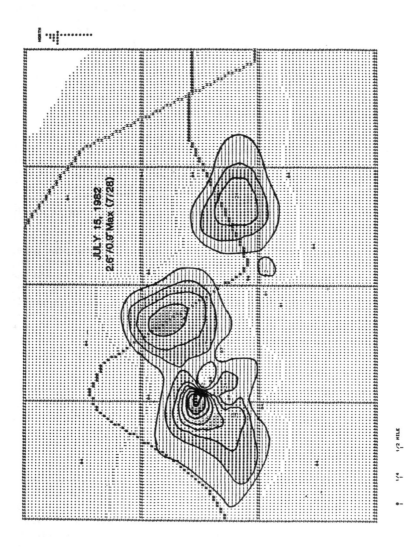

Figure 17.3:   Study Area Ground Water Nitrate Contours (July 15, 1982)

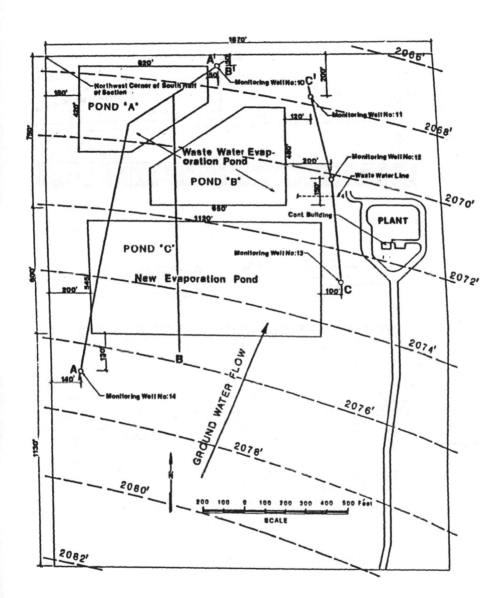

Figure 17.4:   General Layout of the Fertilizer Plant

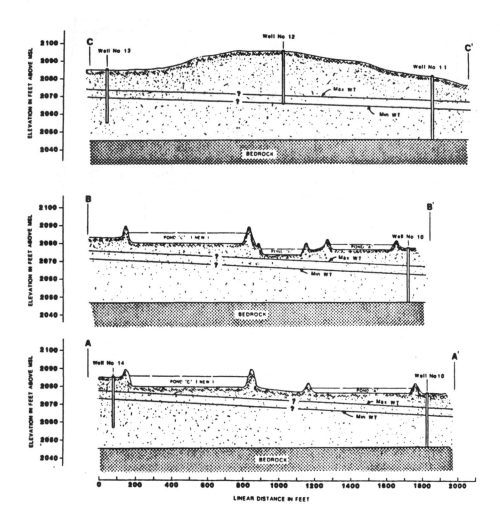

Figure 17.5: Vertical Cross-Section AA', BB', and CC' in the Vicinity of the Plant

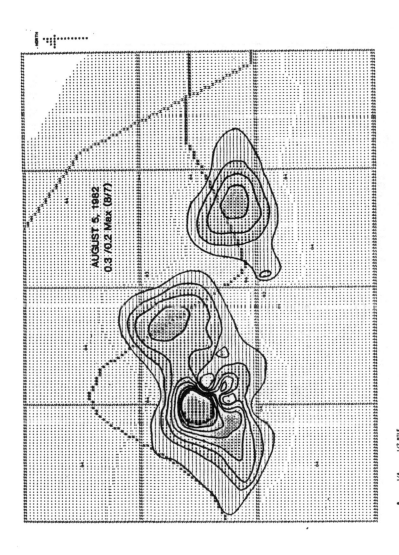

Figure 17.6: Study Area Ground Water Nitrate Contours (August 5, 1982)

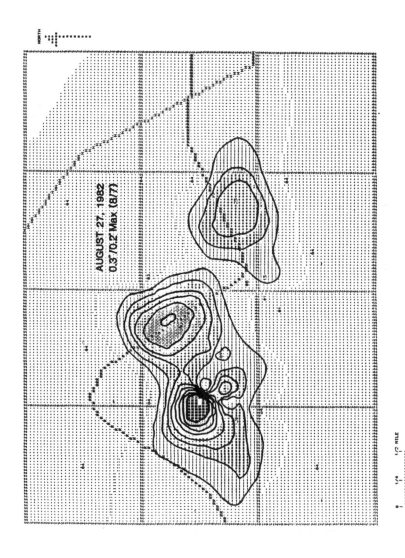

Figure 17.7: Study Area Ground Water Nitrate Contours (August 27, 1982)

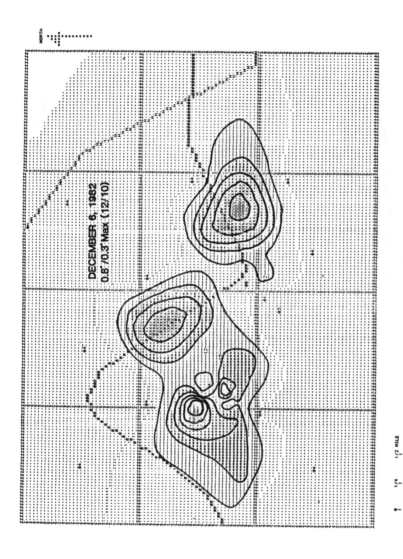

Figure 17.8: Study Area Ground Water Nitrate Contours (December 6, 1982)

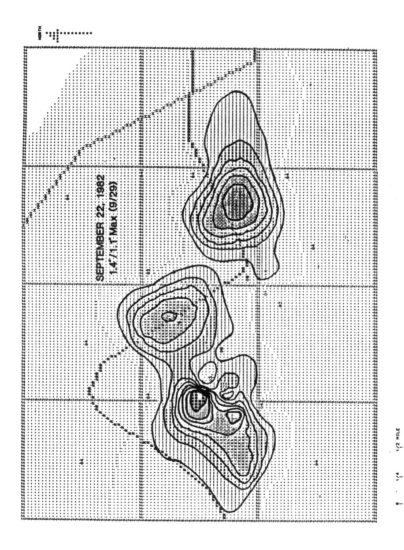

Figure 17.9:   Study Area Ground Water Nitrate Contours (September 22, 1982)

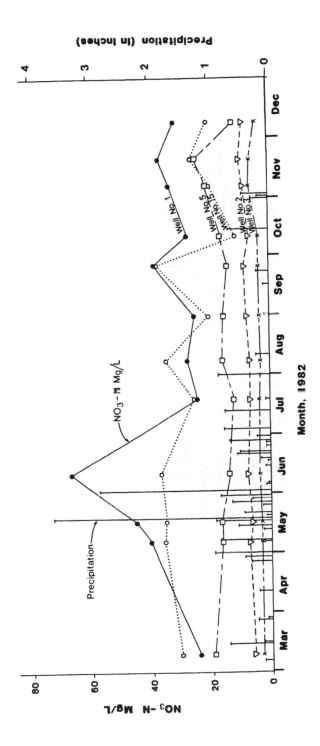

Figure 17.10: Relationship Between Precipitation and NO$_3$-N Levels in Wells

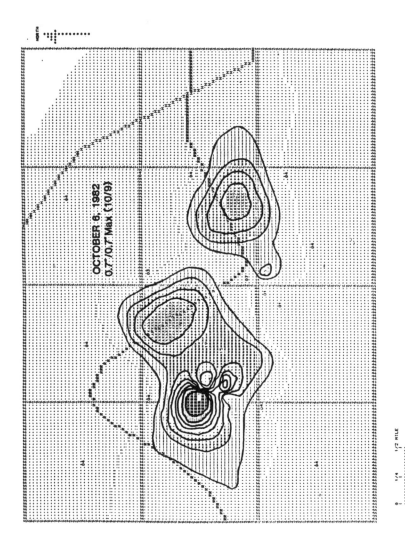

Figure 17.11: Study Area Ground Water Nitrate Contours (October 6, 1982)

aquifers is clearly seen from the increase in nitrogen concentration induced by infiltrating rainwater.

SUMMARY

Analysis of 16 wells in Zone I, over the nine-month study period revealed fluctuations of nitrate concentrations in ground water throughout the year. These fluctuations were of three types:

(1) Large and frequent changes in nitrate concentrations between monthly samples in well numbers 10 and 11, due to the introduction of industrial wastewater in the vicinity of a new pond, which is upgradient from the wells.

(2) Patterns similar to those of seasonal variation, i.e., usually high in the spring and decreasing during the summer and increasing again in the fall were observed in well #1, due to presence/absence of animal waste and fertilizer application, and recharge patterns around the well.

(3) A steady increase in nitrate concentrations throughout the sampling period in well #13 due to a past activity which resulted in N-- enrichment in the soil. In March, 1986, the nitrate levels had decreased to 20 mg/L as nitrogen.

The nitrate problem in the majority of Woodward wells sampled in 1982 was created during rainfall events by nitrate sources leaching into the highly permeable soil. It is recommended that careful attention and planning be given to the proper agricultural practices involving the use of fertilizers, irrigation practices, livestock handling operations, and proper handling of industrial wastes. SYMAP has proved to be an efficient tool for observing spatial water quality trends in water table aquifers.

SELECTED REFERENCES

American Water Works Association, Standard Methods for the Examination of Water and Wastewater--15th Edition, American Public Health Association, 1981, Washington, D.C., p 370.

DeWalle, F.B., and Schaff, R.M., "Groundwater Pollution by Septic Tank Drainfields," Journal Environmental Engineering Division, Vol. 106 , No. EE3, 1980, pp 631-648.

Dudnik, E.E., "SYMAP--User's Reference Manual for Synagraphic Computer Mapping," Report No. 71-1, 1971, Department of Architecture, College of Architecture and Art, University of Illinois, Chicago, Illinois.

Enberg, R.A., "The Nitrate Hazard in Well Water with Special Reference to Holt County, Nebraska," Nebraska Water Survey Paper 21, 1967, University of Nebraska Conservation and Survey Division, Lincoln, Nebraska.

Feth, J.H., "Nitrogen Compounds in Natural Water--A Review," Water Resources Research, No. 2, 1966, pp 41-58.

Gopal, B.K., "Alternate Test Procedure for Measuring Nitrate-N in Well Water--A Comparability Testing Study," A Working Paper, 1981a, Oklahoma Water Resources Board, Oklahoma City, Oklahoma.

Gopal, B.K., "208 Task 861, Groundwater Prototype Project--Nitrate Contamination in Groundwater, Development of an Experimental Design Program for the Problem," A Working Paper, 1981b, Oklahoma Water Resources Board, Oklahoma City, Oklahoma.

Gopal, B.K., "Groundwater Prototype Project--Nitrate Contamination in Groundwater," 208 Work Task 861, A Final Report, 1982, Oklahoma Water Resources Board, Oklahoma City, Oklahoma.

Peele, T.C., and Gillingham, J.T., "Influence of Fertilization and Crops on Nitrate Content of Groundwater and Tile Drainage Effluent," Report No. 33, 1972, Water Resources Research Institute, Clemson University, Clemson, South Carolina, p 19.

U.S. Soil Conservation Service, "Soil Survey--Woodward County, Oklahoma," Series 1960, No. 6, 1963, U.S. Department of Agriculture, Washington, D.C.

CHAPTER 18

SALINE SEEP ON WHEATLAND IN NORTHWEST OKLAHOMA

by

W.A. Berg,
C.R. Cail,
D.M. Hungerford,
J.W. Naney, and
G.A. Sample

Dryland-farm saline seeps have emerged over the past 10 to 40 years in the northern Great Plains (Brown et al., 1983) and in Texas (Neffendorf, 1978) to become major management problems. Changes in land use from native range to dryland farming, improved farming technology which increases soil water recharge, water moving through the soil profile, geology of the area, and climate all contribute to saline seep (Brown et al., 1983). Dryland-farm saline seep is recognized worldwide (Vander Pluym, 1978); however, the problem is not always distinctly separated from saline seep induced by irrigation or naturally occurring saline seep on land in native vegetation.

The purpose of this chapter is to report observations on dryland-farm saline seeps which have developed over the past 5 to 20 years in western Oklahoma. All references to saline seep that follow are associated with dryland farming operations.

SETTING

Most of the observations and all the data presented are from Harper County in northwest Oklahoma. Very gently sloping (1-3%) to undulating (3-8%) uplands underlain by Permian redbeds are farmed for production of winter wheat. Wheat is seeded in the fall and matures in early June. After harvest the land is cultivated to control weeds until reseeded to wheat in the fall. Stubble mulch and terraces are common practices used to control erosion. The usual terrace construction is level with open ends to allow some surface drainage.

Average annual precipitation at Buffalo in central Harper County over the 1950-1985 period was 690 mm. Over this period the least precipitation in one calendar year was 282 mm, and the most was 1205 mm. About 62 percent of the precipitation is received from May through September. Summer daytime high temperatures are often in the 35 to $40°C$ range. Winter low temperatures are about $-18°C$. The frost-free period is usually from mid-April until mid-October. Strong winds are common in March, April, May, and June.

Quinlan soils (loamy, mixed, thermic, shallow Typic Ustochrepts) are shallow (< 50 cm) over redbeds and occur on ridges and upper slopes. Woodward soils (coarse-silty, mixed, thermic Typic Ustochrepts) are moderately deep (50-100 cm) over redbeds and are common on mid-slope positions. Carey (fine-silty, mixed, thermic Typic Argiustolls) and St. Paul soils (fine-silty, mixed, thermic Pachic Argiustolls) are deep (>100 cm) over redbeds on very gentle slopes on lower landscape positions (Nance et al., 1960).

Most of the rocks immediately below the soil mantle are within the Whitehorse Group which ranges from soft, fine-grained, poorly-cemented sandstone (Marlow Formation) to interbedded sandstones, shales, and gypsum (Rush Springs Sandstone) and are subunits of the Permian System (Myers, 1959). Older Permian redbeds which include some finer textured strata (Dog Creek Shale and Blaine Formation) underlie soils in localized areas of the central and eastern parts of the county, respectively.

The rooting zones in soils, alluvium and weathered redbeds have relatively higher hydraulic conductivities than the underlying consolidated redbeds. Thus, the bedrock surface forms a lower flow boundary for percolating water and causes downslope ground water movement as shown in Figure 18.1.

SALINE SEEP DEVELOPMENT

Saline seeps in northwest Oklahoma were not mentioned in the 1960 Harper County soil survey report (Nance et al., 1960), or in a 1962 review of saline-alkali soils in Oklahoma (Reed, 1962). In the 1970s farmers became aware of areas in heretofore highly productive fields where wheat stands were thin and weeds, particularly Kochia scoparia, were a problem. Inspection showed a moist soil with a water table at 0.5 to 2 meter depth. With time the affected areas became broader and extended up the drainage. Eventually, the original affected areas became boggy so that they could not be cultivated and a white salt crust formed after periods of several weeks with no precipitation. At this stage the salt-affected areas supported little or no vegetation. Eventually, salt cedar (Tamarix spp.) invaded some of the seeps.

The incidence and extent of saline seep increased over the 1975 to 1985 period. By 1985, 1300 ha of about 65,000 ha of wheatland in Harper County was known to be affected, and the total area affected was estimated to be several times greater. Other counties in northwest Oklahoma where wheat is grown on soils developed by Permian redbeds also have an increasing incidence of saline seep. Saline seep has been reported in northwest Texas under similar land use, precipitation, and geologic conditions (Neffendorf, 1978).

The size of saline seep areas in Harper County range from 5 to 100 ha in the lower landscape positions of 200 to 2000 ha watersheds underlain at shallow depths (1 to 3 m) by Permian redbeds. The greater portion of affected watersheds are in cultivation for dryland wheat. On steeper areas (>8% slopes) native range and old fields seeded to grass occur. The watersheds are usually terraced. After a rain the terraces usually hold some water in low spots and behind small dams formed by siltation.

It can be hypothesized that saline seeps develop over tens of years from small increments of water percolating through cultivated soils and then move over relatively unleached Permian redbeds to lower landscape positions (Figure 18.1). In contrast, under native vegetation most of the precipitation is transpired or evaporated and little, if any, percolates to the redbeds.

Four areas affected by saline seep were sampled in January, 1986,

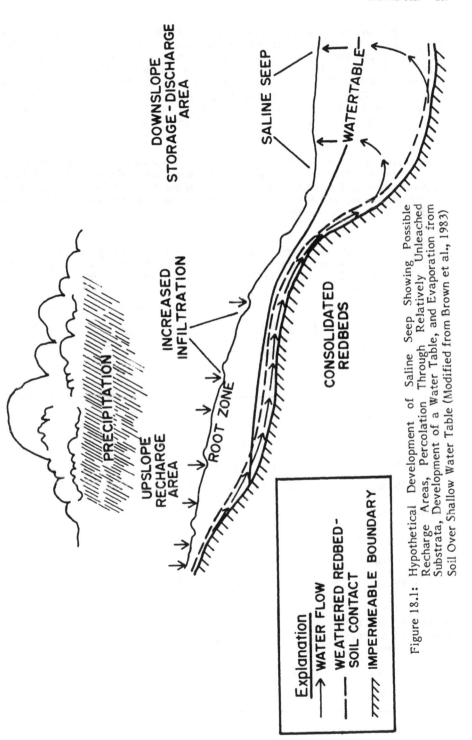

Figure 18.1: Hypothetical Development of Saline Seep Showing Possible Recharge Areas, Percolation Through Relatively Unleached Substrata, Development of a Water Table, and Evaporation from Soil Over Shallow Water Table (Modified from Brown et al., 1983)

following two months with little precipitation. High soluble salt concentrations were present in the surface 5 cm of the salt crusted areas (Table 18.1). At the 5 to 15 cm depth the seep and fringe areas sampled were saline (>4 mmhos/cm; Richards, 1954) but within salinity levels where salt-tolerant species will grow (Bernstein, 1964). The upland soils were not saline. The sodium adsorption ratios of 12 to 29 in salt crusted areas (Table 18.1) indicates that soil dispersion may occur if soluble salts are leached from these soils. The soils were calcareous and were in the pH range of 8.0 to 8.5 as measured in saturated pastes.

Specific problems associated with saline seep include: (1) the deepest, most productive soils are affected, (2) the seep areas dissect fields making cultivation and pest control more difficult, (3) the seeps increase the salt load on local drainages, and (4) water that could be used to produce vegetation is lost by evaporation or drainage.

Table 18.1:   Electrical Conductivity and Sodium Adsorption Ratios of Saturation Extracts from Soil Samples Taken in and Adjacent to Four Dryland Saline Seep Areas in Northwest Oklahoma (January, 1986)

Soil Condition Depth	Electrical Conductivity mmhos/cm		Sodium Adsorption Ratio	
	Mean	Range	Mean	Range
Upland 6 to 20 m from salt-crusted areas				
0 - 5 cm	1.0	0.6- 1.7	2	1- 3
5 - 15 cm	0.8	0.4- 1.2	3	2- 4
Fringe area, discontinuous salt crust, some wheat dead				
0 - 5 cm	12	7 -20	8	6-11
5 - 15 cm	5	4 - 6	4	4- 6
Near bottom of drainage, continuous salt crust, no live wheat				
0 - 5 cm	23	13 -30	22	12-29
5 - 15 cm	8	6 -11	9	6-14

TILE DRAIN SYSTEMS

Tile drain systems have been installed on five saline seep areas in Harper County. This involves 15 km of pipe to drain 300 ha of seep affected area. Four of the five tile drain systems have been effective in ameliorating the saline seep problem. Drainage from three of the systems is monitored for water quantity (Table 18.2) and quality (Table 18.3). Two of the monitored tile drain systems ameliorated the saline seep condition. In these systems the annual drainage through the tile is equivalent to 1.6 to 2.5 cm of precipitation over the watershed (Table 18.2). The monitored tile drain system that did not ameliorate the saline seep condition removed 0.3 cm of water per year on a watershed basis (Table 18.2). This data indicates that the excess water which promotes saline seeps is equivalent to approximately 1.6 to 2.5 cm per year on a watershed basis.

Table 18.2: Quantity of Drainage from Tile Drain Systems Installed in Saline Seeps on Three Watersheds.

| | WATERSHED | | |
OBSERVATIONS	COSBY	MALLORY	HOWARD
Years observed	1980-85	1982-85	1981-85
Watershed area, ha	900	210	2100
Saline seep, ha	105	40	70
Flow from drain, L/minute mean	280	98	115
range	130-380	40-150	76-140
Mean annual drainage ÷ watershed area, cm/yr	1.6	2.5	0.3
Saline seep ameliorated	Yes	Yes	No

The discharge water contains concentrations of soluble salts (Table 18.3, Figure 18.2) which, in general, are greater than the salt concentrations in the local streams (about 2000 umhos/cm, electrical conductivity) receiving the drainage. The local streams eventually drain into the Cimarron River which has electrical conductivity readings of about 4000 umhos/cm in western Harper County (Boyle, 1986). Downstream in eastern Harper County the Cimarron River is usually higher in dissolved salts (Boyle, 1986). The NO_3-N in the tile drainage water is approaching the level (10 mg/L) unacceptable for human consumption. Sulfate is the dominant anion in the drainage. Calcium, magnesium, and sodium are the dominant cations. The relatively low sodium adsorption ratios (3 to 6) indicate that soil irrigated with the water should not become dispersed (sodium affected).

Table 18.3:   Water Quality of Drainage from Tile Drain Systems Installed in Saline Seeps on Three Watersheds (January, 1986)

MEASUREMENT	WATERSHED		
	COSBY	MALLORY	HOWARD
Electrical Conductivity (umhos/cm)	6800	4800	4300
pH	8.2	8.2	8.0
NO_3^--N (mg/L)	7.5	8.6	8.8
$SO_4^=$ (mg/L)	3800	2000	2100
Cl^- (mg/L)	536	450	300
Na^+ (mg/L)	890	530	340
Ca^{++} (mg/L)	540	620	580
Mg^{++} (mg/L)	680	220	260
Sodium Adsorption Ratio	6	3	5

The amount and electrical conductivity of drainage from the tile drain system on the Cosby watershed was monitored over a five-year period (Figure 18.2). The general pattern is that drainage volume is less in the summer and fall. The electrical conductivity of the drainage water appears to increase after major precipitation periods; an exception was in 1985, when precipitation was 180 percent of the long-term annual average.

INFORMATION NEEDS AND MANAGEMENT SUGGESTIONS

The saline seep problem in northwest Oklahoma appears similar to the extensive saline seep problem in the northern Great Plains. Thus, many of the management approaches for ameliorating saline seep in the northern Great Plains (Brown et al., 1983; and Vander Pluym, 1978) should apply to the southern plains. However, there are land management differences between the two regions: (1) the affected areas in Oklahoma are usually cropped to wheat annually; whereas, in the northern plains the land management is commonly a year of fallow followed by wheat; and (2) the Oklahoma wheatlands are usually terraced; those in the northern plains are not.

The immediate need in Oklahoma is to identify recharge areas and to manage these areas to use or discharge the precipitation before it moves through the root zone. Sandier soils, terraces, and farm ponds are probable recharge areas. Establishing perennial forage species or possibly farming warm-season annuals such as sorghum or sunflower should use more of the late spring and summer rainfall than winter wheat. Compacted tillage pans at depths of 20 to 25 cm are common on affected watersheds. Few wheat roots

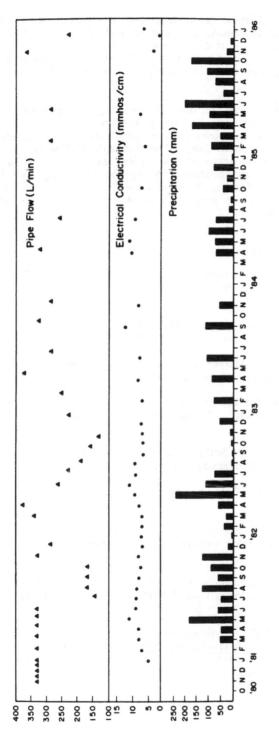

Figure 18.2: Quantity and Electrical Conductivity of Water from Tile Drainage
System Installed in 1980 on Cosby Watershed as Related to
Monthly Precipitation Over Five Years

penetrate into or through these pans. Mechanically breaking the pans or using species better adapted to rooting into and through the pans should result in greater use of soil water. Or, it can be argued that tillage pans act to restrict percolation since the infiltration rate is slow. If terraces are a major recharge source, then use of graded terraces and pipes to drain level terraces may be needed.

As covered in an earlier section, tile drainage systems can ameliorate individual saline seeps. However, installing drainage is expensive and disposal of the saline drainage water poses another problem. Use of perennial salt-tolerant forage species which are adapted to sites with high water tables should result in beneficial use of some of the saline seep areas. Alkali sacaton (Sporobolus airoides) and tall wheatgrass (Agropyron elongatum) have been used on similar saline sites in Kansas, Oklahoma, and Texas. However, drainage or use of salt-tolerant species on seeps is a treatment of symptoms rather than prevention of the problem.

SELECTED REFERENCES

Brown, P.L. et al., "Saline-seep Diagnosis, Control, and Reclamation," Conservation Report No. 30, 1983, U.S. Department of Agriculture, Washington, D.C.

Bernstein, L., "Salt Tolerance of Plants," Information Bulletin No. 283, 1964, U.S. Department of Agriculture, Washington, D.C.

Myers, A.J., "Geology of Harper County, Oklahoma," Bulletin No. 80, 1959, Oklahoma Geological Survey, Norman, Oklahoma.

Nance, E.C. et al., "Soil Survey of Harper County, Oklahoma," Series 1956, No. 8, 1960, U.S. Department of Agriculture, Soil Conservation Service, Washington, D.C.

Neffendorf, D.W., "Statewide Saline Seep Survey of Texas," Master of Science Thesis, 1978, Texas A&M University, College Station, Texas.

Personal Communication, Dale Boyle, U.S. Geological Survey, Woodward, Oklahoma, 1986.

Reed, LW., "A Study of Saline-alkali Soils in Oklahoma," Processed Series, P-430, 1962, Oklahoma State University Experiment Station, Stillwater, Oklahoma.

Richards, L.A. (ed)., "Diagnosis and Improvement of Saline and Alkali Soils," Agriculture Handbook 60, 1954, U.S. Department of Agriculture, Washington, D.C.

Vander Pluym, H.S.A. (ed), "Dryland-saline-seep Control," Proceedings Meeting of the Subcommission on Salt-affected Soils at the 11th International Soil Science Society Congress, 1978, Plant Industry Division, Agriculture Center, Lethbridge, Alberta, Canada.

CHAPTER 19

A NATIONAL ASSESSMENT OF GROUND WATER CONTAMINATION
FROM PESTICIDES AND FERTILIZERS

by

Deborah M. Fairchild

Nonpoint pollution sources are being increasingly addressed in Federal, state, and local ground water quality protection programs. Ground water pollution from prior and/or current usage of pesticides and fertilizers in agricultural operations is being routinely documented. Synthetic organic pesticides are of concern due to their environmental persistence and potential health effects. Nitrates in ground water from commercial fertilizer applications have long been recognized as an indicator of nonpoint pollution. This chapter summarizes current information on ground water contamination in the contiguous United States from the usage of pesticides and fertilizers.

DESCRIPTION OF STUDY

The Soil Conservation Service (SCS) within the U.S. Department of Agriculture was charged with the preparation of the 1985 Resource Conservation Assessment. This Assessment is an outgrowth of the 1977 Soil and Water Conservation Act and includes topics associated with ground water quantity and quality. Of particular importance is the requirement to address the issue of ground water pollution from agricultural chemicals as it relates to future uses of ground water resources for agricultural and other purposes. The objective of a study conducted by the Environmental and Ground Water Institute (O'Hare et al., 1985) was to summarize ground water pollution occurrences in the contiguous United States resulting from the usage of agricultural chemicals. Agricultural chemicals as defined herein include pesticides and nutrients, primarily nitrogen, resulting from either fertilizer or animal waste applications. This study included the conduction of a literature search on ground water pollution from agricultural chemicals and the conduction of a telephone survey.

The information gathering work task included the conduction of a computer-based literature search using the Dialog System operated by Lockheed Corporation in Palo Alto, California, and the National Ground Water Information Center operated by the National Water Well Association; screening and categorizing of the procured abstracts from the computer-based searches; and library searches. Even though over 300 data bases are included in Dialog, only ten were considered relevant after a review of the type and quantity of information available in each data base. Key descriptor words utilized in the search were pesticides, nutrients, herbicides, insecticides, fertilizers, and ground water. A total of 1,317 abstracts were procured. The search of the National Ground Water Information Center was conducted to supplement and confirm the results of the Dialog search. Each of the procured abstracts were read and duplications were eliminated along with abstracts deemed to be only minimally relevant. A total of 239 abstracts were relevant, and they were sorted into 14 subject categories.

The libraries at The University of Oklahoma (Norman), Oklahoma State University (Stillwater), and the Robert S. Kerr Environmental Research Laboratory of EPA (Ada, Oklahoma) were searched to identify pertinent references. The subject card catalog and government documents sections of The University of Oklahoma's main library were searched in addition to the subject card catalog and states index of the university's geology library. In addition, a reference search was also made at The University of Oklahoma's Engineering Library and the libraries of the Science and Public Policy Program and the Environmental and Ground Water Institute.

The second work task was to call federal and state agencies to obtain information (verbally) and request written reports, maps, or other published items that would be useful. A standard telephone protocol was developed to achieve consistency in obtaining the requested information. Each telephone contact was recorded on a standard form that listed the date, person contacted, organization, telephone number, and key discussion points such as current efforts to evaluate or monitor ground water pollution from agricultural chemicals. In total, data sheets were prepared for 55 telephone calls.

Table 19.1 illustrates the sources of information utilized for the analysis of each state's assessment of ground water contamination from pesticides and fertilizers. An "x" in the "abstracts" column indicates that one or more abstracts containing information on agricultural chemicals and ground water in that state were obtained through the Dialog search. An "x" in the "telephone survey" column indicates that pertinent information was obtained from that state's telephone contact. A blank in the telephone survey column means (1) the contact either preferred receiving the questionnaire by mail and no reply was received or (2) promised to check his/her agency's publications and send these by mail. If a promised mailing was received or if locally available reports were acquired, an "x" was placed in the last column. The information listed in Table 19.1 was used to compile this chapter. In this national survey, unavoidable limitations were identified and include:

(1) The inherent limitations of a telephone survey are that it may be difficult to know if the correct contact has been made; there may be no way to ascertain if the contact has been able to supply complete and accurate information; and it is not possible to talk to everyone who has valid facts/opinion or knowledge of the subject.

(2) Often the information obtained via a telephone survey is vague.

(3) There are inherent problems with computer searches of literature, including the fact that entries may be months behind.

(4) Important sources of information may be completely bypassed.

(5) Detection limits on some pesticides may not have allowed for proper analysis of ground water contamination.

OVERVIEW OF RESULTS

Ground water contamination resulting from pesticide usage is addressed

Table 19.1: Information Sources

State	Abstracts	Telephone Survey	Reports Promised by Phone Contact	Reports Received or Acquired*
Alabama	X	X		X
Arizona	X	X		X
Arkansas		X		X
California	X		X	X
Colorado	X	X	X	X
Connecticut	X	X	X	
Delaware		X	X	0
Florida	X	X	X	X
Georgia	X	X		
Idaho	X	X		
Illinois	X	X		
Indiana	X	X		
Iowa	X	X	X	X
Kansas	X	X		
Kentucky	X	X		
Louisiana	X	X		
Maine		X		
Maryland			X	0
Massachusetts	X	X		
Michigan	X	X	X	X
Minnesota	X	X		
Mississippi	X	X		
Missouri	X	X		
Montana	X	X	X	X
Nebraska	X	X	X	0
Nevada		X	X	0
New Hampshire		X		
New Jersey	X	X	X	X
New Mexico	X	X		
New York	X	X		
North Carolina	X	X	X	0
North Dakota		X		
Ohio		X		
Oklahoma	X	X		
Oregon	X	X		

Table 19:1:   (continued)

State	Abstracts	Telephone Survey	Reports Promised by Phone Contact	Reports Received or Acquired*
Pennsylvania	X	X		
Rhode Island		X		X
South Carolina	X	X	X	X
South Dakota		X	X	X
Tennessee		X		
Texas	X	X	X	X
Utah		X		
Vermont		X	X	X
Virginia	X	X		
Washington	X	X		
West Virginia	X	X		
Wisconsin	X	X	X	
Wyoming	X	X		X

*An "O" indicates that the promised report(s) were not received but a supplemental publication was located to aid in analysis.

by state in Table 19.2. It may be particularly interesting to note those states which have identified contamination but are not yet taking any action toward control due to either a lack of interest or necessary funding. Four of these states--Florida, New York, Rhode Island, and Texas--have extensive (regional) problems which have resulted in institutional control measures while California has regional problems with extensive monitoring being undertaken. Arizona, Delaware, Kansas, and Montana have localized problems that were serious enough to warrant institutional measures.

Table 19.3 is a summary of the specific pesticides responsible for ground water contamination in each state. Those additional states which claim pesticide contamination but did not distinguish individual contaminants are also listed. Table 19.4 is a list of high priority pesticides developed by the U.S. Environmental Protection Agency (EPA).

Table 19.5 illustrates the extent of ground water contamination resulting from fertilizer usage (mostly nitrates). Five states (Delaware, New York, Oklahoma, South Dakota, and Texas) have regional problems and institutional control measures for fertilizers.

Table 19.2:   Summary of Ground Water Contamination from Pesticide Usage

State	R	L	S	N	NA	1	2	3	4*
Alabama		X	X				X		
Arizona		X				X			
Arkansas			X				X		
California	X						X		X
Colorado		X					X		
Connecticut	X						X		
Delaware		X				X	X		
Florida	X					X			
Georgia		X					X		
Idaho		X	X				X		X
Illinois		X	X						X
Indiana		X							X
Iowa		X							X
Kansas		X				X			
Kentucky			X					X	
Louisiana		X						X	
Maine	X								X
Maryland		X					X		
Massachusetts	X						X		
Michigan		X						X	
Minnesota		X	X				X		
Mississippi				X			X		
Missouri		X						X	
Montana		X				X			
Nebraska	X						X		
Nevada				X				X	
New Hampshire				X				X	
New Jersey	X							X	
New Mexico			X			X	X		
New York	X					X			
North Carolina					X	X			
North Dakota				X				X	
Ohio		X						X	
Oklahoma					X	X			
Oregon				X			X		

Table 19.2: (continued)

State	R	L	S	N	NA	1	2	3	4*
Pennsylvania		X					X		
Rhode Island	X					X			
South Carolina		X					X		
South Dakota	X					X			
Tennessee		X					X		
Texas	X	X				X	X		
Utah		X					X		
Vermont		X					X		
Virginia			X				X		X
Washington	X						X		
West Virginia		X						X	
Wisconsin		X					X		
Wyoming			X					X	
TOTALS	12	25	8	6	2	12	25	11	7

Contamination Extent:

R = regional contamination documented (via ground water monitoring, published studies, etc.)

L = localized contamination documented

S = regional or localized contamination suspected

NA = contamination exists but is due to handling or nonagricultural activities

N = no significant problems revealed at present time

Status of State Awareness:

1 = specific institutional control measures implemented (state-wide monitoring for specific ag-chemical contaminants, control on application rates, etc.)

2 = evaluation of issue taking place or monitoring systems in planning stage

3 = no actions at present time (due to lack of funds or interest)

4 = need has been expressed for revision of present institutional control measures (e.g., need to update allowable concentrations of certain chemicals or types of chemicals that can be legally applied, etc.)

Table 19.3:   Specific Pesticide Contaminants (includes only
documented contaminants from information sources)

EDB	Arizona, Florida, Georgia, Massachusetts, South Carolina, Washington
aldicarb (Temik)	Maine, Massachusetts, New Mexico, New York, Rhode Island, Wisconsin, Florida
DDT	Idaho, Louisiana, Wisconsin
atrazine	Delaware, Kansas, Nebraska
alachlor	Kansas, Nebraska
chlordane	Oklahoma, Virginia
tebuthiuron	Indiana
phorate	Indiana
carbaryl	Indiana
toxaphene	Louisiana
endrin	Louisiana
pordon	West Virginia
allanine	Delaware
lindane	Wisconsin
arsenic	Texas

*Note: Not all state information sources revealed specific contaminants; those which cited undistinguished pesticides include: Alabama, California, Illinois, Kentucky, Michigan, Minnesota, Missouri, Utah, Vermont, North Carolina, Ohio, South Carolina, Tennessee.

Table 19.6 is a listing of those states which have displayed, through the information sources, some form of control over pesticide and nutrient usage through licenses or legislation which require limitations on the amount of agricultural chemicals utilized and/or an evaluation of the extent of possible ground water contamination. As far as this survey has revealed, these states exert a legal influence over agricultural chemical usage in excess of federal regulation, and/or they are taking steps to systematically evaluate pesticide and nutrient contamination of ground water in their state. These are only the states for which their contact gave a positive response--other states may indeed have such legislation without an individual contact's knowledge.

Table 19.4:   Agricultural Pesticides of Highest Priority

Common Name	Trade Name

HALOGENATED ORGANICS

D-D Mixture
(1,2-Dichloropropane, 1,3-Dichloropropene
and Trichloropropanes)

DBCP
(Dibromochloropropane)

EDB
(Ethylene Dibromide)

Methyl Bromide
(Bromomethane)

Chloropicrin
(Trichloronitromethane)

Endosulfan and Breakdown Product (Endosulfan Sulfate)	Thiodan
PCNB (Pentachloronitrobenzene)	Terraclor
Dicofol (1,1-Bis (chlorophenyl)-2,2,2-trichloroethanol)	Kelthane

Dimethyltetrachloroterephthalate

ORGANO PHOSPHORUS

Parathion (Ethyl Parathion)

Dimethoate	Cygon

Diazinon

Acephate	Orthene
Disulfoton	Di-Syston
Phorate	Thimet
Fenamiphos	Nemacur*
Metbamidophos	Monitor
Chlorpyrifos	Dursban, Lorsban

Table 19.4:   (continued)

Common Name	Trade Name

CARBAMATES

Maneb
and Breakdown Product (Ethylene Thio Urea)

Ziram
and Breakdown Product (Ethylene Thio Urea)

Carbaryl — Sevin

Methomyl — Lannate, Nudrin

Carbofuran — Furadan

Aldicarb — Temik

Oxamyl — Vydate

Benomyl
and Breakdwon Product
(Methyl 2-benzimidazolecarbamate)

TRIAZINES

Atrazine

Simazine

Prometryn — Caparol

Cyanazine — Bladex*

DINITRO PHENOLS

DNBP (Dinoseb)

DNOC (Dinitrocresol)

UREA DERIVATIVE

Diuron — Karmex, Krovon

DICARBOXIMIDES

Captan

Table 19.4:   (continued)

Common Name	Trade Name
BENZONITRILE	
Chlorothalonil	Bravo
ACETANILIDE	
Alachlor	Lasso
BIPHYRIDYL	
Paraquat	
ACETAMIDE	
Diphenamid	Dymid, Enide
SUBSTITUTED URACIL	
Bromacil	Hyvar
PHTHALATE	
Endothall	

*Use reported by Department of Food and Agriculture under trade name.

STATE-BY-STATE DISCUSSION

This section contains short, written summaries on selected states' agricultural chemical ground water contamination problems. The sources of the information utilized to compile these summaries were listed in Table 19.1. Eleven states were chosen because they had significant regional and/or local pesticide or fertilizer contamination which warranted state licensing, legislation, or other control measures. Complete information on all the states was included in the study report (O'Hare et al., 1985).

Arizona

In the State of Arizona, nitrates occur naturally at concentrations as high as 400 ppm. Fluorides and borons are also naturally occurring. An investigation in Paradise Valley, Arizona, revealed areas of ground water with up to 132 ppm nitrate (Silver and Fielden, 1980). Two sources were suggested: leaching from

Table 19.5: Summary of Ground Water Contamination from Nutrient (Fertilizer) Usage

State	R	L	S	N	NA	1	2	3	4*
Alabama		X	X				X		
Arizona					X			X	
Arkansas			X				X		
California	X						X		X
Colorado		X					X		
Connecticut	X						X		
Delaware	X					X	X		
Florida	X							X	
Georgia	X						X		
Idaho					X		X		
Illinois		X	X					X	
Indiana		X							X
Iowa	X								X
Kansas		X				X			
Kentucky			X					X	
Louisiana				X				X	
Maine	X								X
Maryland		X					X		
Massachusetts				X			X		
Michigan		X						X	
Minnesota					X		X		
Mississippi				X			X		
Missouri		X						X	
Montana				X		X			
Nebraska	X				X		X		
Nevada					X			X	
New Hampshire				X				X	
New Jersey	X							X	
New Mexico		X					X		
New York	X					X			
North Carolina		X					X		
North Dakota				X				X	
Ohio		X						X	
Oklahoma	X					X			
Oregon		X					X		

Table 19.5: (continued)

State	R	L	S	N	NA	1	2	3	4*
Pennsylvania		X			X		X		
Rhode Island				X			X		
South Carolina					X		X		
South Dakota	X				X	X			
Tennessee		X					X		
Texas	X	X				X	X		
Utah			X				X		
Vermont		X					X		
Virginia			X				X		
Washington	X						X		
West Virginia				X				X	
Wisconsin		X					X		
Wyoming			X				X		
TOTALS	14	17	6	9	8	7	27	13	4

Contamination Extent:

R = regional contamination documented (via ground water monitoring, published studies, etc.)
L = localized contamination documented
S = regional or localized contamination suspected
NA = contamination exists but is due to handling or nonagricultural activities
N = no significant problems revealed at present time

Status of State Awareness:

1 = specific institutional control measures implemented (statewide monitoring for specific ag-chemical contaminants, control on application rates, etc.)
2 = evaluation of issue taking place or monitoring systems in planning stage
3 = no actions at present time (due to lack of funds or interest)
4 = need has been expressed for revision of present institutional control measures (e.g., need to update allowable concentrations of certain chemicals or types of chemicals that can be legally applied, etc.)

Table 19.6: States Exhibiting Legislative Control or Interest in Agricultural Chemical Contamination of Ground Water*

Arkansas	New Mexico
California	New York
Connecticut	North Carolina
Florida	Oklahoma
Idaho	Rhode Island
Illinois	South Dakota
Indiana	Texas
Iowa	Vermont
Maine	Virginia
Montana	West Virginia

*Examples may include applicator licenses, review councils and, systematic evaluation. State listing does not necessarily indicate complete awareness or control of issue.

nearby formations or infiltration due to the use of nitrogenous fertilizers and disposal of treated wastewater.

Recently, the cancer-causing compound EDB (ethylene dibromide) was found in a water supply well in Cashion, a suburb of Phoenix. The source of this contamination is not known, but the chemical had been used as a pesticide until banned in 1983. As a result of the incident, the Department of Health Services compiled new guidelines for public drinking water (U.S. Water News, 1985a).

The primary ground water contamination concerns resulting from agricultural practices in Arizona have been increasing salinity problems associated with irrigation, especially in the Yuma area and around Heala. There is a State Board of Pesticide Control, but it has no ground water monitoring program.

California

The State of California is a large user of ground water and agricultural chemicals. The U.S. Water News reports that one out of every four wells in California has water contaminated by industrial chemicals or pesticides (U.S. Water News, 1985). A 1980 estimate of the total annual use of pesticides in California was 121 million pounds; due to gaps in reporting information, this number may be actually as much as 300 percent higher (Litwin et al., 1983). Litwin et al. (1983) found more than 50 different pesticides in ground water from 23 California counties. A majority of these incidents resulted from pesticide application, while some were caused by local spills at pesticide manufacturing, handling and disposal sites. Findings to date do not fully represent the extent of pesticide contamination because of the limited scope of monitoring. The study concluded that coordination among state and local agencies responsible for ground water monitoring of pesticides is inadequate.

Also of great concern is the contamination by nitrate-N from fertilizer application in the Santa Maria Valley, the Fresno-Clovis metropolitan area, the Redlands area, and the Upper Santa Ana Basin area (Hughes, 1977; Schmidt, 1972; and Eccles and Bradford, 1977). Nitrate concentrations have reached as much as 50 mg/l in the Santa Maria Valley ground water basin (Hughes, 1977). An investigation on a 960-acre citrus watershed revealed that 45 percent of the nitrogen applied per year was lost to drainage, providing a serious threat to ground water (Bingham, Davis and Shade, 1971). Several studies have examined the development of nitrogen fertilizer application management practices to reduce nitrate contamination of ground water without affecting crop yield and productivity (Pallares, 1978; Nightingale, 1972; and Embleton et al., 1979).

Delaware

In the U.S. Environmental Protection Agency's "State Ground-Water Program Summaries" (1985), Delaware ranked agriculture as one of its major sources of contamination. Nitrates from both fertilizers and septic tanks are a problem in the excessively well drained, sandy aquifers. Poultry wastes have also caused some contamination problems. Low concentrations of atrazine have been found in the potato growing areas; however, these areas have been tested for aldicarb with negative results.

There is an existing policy for protecting ground water quality in the general state statutes and the Department of Natural Resources and Environmental Conservation is currently developing a ground water quality monitoring program and an aquifer classification system. The Department of Agriculture has been delegated the task of pesticide enforcement (U.S. Environmental Protection Agency, 1985).

There are occasional studies on nitrates and pesticides in ground water, and, at this time, the University of Delaware is involved in a regional study of pesticide leaching in ground water with Penn State and Cornell University.

Florida

Florida is one of the most aware states in the nation regarding agricultural chemical contamination of ground water. Reasons for this may include: 92% of Florida's population relies on ground water for its drinking water supplies; the state uses more pesticides than many other states (while it is 33rd in the amount of planted acreage), and ethylene dibromide (EDB) has been found in 722 drinking water wells in the state as of November 28, 1984 (Florida Department of Environmental Regulation (DER), 1984). Interest in the agricultural chemical issue has developed since the discovery of aldicarb (Temik) contamination in an aquifer beneath a southern Florida citrus field in 1982. The DER Pesticide Review Section has prepared a summary of aldicarb (Temik) studies that have taken place in Florida since that date (Pesticide Review Section, 1984). Also investigated have been levels of DDT (Pickrell, Dwinell and Vogel, 1984), methyl bromide and chloropicrin (Pickrell, Dwinell and Tterlikkus, 1985), and ortho diquat herbicide (EDB) (Pickerell, 1984).

Chapter 487 of the Florida Statues defines the Florida Pesticide Law, which includes requirements for registration of pesticides with the Florida Department of Agriculture and Consumer Services, installation of antisyphon devices for irrigation systems, and the creation of the Pesticide Review Council to evaluate the issue. As part of the Florida Groundwater Protection Program, an ambient ground water quality monitoring program under DER began in 1985 and will continue for four years. The program includes drilling up to 1,600 new monitoring wells and utilizing approximately 1,000 other wells (Florida Department of Agriculture and Consumer Services, 1985).

Chapter 5E-2 in the Rules of the Department of Agriculture and Consumer Services refers to the use of pesticides, and their restricted use. Forty-five different active ingredients are restricted in formulation and use and may only be applied under the direct supervision of a licensed applicator. Chapter 5E-9 designates requirements for pesticide applicators (Florida Department of Agriculture and Consumer Services, 1985).

Recently, the Pesticide Review Section of DER compiled a list of 18 pesticide-active ingredients which are suspected ground water contaminants (Table 19.7). They are requesting that the manufacturers supply more information on these products. In addition, the various water management districts in Florida have compiled lists of pesticides for their areas. These lists include pesticides used and found in ground water.

At present, great interest is focused on EDB (750 ppb) and aldicarb (1200 ppb) along the Florida Ridge (a sandy ridge, extending north-south down the center of the state). Numerous nitrate occurrences are common throughout the state but this problem tends to be overlooked.

Table 19.7:   Pesticide Active Ingredients Suspected as Ground Water Contaminants in Florida (Pesticide Review Section, 1984)

Atrazine	Hexazinone
Bromacil	Mancozeb
Carbofuran	Metalaxyl
Chloropicrin	Metam-sodium
Chlorpyrifos	Methyl Bromide
1,3 Dichloropropene	Methyl Isothiocyanate
Ethoprop	Methomyl
Fenamiphos	Oxamyl
Fosetyl-al	Simazine

Kansas

Ammonia, atrazine, and alachlor have been found in separate incidents over the past 10 to 15 years in public supply wells near co-op storage areas and in association with spills.   These wells are usually completed into shallow localized aquifers.

Two studies have been conducted on nitrates in Kansas ground water. In one study (Murphy and Gosch, 1970), large amounts (5000 kg/ha) of nitrate-nitrogen were found in the top 4 m of soil under feedlots, with the amount increasing with the age of the lot. Also in this study, wells in different areas receiving uniform high rates of applied nitrogen showed widely varying amounts of nitrate-nitrogen with no indications of definite trends. In the second study, nitrates observed in farm wells were found to be related to improper location and construction of wells and not to fertilizer application (Ridder, Oehme and Kelley, 1974). Animal health problems could not be correlated to either the contaminated wells or to the fertilizer application; however, it was noted that the complexity of the problem made direct correlations difficult.

The Kansas Water Resources Board has studied the effects of irrigation on water quality (Balsters and Anderson, 1979). The chemistry of the ground water is discussed as well as the factors that alter the quality, including salt water intrusion, leaching, and added chemicals.

Ground water quality is monitored across the state by the Department of Health and Environment in conjunction with the U.S. Geological Survey. Once the Department of Health and Environment has delineated a problem, the Department of Agriculture traces the source and cause of the pollution. The Department of Agriculture is also responsible for licensing for application and storage of agricultural chemicals.   Applications are monitored and label directions and cautions are strictly enforced. Also, a chemigation law has been passed recently (U.S. Water News, 1985c).

Montana

Very low levels of phenoxy herbicides and Tordon were found in the ground water in Teton County in a 1984 survey of three agricultural production areas by the Environmental Management Division of the Montana Department of Agriculture (supported by a grant from the U.S. EPA). A chlorinated hydrocarbon scan in the same area identified no detectable amounts. Other areas included in the study were in Flathead, Lake, and Beaverhead Counties. Neither chlorinated hydrocarbons nor phenoxy herbicides were found in these areas.

In 1975, Montana was included along with Colorado, Oregon, Idaho, Washington, and Wyoming in a six-state ground water evaluation that listed irrigation return flow and dryland farming as second in severity of man-made ground water quality problems. Discharge of effluent from septic tanks and sewage treatment plants was rated most severe (van der Leeden, Cerrillo and Miller, 1975). Feedlots and application of fertilizers and pesticides were listed as less important, but still considerable.

Montana has enacted a Pesticide Control Act that very closely follows the Federal Insecticide, Fungicide, and Rodenticide Act.   Some ground water quality is monitored cooperatively by the U.S. Geological Survey and the Montana Department of Agriculture, but pesticides are not analyzed; and a comprehensive plan has not been established (U.S. EPA, 1985). Contamination problems are handled by the Environmental Management Division of the Department of Agriculture.

New York

Aldicarb has contaminated the ground water in Ft. Edward, Washington County, and in the upper two sandstone aquifers in eastern Long Island. In a tentative settlement, the Union Carbide Corporation has agreed to provide $100 apiece to some 2,400 property owners and to maintain water filtration systems for an indefinite period of time.

Nitrate contamination has also been a problem in Rensselaer, Cattaraugus, Cortland, Nassau, and Suffolk Counties and have all been related primarily to fertilizers (Butler, Nichols, and Harsh, 1978; Porter, 1980; and Soren, 1977). Virtually no ground water contamination from pesticides was found in Suffolk County in 1977 (Soren, 1977).

Contamination problems are generally addressed by the Department of Environmental Conservation. If drinking water is contaminated, the Health Department is involved. Chemical application rates are monitored by the Bureau of Pesticides, and ground water is monitored when possible contamination incidents are reported. Chlordane has been banned and the use of aldicarb has been restricted.

Oklahoma

High nitrate levels have been found in the shallow aquifers in Custer, Payne, Woods, and Tillman Counties. Chlordane has been reported in isolated instances in private wells after termite extermination. Task 1401, a three-year study, has been conducted in cooperation with the U.S. EPA to monitor contamination resulting from the land application of animal wastes. Soil, surface and ground water have been tested. Results indicate no increasing trend at this time; however, the state is continuing the monitoring program.

The pesticide control legislation in the state is similar to the national laws. Fertilizer rates are suggested by the County Agent. Ground water contamination problems are referred to the Pesticide Management Section of the Plant Industry Division of the Oklahoma Department of Agriculture.

Ambient ground water quality has been monitored for three years. A pesticide monitoring program for ground water in the bottom lands and alluvial soils is in the planning stages.

Rhode Island

Aldicarb contamination has occurred in the potato growing areas in the shallow outwash and till aquifers. This problem has been known for over a year and a half, and Union Carbide has put filters on the wells that have been contaminated. Ground water quality is being monitored in all areas known to be contaminated and these areas are being mapped at this time. A study of private wells in different land use areas is in the planning stages. Some nitrate contamination has also been reported (U.S. EPA, 1985).

The State Department of Environmental Management is responsible for source identification and law enforcement, but the Department of Health is often also involved. A Pesticide Advisory Board has been reestablished. Aldicarb has been banned and legislation is pending to increase licensing fees and/or taxes on pesticides in order to fund the needed studies and responses to the problem. All the major aquifers in the state are gravel and sand. A total of 300 private and 7 public wells has been contaminated by various sources across the state.

South Dakota

Some pesticide and nitrate contamination due to fertilizers, animal waste, and septic tanks has been recognized for several years in the Big Sioux Aquifer in the eastern part of the state. This, in part, has led to the Oakwood Lakes-Poinsett Rural Clean Water Project to improve and protect the surface and ground water quality of the project area by the application of selected Best Management Practices. As a preliminary activity, the "208" Big Sioux Aquifer Study is currently underway to assess the magnitude, areal extent and source of water quality problems within the Big Sioux Aquifer (South Dakota State Coordinating Committee, 1982). The Department of Water and Natural Resources is the agency in South Dakota that responds to ground water contamination problems.

Texas

According to the "State Ground-Water Program Summaries" (U.S. EPA, 1985), the State of Texas is third in the nation in agricultural production and first in the use of agricultural chemicals. Numerous isolated incidents of ground water contamination by pesticides have been documented.

Arsenic contamination of a domestic well in 1983 near the Town of Knott in Howard County led to an extensive investigation and the conclusion that the source was an arsenic-based defoliant that had been applied to the cotton fields at harvest since the 1930's. The arsenic had become particularly concentrated in the gin areas (Texas Department of Water Resources, 1984).

Other agriculture related contamination, according to The State of Texas Water Quality Inventory, (1984), includes high nitrate concentrations, mostly in the Low Rolling Plains (Abilene and Wichita Falls area) and the High Plains (Midland, Odessa, Lubbock, and Amarillo area), but also individual incidents in the Lockhart and Taylor alluvial aquifers in central Texas and Hidalgo County.

In 1982, the Texas Department of Water Resources published "Pesticide and PCB Concentrations in Texas Water, Sediment and Fish Tissue" (Report 264). Although it does not include ground water concentrations per se, it does include sediment samples associated with river basins. Perennial rivers in arid climates (most of Texas) are usually losing rivers; therefore, if the river and the sediments under it are contaminated, the associated alluvial aquifer is contaminated or soon will be. The study also addresses the characterization, sources and uses, metabolism and toxicity, and water quality criteria as established by the U.S. EPA. The 17 compounds addressed are organized into four groups: (1) chlorophenoxy herbicides--2,4-D, 2,4,5-T, and Silvex; (2) chlorinated hydrocarbons--heptachlor, heptachlor epoxide, lindane, methoxychlor, aldrin and dieldrin, endrin, chlordane, toxaphene, and DDT; (3) organophosphate pesticides--malathion, parathion and methyl parathion, and diazinon; and (4) PCBs. The Trinity, San Jacinto, and San Antonio River Basins show significant sediment contamination by several pesticide compounds, while the Sulpher Cypress Creek, Sabine, Brazos, Colorado, and Guadalupe Basins have a large number of positive determinations for only one or two compounds. The most frequently detected compounds were PCB's and DDT breakdown products.

Contamination problems and complaints are reported to the Texas Department of Water Resources and the Department of Agriculture. A ground water quality monitoring network consisting of 5,827 wells is maintained by the Department of Water Resources. Approximately 750 wells are sampled annually for common constituents such as calcium, sodium, chlorides, and nitrates. Special constituents such as pesticides and hydrocarbons are analyzed on the basis of citizen complaints or special projects (Department of Water Resources, 1984). The Department of Agriculture is working with the Department of Water Resources to identify areas of potential contamination and to initiate testing to assess the seriousness of the problem (U.S. EPA, 1985).

Chapter 75 of the Texas Agriculture Code is concerned with herbicides and Chapter 76 is concerned with pesticides. The Department of Agriculture has the delegated enforcement responsibilities (U.S. EPA, 1985).

CONCLUSIONS

The following conclusions may be drawn about the state of the nation regarding agricultural chemical contamination of ground water and acknowledgment of this problem:

(1)   Many states have only limited ground water information on pesticides and nutrients; however, the interest in these contaminants is increasing and many efforts are underway to gather data. Funding for monitoring programs appears to be one of the major causes for the lack of attention to pesticides and nutrients in ground water.

(2)   A large number of states do not presently have laws or regulations that relate to pesticide or fertilizer applications; many existing laws and regulations need updating.

(3)   In some states the institutional authority for addressing pesticide contamination of ground water is still unclear.

(4)   Twenty-six states with identified or suspected pesticide contamination of ground water were found to have adequate legislative awareness of the pesticide issue or were taking steps to provide evaluation of the pesticide threat to ground water. Fourteen states with identified or suspected pesticide contamination

were found to be either taking no actions due to a lack of interest or funds or were in need of changes in their legal controls. Two of those states needing changes are beginning an evaluation of the situation.

In the case of nutrient contamination, 23 states with identified or suspected contamination demonstrate adequate legislative awareness or are taking steps to evaluate the issue. Of eleven more states claiming an identified or suspected nutrient contamination, ten are either taking no action or express a need for revision of present controls; and only one was found to be taking steps to evaluate the issues.

(5)   Many field studies have been conducted across the nation on agricultural contamination of ground water, the majority of which focus on nutrients (mostly nitrates), with lesser attention given to pesticides.

(6)   Research on the fate and transport of pesticides in ground water is presently gaining momentum; many studies are in progress.

(7)   Thirty-three states reported pesticides as a confirmed or suspected ground water contaminant.

(8)   Nitrates in ground water are regulated under stricter standards than pesticides; the lack of pesticide standards hinders the analysis of pesticide contamination.

(9)   Information is steadily being accumulated on management practices to reduce inefficient use (loss to ground water) of nitrate fertilizers. It appears that lack of technology to reduce contamination of ground water from the use of agricultural chemicals is not the restricting factor.

(10)   There is a lack of publicly available data on agricultural chemical contamination of ground water health effects, as well as political and proprietary constraints.

(11)   There is a lack of communication between separate agencies handling ground water contamination and data gathering.

SELECTED REFERENCES

Balsters, R.G. and Anderson, C., "Water Quality Effects Associated with Irrigation," Kansas Water News, Vol. 22, No. 1 and 2, Winter 1979, pp. 14-22.

Bingham, F.T., Davis, S. and Shade, E., "Water Relations, Salt Balance, and Nitrate Leaching Losses of a 960-Acre Citrus Watershed," Soil Science, Vol. 112, No. 6, 1971, pp. 410-417.

Butler, W., Nichols, W.J., and Harsh, J.F., "Quality and Movement of Ground Water in Otter Creek-Dry Creek Basin, Cortland County, New York," Water Resources Investigations 78-3, 1978, U.S. Geological Survey, Albany, New York.

Eccles, L.A., and Bradford, W.L., "Distribution of Nitrate in Ground Water, Redlands, California," USGS/WRI-76-117, March, 1977, Water Resources Division, U.S. Geological Survey, Menlo Park, California.

Embleton, T.W. et al., "Nitrogen Fertilizer Management of Vigorous Lemons and Nitrate-Pollution Potential of Groundwater," Technical Completion Report Contribution 182, December, 1979, California Water Resources Center, University of California, Davis, California.

Florida Department of Agriculture and Consumer Services, "Florida Pesticide Law and Rules," April, 1985, Tallahassee, Florida.

Florida Department of Environmental Regulation, "Florida's Ground Water Protection Program," Fact Sheet, December, 1984, Tallahassee, Florida.

Hughes, J.L., "Evaluation of Ground Water Quality in the Santa Maria Valley, California," Water-Resources Investigations 76-128, July, 1977, U.S. Geological Survey, Menlo Park, California.

Litwin et al., "Groundwater Contamination by Pesticides: A California Assessment," Publication No. 83-45P, June, 1983, State Water Resources Control Board, Sacramento, California.

Murphy, L.S., and Gosch, J.W., "Nitrate Accumulation in Kansas Groundwater," Project Completion Report, March, 1970, Kansas Water Resources Research Institute, Kansas State University, Manhattan, Kansas.

Nightingale, H.I., "Nitrates in Soil and Ground Water Beneath Irrigated and Fertilized Crops," Soil Science, Vol. 114, No. 4, 1972, pp. 300-311.

O'Hare, M. et al., "Contamination of Ground Water in the Contiguous United States from Usage of Agricultural Chemicals (Pesticides and Fertilizers)," July, 1985, Environmental and Ground Water Institute, University of Oklahoma, Norman, Oklahoma.

Pallares, C., "Nitrogen Fertilizer Management of a Lemon Orchard as Related to Nitrate-Pollution Potential of Ground Water," OWRT-A-057-CAL(1), January, 1978, Office of Water Research and Technology, U.S. Department of the Interior, Washington, D.C.

Pesticide Review Section, "Summary of Florida Aldicarb Studies," August, 1984, Department of Environmental Regulation, Tallahassee, Florida.

Pickrell, S., "Fate of EDB from the Use of Ortho Diquat Herbicide," H/A Report of Investigation, May, 1984, Department of Environmental Regulation, Pesticide Review Section, Tallahassee, Florida.

Pickrell, S., Dwinell, S., and Vogel, D., "Environmental Levels of DDT Associated with Dicofol Use in Florida Citrus," Report PRS-84-03, September, 1984, Tallahassee, Florida.

Pickrell, S., Dwinell, S., and Tterlikkis, D., "Chemical Residues in Freshwater Resulting from Soil Fumigation by Methyl Bromide and Chloropicrin," Report PRS 85-01, March, 1985, Tallahassee, Florida.

Porter, K.S., "An Evaluation of Sources of Nitrogen as Causes of Ground-Water Contamination in Nassau County, Long Island," Ground Water, Vol. 18, No. 6, November-December, 1980, pp. 617-625.

Ridder, W.E., Oehme, F.W., and Kelley, D.C., "Nitrates in Kansas Groundwaters as Related to Animal and Human Health," Toxicology, Vol. 2, No. 4, 1974, pp. 397-405.

Schmidt, K.D., "Nitrate in Ground Water of the Fresno-Clovis Metropolitan Area, California," Ground Water, Vol. 10, No. 1, January-February, 1972, pp. 50-61.

Silver, B.A. and Fielden, J.R., "Distribution and Probable Source of Nitrate in Ground Water of Paradise Valley, Arizona," Ground Water, Vol. 18, No. 3, May-June 1980, pp. 244-251.

Soren, J., "Ground Water Quality Near the Water Table in Suffolk County, Long Island, New York," Long Island Water Resources Bulletin LIWR-8, 1977, Water Resources Division, U.S. Geological Survey, Mineola, New York.

South Dakota State Coordinating Committee, "Comprehensive Monitoring and Evaluation Plan for the Oakwood Lakes-Poinsett Rural Clean Water Program," July, 1982, South Dakota.

Texas Department of Water Resources, "Pesticides and PCB Concentrations in Texas-Water Sediment, and Fish Tissue," January, 1982, Texas.

Texas Department of Water Resources, "The State of Texas Water Quality Inventory," 7th Ed., December, 1984, Texas.

U.S. Environmental Protection Agency, "State Ground-Water Program Summaries," Vol. II, March, 1985, Washington, D.C.

U.S. Water News, "Arizona Enacts Ban on EDB," March, 1985a.

U.S. Water News, "One in Four California Wells is Tainted by Chemicals," March, 1985b.

U.S. Water News, "Laws Protect Groundwater from Ag Chemical Storage," March, 1985c.

van der Leeden, F., Cerrillo, L.A., and Miller, D.W., "Ground Water Pollution Problems in the Northwestern United States," EPA-660/3-75-018, May, 1975, U.S. Environmental Protection Agency, Corvallis, Oregon.

CHAPTER 20

QUANTITATIVE STUDIES OF BIODEGRADATION OF PETROLEUM AND SOME MODEL HYDROCARBONS IN GROUND WATER AND SEDIMENT ENVIRONMENTS

by

Fu-Hsian Chang,
Marc Hult, and
Nancy N. Noben

The biodegradation of waste oils or accidentally spilled crude oil in terrestrial and aquatic environments has been reviewed by Atlas (1977), Bartha and Atlas (1977) and Colwell and Walker (1977). Extensive and careful study of oil biodegradation in soil reported by Raymond, Hudson, and Jamison (1976) and Dibble and Bartha (1979) was conducted in the field under local weather and precipitation conditions. It was reported that oil was decomposed by microbial activity under anaerobic conditions utilizing the oxygen in sulfates and nitrates. This activity resulted in the formation of hydrogen sulfide, nitrite and free nitrogen (Bastin, 1926; Tauson and Aleshina, 1932; and Tauscon and Veselow, 1934). Microbial degradation of oil is a major process through which hydrocarbons are removed from the sediment environment (ZoBell, 1964).

The ecological importance of hydrocarbon biodegradability is the persistence of crude oil components in the environment which are less biodegradable and more toxic. The biodegradability of petroleum is highly dependent on its molecular configuration and composition of its hydrocarbon components (Atlas, 1975; and ZoBell, 1969). The rate at which microbial degradation takes place decreases from normal alkanes to isoalkanes and cycloalkanes to aromatics (Blumer and Sass, 1972).

Microbial oxidation of various types of oil under enriched culture conditions have been conducted by Chang et al., (1985), Dibble and Bartha (1979), Gibbs (1975), and ZoBell (1972), at temperatures higher than those in nature. Cook and Westlake (1974) and Parkinson (1973) studied microbiological degradation of northern crude oils in a colder climate and found that spilling a crude oil stimulated the growth of microflora. However, the crude oil-degrading microorganisms were not distinguished from the nonoil-degrading microorganisms in their studies. Little work has been done under conditions resembling the natural ground water environment, especially related to quantitative rate measurements.

Many studies of degradation of crude oil by microorganisms have dealt with a single microorganism degrading a single hydrocarbon. The effect of mixed populations on a single hydrocarbon and the effect of single microorganisms on multiple hydrocarbons are poorly understood. Some studies have demonstrated that oil biodegradation depends on various environmental parameters despite the abundance of hydrocarbon-degrading microorganisms (Atlas and Bartha, 1972a, 1972b; Gibbs, 1975; and Gibbs, Pugh, and Andrews, 1975). The rates of crude oil and model hydrocarbon biodegradation were stimulated by an increase in temperature, inorganic nutrient concentration, and oxygen availability (Chang et al., 1985).

The metabolism of $^{14}C_1$-hexadecane by samples of lake water conducted by Caparello and LaRock (1975) demonstrated a correlation between the number of hydrocarbon-oxidizing bacteria present and the lag before hexadecane metabolism.

In view of the occasional accidental ground water pollution by crude oil and because the role of microorganisms and their degradative potential have not been clearly established, the impact of various environmental parameters on biodegradation of crude oil and model hydrocarbons in simulated subsurface environments was studied.

MATERIALS AND METHODS

Field Sampling

Samples used in this study were collected from the following localities: (1) control samples of uncontaminated outwash sand from both water unsaturated and saturated zones near the oil-spill site, (2) uncontaminated ground water 4 ft below the water table, and (3) contaminated samples from both unsaturated and saturated zones at the oil-spill site near Bemidji, Minnesota.

Sediment samples were taken at several sites near the spill and at a control site away from the spill. Samples were recovered with a split spoon sampler that was lowered through a hollow-stem auger. The auger flights were steam cleaned before use. The split-spoon sampler was lined with methanol-washed brass liners to minimize organic contamination. Sediment samples were put in double-layer plastic bags in a storage tank and stored in a $4°C$ walk-in refrigerator before use. Water samples were collected by pumping from individually-cased wells. Each well was screened for about one meter. A minimum of three casing volumes was removed and the pumping was continued until the specific conductance and temperature of the ground water were constant before samples were collected. Ground water samples were put in storage tanks and stored in a $4°C$ walk-in refrigerator before use.

Chemicals

Crude oil was collected from the contaminated site. Pesticide-grade solvents, acetone and hexane (Fisher Scientific Co., Minneapolis, MN) were used in the preparation of glassware and sample collecting tools. Certified reagent-grade cyclohexane and hexadecane (American Scientific Products, Minneapolis, MN), phenanthrene, 1-naphthol, and pyrene (Eastman Kodak Co., Rochester, NY) and n(1-^{14}C) hexadecane (250 μ Ci/mMole) (Amersham/Searle, Arlington Heights, IL) were used as model hydrocarbons for biodegradation studies.

Organisms and Media

Eight bacteria and four fungi were isolated from ground water and sediment from the oil-spill site. These microorganisms were tested in experiments involving pure culture or mixed cultures. Fungal cultures were grown on Sabouraud Dextrose Agar (Difco) and bacteria were maintained on Tryptic Soy Agar (Difco). Inoculum was prepared by washing the cells from the surface of the slant with sterile basal or Tauson broth medium.

The media used in this study for isolation of heterotrophic and crude oil-degrading microorganisms have been described elsewhere (Chang and Ehrlich, 1984; and Chang et al., 1985). To measure microbial degradation of model hydrocarbons, triplicated 250 ml flasks containing 125 ml basal broth or Tauson's broth were inoculated with sediment or ground water as soon as possible (usually less than 24 hours) after sample collection. The inoculated flasks were incubated on a rotary shaker (100 rpm) at 22°C and 6°C for 14 days. Subcultures were prepared once every two days by plating on Tauson's agar or basal agar and examining for growth.

In order to differentiate fungal and yeast activity from bacterial activity, the medium used to support fungi was acidified to pH 5.5 with H_2SO_4, and a bactericide (chloramphenicol or terramycin) was added. The pH of the medium used to culture bacteria was adjusted to 7.0 by dilute KOH solution, and a fungicide (lotrimin or nystatin) was added. For mixed isolate cultures of bacteria, fungi, and yeasts, the pH of the medium was adjusted to 7.0 and no antibiotics or fungicides were added. In all cases, the media were mixed with sterile model hydrocarbons (v/v = 1%) prior to inoculation with microorganisms.

Microcosms

As shown in Figure 20.1, five 20-gallon aquarium tanks were used to set up the microcosm system. One of the tanks was used as a reservoir which was filled with 15 percent glycol solution. When the pump was turned on, the glycol solution in the reservoir was cooled and the temperature of the solution was calibrated at -5°C. The submersive pump pumped the -5° glycol solution through the cooling coils inside each aquarium tank and returned it to the reservoir. The length of the cooling coil in each tank was calculated and the flow rate to each tank was calibrated by adjusting the open/close valve so that the water temperature of each tank was about $1-2^{\circ}$C below the desired temperature. One circulator with a heating element was set on each tank to control the temperature at a constant level ($\pm0.2^{\circ}$C). The mechanical pump automatically shut off when the glycol solution reached -5°C during the circulation. It would also shut off automatically in the event that the pump or compressor overheated.

Cylinders of polyvinyl chloride (PVC) columns (4" I.D., 15" H) and glass graduated cylinders (250 ml) with caps were filled with 1000 g sediment and filled with ground water. These cylinders were immersed in water inside the aquarium tanks. The PVC cylinders were covered with PVC caps and sealed with PVC glue, and the graduated cylinders were capped with neoprene stoppers in order to avoid gas or water exchange inside the microcosms. Glass graduated cylinders were used as a control and compared with the PVC cylinders for chemical analysis of hydrocarbons.

A. Pump
B. Compressor
C. Automatic shut-off switch
D. Evaporator
E. Submersive pump
F. 15% glycol solution
G. ½" copper tubing
H. Switch-control flow rate
I. Thermometer
J. Cooling coil
K. Circulator with heating element and temperature dial
L. 7/16" Tygon tubing
M. Styrofoam insulated 20-gallon aquarium tank
N. 5/16" I.D. Tygon tubing
O. Reservoir, 20-gallon aquarium tank insulated with styrofoam

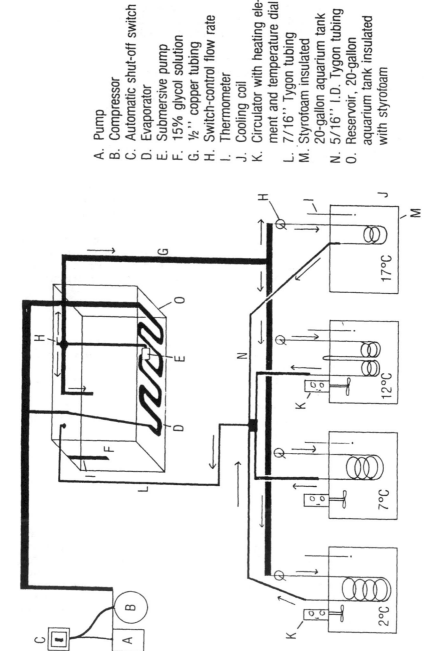

Figure 20.1: Schematic of Microcosms

Biodegradation Assessment of Crude Oil and Model Hydrocarbons

Cylinders in the microcosms were filled with sediment and ground water contaminated by crude oil, or with uncontaminated sediment and ground water mixed with one of the following model hydrocarbons: cyclohexane, hexadecane, 1-naphthol, phenanthrene, pyrene or crude oil. The mixtures were inoculated with mixed indigenous microbial cultures that were isolated from the spill site and were known to degrade crude oil (Chang and Ehrlich, 1984). Some sediment/ground water samples were amended with 20-200 mg/l NH_4NO_3, 50-200 mg/l KH_2PO_4 + K_2HPO_4, 50-500 mg/l $MgSO_4$, and 20-100 mg/l $CaCl_2$ · $2H_2O$. The microcosms were incubated for 3-6 months at $2°$, $7°$, $12°$, and $17°C$. Samples for chemical and biological analyses were initially collected at intervals of 3 to 10 days, then increased to triweekly. In some experiments, fungicide (lotrimin or nystatin) and bactericide (chloramphenicol or terramycin) were added to distinguish between the degradation by bacteria and fungi, respectively. Nonnutrient-treated or sterilized samples were used as controls in all the experiments.

In other sets of study, 500 ml widemouth jars were filled with 100 g oil-contaminated topsoil (or oil-contaminated aquifer sediment) and 20 ml distilled water (or oil-contaminated ground water). Sediment and ground water collected from an upgradient uncontaminated site was mixed with cyclohexane, hexadecane, 1-naphthol, phenanthrene, pyrene or crude oil for the study of biodegradation rates of model hydrocarbons. Mineral nutrients at different concentrations, fungicide and bactericide were also added to the top soil/distilled water or sediment/ground water mixtures and incubated in the dark at $6°$ and $22°C$ for 2-4 months.

Total numbers of viable heterotrophic bacteria and hydrocarbon-degrading microorganisms in each microcosm were determined at different incubation times by the most-probable-number technique and spread plate count (Chang and Ehrlich, 1984).

The rate of microbial degradation of a model hydrocarbon ($1-^{14}C$-hexadecane, 250 μ Ci/mMole) was measured. The evolved CO_2 and $^{14}CO_2$ was trapped in KOH solution and measured by titration and scintillation counting techniques (Chang and Broadbent, 1981; and Chang and Alexander, 1983). The rates of biodegradation of petroleum and hexadecane in a mixture of sediment and ground water were also estimated by measuring oxygen uptake as reported by Chang and Ehrlich (1984).

To test microbial growth on model hydrocarbons, triplicated 50 ml test tubes containing 20 ml Tauson's broth medium and 0.2 ml model hydrocarbon were inoculated with mixed microorganisms or each single isolate. The inoculated tubes were incubated at $6°$ and $22°C$. Growth to turbidity after 10-14 days incubation confirmed the presence of microbes capable of utilizing the model hydrocarbon as a sole source of carbon.

RESULTS AND DISCUSSION

Mineralization of the oil in contaminated topsoil and aquifer sediment by indigenous microorganisms of the oil-spill site at $22°C$ show that both topsoil and aquifer sediment are mineral nutrient deficient, especially in the aquifer sediment (Table 20.1). The stimulatory effect caused by the addition of mineral

Table 20.1: Cumulative Carbon Dioxide Evolution and Carbon Mineralization from Oil-Contaminated Topsoil and Aquifer Sediment

Treatment	Topsoil (59 Days Incubation)		Aquifer Sediment (130 Days Incubation)	
	CO_2 (mg)	Net Carbon (mg)	CO_2 (mg)	Net Carbon (mg)
A. (100 g topsoil + 20 ml distilled water) or 100 g sediment + 20 ml ground water	137	37	55	15
B. (A + mineral nutrients)*	423	115	599	162
C. (B + bactericide)	403	109	526	143
D. (B + fungicide)	356	97	276	75

*Mineral Nutrients = 2 mg N (as NH_4NO_3) + 1 mg P (as wt/wt = 1:1=KH_2PO_4 : K_2HPO_4) + 13 mg S (as $MgSO_4$)

Bactericide = Terramycin

Fungicide = Lotrimin

nutrients was about 3 and 10-fold in topsoil and sediment, respectively. Fungal activity accounted for more degradative potential than bacterial activity in topsoil as well as in aquifer sediment. The rate of biodegradation of oil in topsoil was 1.95 mg/day compared to that in the sediment (1.25 mg/day) indicating that significantly more microbial activity occurred in topsoil than in the aquifer environment (Table 20.1).

Biodegradation of $^{14}C_1$-hexadecane by indigenous microorganisms in uncontaminated sediment at $22^{\circ}C$ is presented in Tables 20.2 and 20.3. There was little measurable $^{14}CO_2$ produced in any treatment after 7 days of incubation. However, microbial activity began to accelerate after 10 days in incubation. This lag phase shows that microbial adaptation occurred when hexadecane was applied to the control sediment. Mineral nutrients availability was a limiting factor for biodegradation of hexadecane since the addition of nutrients at different concentrations promoted significant ($p \leq 0.05$) degradation. Fungi were more sensitive to hexadecane than bacteria, and it took a longer time for fungi to adapt to the contaminated environment (compare treatments 6 and 7, 8 and 9). Nutrient availability not only stimulated rates of biodegradation, but also shortened the adaptation period (compare treatments 6 and 8). Fungal and bacterial activities reached approximately equal levels after about 110 days of incubation, then fungal activity predominated in hexadecane decomposition for the remainder of the incubation period (Tables 20.2 and 20.3). The maximum degradative potential (1.12 mg/day) of hexadecane by indigenous microorganisms at the control site was less than that for crude oil (1.25 mg/l) in the contaminated aquifer sediment (Tables 20.1 and 20.3). A possible explanation is that the microbial community at the contaminated site had grown and stabilized since the oil-spill occurred five years before the samples were collected. Higher bacterial and fungal degradative potential was found in $^{14}C_1$-hexadecane amended sediment (0.71-1.19 mg/day) than in oil-contaminated sediment (0.58-1.10 mg/day).

Figure 20.2 shows the biodegradation of model hydrocarbons by mixed isolate cultures and extracted indigenous cultures in control sediment and ground water samples that were treated with mineral nutrients and incubated at $6^{\circ}C$ under aerobic conditions. It is apparent that hexadecane is most susceptible to biodegradation followed by crude oil, phenanthrene and cyclohexane, with 1-naphthol the least susceptible to biodegradation. The degradative potential of crude oil and hexadecane by indigenous microorganisms at $6^{\circ}C$ were 0.37-0.40 and 0.68-0.75 mg/day, respectively. These rates were significantly lower than for those samples incubated at $22^{\circ}C$. Degradation rates of phenanthrene by extracted indigenous cultures exceeded that of cyclohexane, crude oil and hexadecane after incubation periods of 28 days, 54 days and 115 days, respectively. Extracted indigenous mixed cultures were more active in degrading all (with the exception of hexadecane) the model hydrocarbons than mixed isolated cultures. The isolated cultures could have lost their full degradative potential as a consequence of having been stored on Tauson's agar with crude oil as sole source of carbon for more than 12 months with routine subcultures at 2 month intervals. Alternatively, antagonism between species of bacteria and fungi may have occurred as a consequence of the production of microbicidal agents by one or more of these microbes.

The rates of crude oil biodegradation calculated by oxygen uptake from microcosms that differ by $10^{\circ}C$ (Q_{10}) are illustrated in Table 20.4. Q_{10} values ranged from 2.18-2.56 for $2^{\circ}C$ and $12^{\circ}C$, and 2.38-2.73 for 7° and $17^{\circ}C$. In general, Q_{10} values decreased with increasing temperatures. Temperature

Table 20.2:   Cumulative CO_2 Evolution (mg)* by Indigenous Microbes in $^{14}C_1$-Hexadecane Treated Sediment (Up-Gradient Sediment) (Total CO_2)

Treatment**	Incubation Time (Days)						
	20	35	49	82	95	110	124
#2	4.96	6.54	8.07	13.51	15.65	17.00	18.00A
#3	65.15	91.37	108.57	181.69	210.59	222.15	231.15B
#4	74.23	100.55	127.46	213.30	247.12	272.68	291.98C
#5	78.92	148.93	201.74	337.61	391.13	482.61	509.81F
#6	5.28	24.64	91.00	188.55	226.98	296.94	352.44D
#7	69.90	110.57	139.75	233.86	270.93	301.87	326.57D
#8	18.46	81.24	155.16	310.70	371.97	457.62	541.52G
#9	82.04	141.33	204.43	342.11	396.35	441.64	472.74E

*Values are means of 3 replicates

**#2 = 100 g sediment + 20 ml ground water (both were control) + $1 \mu Ci ^{14}C_1$-hexadecane + 1 ml hexadecane

#3 = #2 + 10ppm $MgSO_4$ + 2ppm NH_4NO_3 + 0.5ppm KH_2PO_4 + K_2HPO_4

#4 = #2 + 50ppm $MgSO_4$ + 10ppm NH_4NO_3 + 5ppm KH_2PO_4 + K_2HPO_4

#5 = #2 + 500ppm $MgSO_4$ + 20ppm NH_4NO_3 + 10ppm KH_2PO_4 + K_2HPO_4

#6 = #4 + 10 mg antibacterial chloramphenicol

#7 = #4 + antifungal nystatin (10 mg)

#8 = #5 + 10 mg chloramphenicol (antibacterial agent)

#9 = #5 + 10 mg Nystatin

***Values followed by different letters are different at 5% level.

coefficients (Q_{10}) were also calculated by CO_2 evolution in oil-contaminated sediment samples before and after the temperature was changed from $22°C$ to $12°C$ under various nutrient concentrations (Table 20.5). Again, Q_{10} values increased with decreasing nutrient concentrations despite the rates of CO_2 production being significantly lower in low nutrient treated sediment samples. It is interesting to observe that antibacterial drug (terramycin)-treated sediment resulted in lower Q_{10} values than antifungal drug (lotrimin)-treated sediment despite higher CO_2 evolution in terramycin-treated sediment (Table 20.5).

Percentage labelled isotope recovery from $^{14}C_1$-hexadecane degradation by mixed isolate culture expressed by the Q_{10} value was affected by three environmental factors;   temperature, nutrient concentration and oxygen

Table 20.3: Cumulative Carbon Mineralization (mg)* of $^{14}C_1$-Hexadecane by Indigenous Microbes in Control Sediment and Ground Water Mixture (Total Carbon)

Treatment**	Incubation Time (Days)						
	20	35	49	82	95	110	124
#2	1.35	1.78	2.20	3.68	4.27	4.64	4.88 A***
#3	17.77	24.92	29.61	49.55	57.40	60.58	62.98 B
#4	20.24	27.42	34.76	58.17	67.39	74.36	79.66 C
#5	21.52	40.62	55.01	92.08	106.68	131.63	139.03 F
#6	1.44	6.72	24.82	51.42	61.90	80.98	96.08 D
#7	18.52	29.61	37.57	63.24	73.35	81.78	88.48 E
#8	5.04	22.18	42.32	84.74	101.45	124.81	147.71 G
#9	22.37	38.55	55.75	93.30	108.09	120.44	128.94 E

*Values are means of 3 replicates

**#2 = 100 g sediment + 20 ml ground water + 0.5 ml $^{14}C_1$-hexadecane ($1_{\mu}C_i$) + 1.0 ml hexadecane

#3 = #2 + 10ppm $MgSO_4$ _ 2ppm NH_4NO_3 + 0.5ppm KH_2PO_4 + K_2HPO_4

#4 = #2 + 50ppm $MgSO_4$ + 10ppm NH_4NO_3 + 5ppm KH_2PO_4 + K_2HPO_4

#5 = #2 + 500ppm $MgSO_4$ + 20ppm NH_4NO_3 + 10ppm KH_2PO_4 + K_2HPO_4

#6 = #4 + 10 mg antibacterial chloramphenicol

#7 = #4 + 10 mg antifungal nystatin

#8 = #5 + 10 mg chloramphenicol

#9 = #5 + 10 mg nystatin

***Values followed by different letters are different at 5% level.

availability (Table 20.6). Mixed cultures of bacterial, fungal and yeast isolates yielded the highest recovery percentage (as $^{14}CO_2$). Maximum rates of recovery at $6°$ and $22°C$ ranged from 26-35% and 72-80%, respectively, under aerobic conditions (dissolved oxygen >0.9 mg/l) after 92 days incubation. However, the recovery rates were lower under microaerobic conditions ($0.2 \leq$ D.O. ≤ 0.9 mg/l) than aerobic conditions. The percent $^{14}CO_2$ recovery ranged from 13-22% and 5-10% for $22°$ and $6°C$, respectively. Fungal and yeast activity yielded higher percent recoveries than bacterial activcity under aerobic conditions. However, opposite results were found under microaerobic conditions. Our results demonstrate that slow oxidation of $^{14}C_1$-hexadecane to $^{14}CO_2$ occurs under microaerobic conditions by mixed isolate cultures indigenous to the aquifer environment of the oil-spill site. This agrees with the well-accepted theory of the environmental limitation on the rate of hydrocarbon biodegradation in the absence of oxygen (Ward and Brock, 1978).

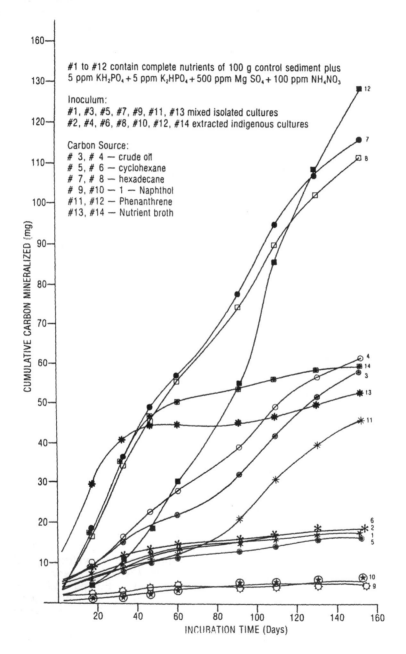

Figure 20.2: Biodegradation of Model Hydrocarbons

Table 20.4: Mean* Net Rates of Oxygen Uptake (mg O_2/week) Before and After Temperature Change, and Values for Q_{10}

Incubation Temp.	A	B	C	D	E	F	G	H	I
					BOD Bottle Set Number**				
12°C	0.74	8.49	10.85	14.96					0.16
2°C	0.29	3.37	4.54	6.86					0.07
Q_{10}	2.56	2.52	2.39	2.18					2.28
17°C					0.92	9.51	11.90	17.38	0.26
7°C					0.34	3.59	4.72	7.30	0.09
Q_{10}					2.73	2.65	2.52	2.38	2.88

*Values are means of three replicates

**A and E: ground water + sediment 30g + 0.1% crude petroleum

B and F: A or E + 0.5ppm P + 2ppm N + 10ppm $MgSO_4$

C and G: A or E + 5ppm P + 10ppm N + 50ppm $MgSO_4$

D and H: A or E + 10ppm P + 20ppm N + 500ppm $MgSO_4$

I: Control, no inorganic nutrient or crude petroleum was added

Table 20.5:   Mean*Carbon Dioxide Production (mg CO2/week) Before and After Temperature Change in Oil Contaminated Sediment and Values for Q10

Incubation Temp.	Treatment**								
	A	B	C	D	E	F	G	H	I
22°C	32.3	16.8	9.74	2.72	1.71	28.3	14.9	23.2	3.60
12°C	12.8	5.94	3.29	0.89	0.55	10.8	5.21	8.59	1.20
Q_{10}	2.53	2.83	2.95	3.04	3.09	2.61	2.86	2.70	2.99

*Values are means of 3 replicates

**A = 100 g oil contaminated sediment + 500ppm S + 20ppm N + 10ppm P

B = 100 g contaminated sediment + 50ppm S + 10ppm N + 5ppm P

C = 100 g contaminated sediment + 10ppm S + 2ppm N + 1ppm P

D = 100 g contaminated sediment only

E = 100 g control sediment only

F = 100 g contaminated sediment + 50ppm S + 10ppm N + 5ppm P + 0.5 ml terramycin

G = 100 g contaminated sediment + 50ppm S + 10ppm N + 5ppm P + 1.0 ml lotrimin

H = 100 g contaminated sediment + 0.5 ml terramycin (antibacterial)

I = 100 g contaminated sediment + 1.0 ml lotrimin (antifungal)

Table 20.6: Percentage* Isotope-Recovery (as $^{14}CO_2$) of $^{14}C_1$-Hexadecane Biodegradation Under Various Environmental Conditions (92 Days Incubation)

| Culture and Nutrient Addition[†] | | Oxygen Availability | | | | | |
| | | Aerophilic | | | Microaerophilic | | |
		6°C	22°C	Q_{16}**	6°C	22°C	Q_{16}
Bacteria, Fungi	1	5.1	12.5	2.45	2.0	4.6	2.30
and Yeasts	2	24.3	56.9	2.34	4.9	11.1	2.27
	3	35.4	90.3	2.27	10.2	21.8	2.14
Bacteria	1	3.7	10.9	2.95	1.5	4.3	2.87
	2	20.2	49.5	2.45	4.5	10.2	2.27
	3	31.8	72.6	2.28	8.6	18.4	2.14
Fungi and Yeasts	1	3.2	11.5	3.59	0.9	2.8	3.11
	2	16.6	51.6	3.11	2.4	7.0	2.92
	3	26.3	78.8	2.99	4.7	13.3	2.83

*Values are averages of three replicates

**Q_{16} = %^{14}C recovery at 22°C ÷ %^{14}C recovery at 6°C

[†]1 = Control, no mineral nutrient was added

 2 = 2ppm NH_4NO_3 + 0.5ppm ($K_1 + K_2$) HPO_4 + 10ppm $MgSO_4$

 3 = 20ppm NH_4NO_3 + 5ppm ($K_1 + K_2$) HPO_4 + 100ppm $MgSO_4$

Figure 20.3 illustrates the relationship of logarithm of inoculum density as a function of incubation time required for 20 and 50 percent $^{14}CO_2$ recovery from $^{14}C_1$-hexadecane. The length of incubation time required for a certain percentage recovery was, therefore, a measure of the density of the hydrocarbon-degrading microbial population in the original sample. Changing the inoculum density for a given isolate culture did not change the slope of the growth phase for the same percentage ^{14}C recovery.

Mineralization rates for n-$^{14}C_1$-hexadecane by mixed indigenous bacterial and fungal cultures at 6° and 22°C are depicted in Figure 20.4. Total percent recovery was significantly higher at 22°C than at 6°C in all the nutrient-treated samples. The ^{14}C recovery was directly proportional to the inorganic nutrient concentrations added to the samples. In sediment samples that were not treated with mineral nutrients, there was 14 percent and 5 percent of $^{14}C_1$-hexadecane mineralization after 126 days of incubation at 22° and 6°C, respectively. With mineral nutrient treatments the ^{14}C recovery increased up to more than 5 fold of those nonnutrient-treated samples. This indicates that ^{14}C-hexadecane and its degradation products were available for mineralization and did not volatize from the sediment/ground water suspension. Initial lag phases were present in all cases, and this indicates that ^{14}C recovery is mediated by microbial activities.

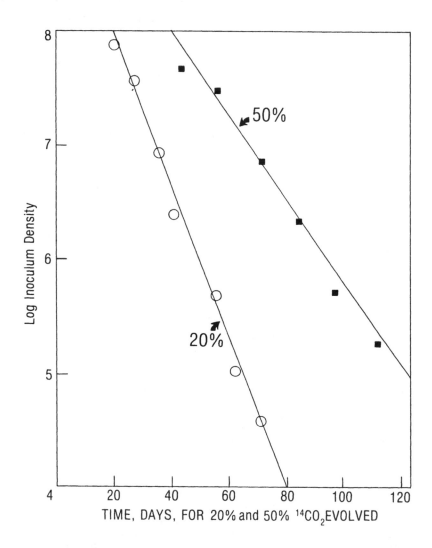

Figure 20.3: Variation of Inoculum Density with Incubation Time

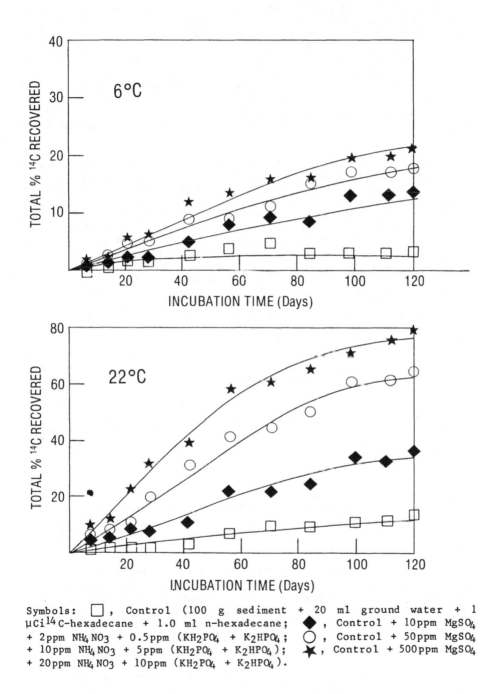

Symbols: ☐ , Control (100 g sediment + 20 ml ground water + 1
μCi^{14}C–hexadecane + 1.0 ml n–hexadecane; ◆ , Control + 10ppm MgSO$_4$
+ 2ppm NH$_4$NO$_3$ + 0.5ppm (KH$_2$PO$_4$ + K$_2$HPO$_4$; ◯ , Control + 50ppm MgSO$_4$
+ 10ppm NH$_4$NO$_3$ + 5ppm (KH$_2$PO$_4$ + K$_2$HPO$_4$); ★ , Control + 500ppm MgSO$_4$
+ 20ppm NH$_4$NO$_3$ + 10ppm (KH$_2$PO$_4$ + K$_2$HPO$_4$).

Figure 20.4: Mineralization Rates for Hexadecane

Figure 20.5 illustrates the relationship between chemical oxygen demand (COD) and time of incubation for control sediment and ground water samples that were treated with crude oil (100 g sediment + 450 ml ground water + 1 ml crude oil) and incubated at $6°$ and $22°$C. Temperature and nutrient availability were two important factors controlling the COD reduction. The reductions in COD after 60 days incubation were 10.7-35.7 percent and 16.0-46.4 percent for samples incubated at $6°$ and $22°$C, respectively. The steady reduction in COD values indicates a progressive loss of organic matter (hydrocarbons) as a result of diffusion and solubilization. The reduction of COD was proportional to the initial nutrient addition (i.e., nutrient availability) in the sediment/ground water suspension.

There were fluctuations in the proportion of crude oil-degrading bacteria and heterotrophic bacteria in both sediment and ground water environments (Figure 20.6). When control sediment or ground water was enriched with crude oil, about 95 percent of the heterotrophic bacteria isolated from inorganic nutrient-treated sediment grew on oil-enriched Tauson's agar after 27 days incubation at $22°$C; and about 88 percent of the bacteria isolated from ground water treated with inorganic nutrients grew on oil-enriched Tauson's agar. A high percentage (98%) of oil-degrading bacterial colony forming units (CFU) was found in sediment and ground water isolates after 33 days of incubation from samples that were not treated with inorganic nutrients. Initial lag phases of oil-degrading bacterial CFU percentages were present in both nutrient-treated and nonnutrient-treated sediment and ground water samples. However, samples that were not treated with inorganic nutrients had a longer lag phase (about 14 days), and it declined to a lower percentage of oil-degrading bacterial CFU (35%) than those of nutrient treated samples (62-87%) after 60 days of incubation. It is apparent from Figure 20.6 that the continued presence of crude oil led to the maintenance of a significant population of oil degrading bacteria in the sediment and ground water.

Experimental results of the degradation of specific model hydrocarbons by 12 microbial isolates are given in Table 20.7. Crude oil was degraded by all 8 bacteria and 4 fungi. Percentages of bacterial isolates that were able to utilize cyclohexane, hexadecane, 1-naphthol, phenanthrene and pyrene as sole sources of carbon were 37.5, 100, 25, 37.5, and 12.5 percent, respectively. Percentages of fungal isolates that were able to degrade cyclohexane, hexadecane, 1-naphthol, phenanthrene and pyrene were 75, 75, 25, 25, and 25 percent, respectively. The polyaromatic hydrocarbon pyrene was the most resistant to biodegradation among the six model hydrocarbons tested. All of the bacteria and 3 of the 4 fungi were able to utilize hexadecane. This suggests that the use of hexadecane as an indicator of the degradation potential of hydrocarbons by microorganisms is quite appropriate.

When 1 ml of each model hydrocarbon was added to 100 g control sediment and ground water suspension and inoculated with extracted mixed cultures of the oil-contaminated sediment, the reduction in biochemical oxygen demand (BOD) and COD was the greatest in hexadecane-treated samples followed by crude oil, cyclohexane, phenanthrene, 1-naphthol, and pyrene, in a decreasing order, at $6°$ and $22°$C (Table 20.8). Samples treated with inorganic nutrients had significantly ($p \leq 0.01$) more BOD and COD reductions than those samples that were not treated with nutrients in all instances. Reductions in BOD, in all cases, were significantly ($p \leq 0.05$) greater than reductions in COD. Chemical oxygen demand reductions compared with BOD reductions in sediment treated with cyclohexane, hexadecane, 1-naphthol, phenanthrene, crude oil and

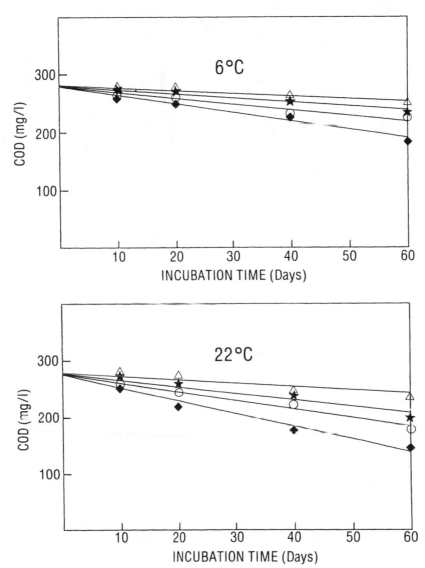

Symbols: △ , Control (100 g sediment + 20 ml ground water + 1.0 ml crude oil; ★ , Control + 10ppm $MgSO_4$ + 2ppm NH_4NO_3 + 0.5ppm $(KH_2PO_4$ + $K_2HPO_4)$; ○ , Control + 50ppm $MgSO_4$ + 10ppm NH_4NO_3 + 5ppm $(KH_2PO_4$ + $K_2HPO_4)$; ◆ , Control + 500ppm $MgSO_4$ + 20ppm NH_4NO_3 + 10ppm $(KH_2PO_4$ + $K_2HPO_4)$.

Figure 20.5: Variation of C O D with Incubation Time

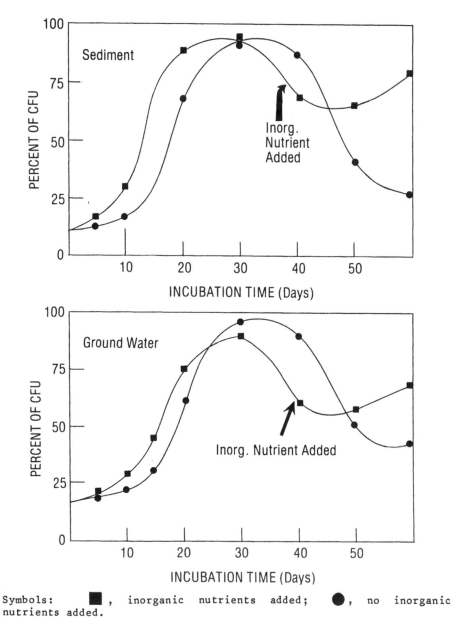

Symbols: ■ , inorganic nutrients added;   ● , no inorganic nutrients added.

Figure 20.6: Variation of Colony Forming Units with Incubation Time

Table 20.7: Degradation of Model Hydrocarbons by Bacteria and Fungi

Isolate Number	Growth On					
	Cyclohexane	Hexadecane	1-Naphthol	Phenanthrene	Crude Oil	Pyrene
Bacteria						
1	+	+	-	+	+	-
2,5,6	-	+	-	-	+	-
4	+	+	+	+	+	-
3	+	+	+	+	+	+
7,8	-	+	-	-	+	-
Fungi						
1,3	+	+	-	-	+	-
2	+	+	+	+	+	+
4	-	-	-	-	+	-

"+" Indicates isolate can utilize that particular hydrocarbon.

pyrene were 64-68 percent, 61-64 percent, 62-29 percent, 61-67 percent, 54-72 percent and 58-63 percent, respectively. These data imply that crude oil contamination in the subsurface environment could be cleaned to a great extent by mixed cultures enriched with mineral nutrients.

Table 20.8: Chemical Oxygen Demand (COD) and Biochemical Oxygen Demand (BOD) Removal (%) in Sediment/Ground Water Suspension Treated with Model Hydrocarbons and Incubated at 6° and 22°C for 40 Days

Model Hydrocarbon	COD				BOD			
	6° C		22° C		6° C		22° C	
	INT*	NINT**	INT	NINT	INT	NINT	INT	NINT
Cyclohexane	15	4.6	28	7.7	22	7.2	44	12
Hexadecane	24	9.8	43	16	38	16	68	25
1-Naphthol	8.3	3.4	16	4.9	12	5.5	24	7.6
Phenanthrene	13	4.1	22	6.1	20	6.5	33	10
Crude Oil	18	5.4	35	9.3	25	8.2	65	14
Pyrene	5.6	2.5	9.9	3.8	9.3	4.0	17	6.0

*INT = Inorganic Nutrients Treated (2 mg N + 1 mg P + 13 mg S)
**NINT = Non-inorganic Nutrient Treated

Our results show that all the model hydrocarbons are subjected to biodegradation in the subsurface aquifer environments by indigenous mixed microbial populations. Some preference was shown for degradation of crude oil and n-hexadecane, although degradation rates were reduced significantly at lower temperatures. Hydrocarbon biodegradation was limited by the nutritional imbalance between the substrate carbon, nitrogen and phosphorus required for microbial growth. This stimulating effect on oil and model hydrocarbon biodegradation by the addition of inorganic nutrients has also been reported by several authors (Jobson et al., 1974; Kincannon, 1972; Atlas and Bartha, 1972b; and Raymond, Hudson, and Jamison, 1976) who found an enhancement of oil biodegradation with mineral fertilizer.

Concentrations of inorganic nutrients in ground water at this particular oil-spill site are extremely low. The laboratory studies suggest that nutrient availability is the major factor limiting microbial activity at the site. However, the total amount of nutrients and microbial biomass may increase by trapping the flow of ground water and from the degradation of the crude oil.

A rise in temperature creates a rise in metabolic activity of microbial communities in both aerobic and microaerobic conditions. When hexadecane and aromatic hydrocarbons were incubated at 6° and 22°C under strict anaerobic conditions, there was no clear evidence for rapid degradation to CO_2 by indigenous microorganisms after 60 days incubation (data not shown). This is supported by Ward and Brock (1978), who reported that low levels of $^{14}CO_2$ detected during anaerobic incubation were not increased by addition of nitrate or in prereduced media containing nitrate and sulfate. Microbial processes in the contaminated aquifer appear to be restricted from microaerobic and anaerobic conditions. Although rates of anaerobic hydrocarbon degradation are slow, they could be ultimately significant as suggested by Davis (1967). The temperatures of $2°$, $7°$, $12°$, and $17°C$ were chosen to cover a wide range of ground water temperatures in most parts of the United States. The Q_{10} values were 2.18 to 3.09, which agrees with the Q_{10} values reported by ZoBell (1964) of slightly higher than 2 for nonproliferating cells, and about 3 for proliferating cells under enriched culture conditions.

Application of the radioisotope $^{14}C_1$-hexadecane to sediment samples collected from the aquifer indicated that hydrocarbon-degrading microorganisms are ubiquitous. They were found in sediment and ground water in both upgradient uncontaminated and down-gradient contaminated areas. The abundance of these organisms was greatest in oil-spill areas where microbial communities have redeveloped and the majority of the microbes have adapted to the oil polluted environment.

Antimicrobial agents provided a means for differentiating the degradation of model hydrocarbons by fungi and bacteria. Fungal activity accounted for the majority of the degradative potential at $22°C$; however, bacterial activity was responsible for the majority of model hydrocarbon degradation at $6°C$.

Microbial degradative potential for crude oil and model hydrocarbons in sediment and ground water samples was a function of the numbers of mixed indigenous isolates present in the aquifer. This is supported by Walker and Colwell (1974), who concluded that utilization of model petroleum at low temperatures is a function of the types and numbers of microorganisms present in an original inoculum taken from the natural environment. However, our results contradict Gibbs (1975), who found plate counts of total heterotrophic bacteria in the water which showed no quantitative relation with the rate of oxidation, and scarcely any difference between the oiled and control vessels. Microbial degradation of model hydrocarbons and crude oil was not completely lacking at low temperatures ($2°$, $6°$, or $7°C$). Indeed, there were about 4.3×10^4 -5.7×10^5 per ml psychrophilic indigenous microorganisms capable of degrading hydrocarbons at low temperatures in ground water environment. This is also in agreement with Walker and Colwell's findings (1974).

Microbial degradation of crude oil and model hydrocarbons was enhanced by an increase in temperature, oxygen availability, and inorganic nutrient concentration as shown by an increase in CO_2 evolution ($^{14}CO_2$ recovery), BOD and COD reduction. The present work indicates that biodegradation of hydrocarbons in a contaminated aquifer may be quite important in affecting the transport of aliphatic and low molecular weight hydrocarbons but is probably less important for the larger polynuclear aromatic compounds (such as pyrenous compounds).

ACKNOWLEDGMENTS

This research was supported by the United States Geological Survey,

Water Resources Division, as part of their national research project on Toxic Waste-Ground Water Contamination study.

The authors are indebted to Janet M. Erickson for her assistance and to the Department of Biology for the use of their Beckman Direct Data Readout Liquid Scintillation System. The authors also thank Charles S. Holt for permission to use his wet laboratory to carry out the microcosm study.

SELECTED REFERENCES

Atlas, R.M., and Bartha, R., "Biodegradation of Petroleum in Seawater at Low Temperatures," Canadian Journal of Microbiology, Vol. 18, 1972a, pp 1851-1855.

Atlas, R.M., and Bartha, R., "Degradation and Mineralization of Petroleum in Seawater: Limitation by Nitrogen and Phosphorus," Biotechnology and Bioengineering, Vol. 14, 1972b, pp 309-318.

Atlas, R.M., "Effects of Temperature and Crude Oil Composition on Petroleum Biodegradation," Applied Microbiology, Vol. 30, 1975, pp 396-403.

Atlas, R.M., "Stimulated Petroleum Biodegradation," Critical Review of Microbiology, Vol. 5, 1977, pp 371-386.

Bartha, R., and Atlas, R.M., "The Microbiology of Oil Spills," Advances in Applied Microbiology, Vol. 22, 1977, pp 225-266.

Bastin, E.A., "The Presence of Sulfate-Reducing Bacteria in Oil Field Water," Science, Vol. 63, 1926, p 21.

Blumer, M., and Sass, J., "Oil Pollution: Persistance and Degradation of Spilled Fuel Oil," Science, Vol. 176, 1972, pp 1120-1122.

Caparello, D.M., and LaRock, P.A., "A Radioisotope Assay for the Quantification of Hydrocarbon Biodegradation Potential in Environmental Samples," Microbiology Ecology, Vol. 2, 1975, pp 28-42.

Chang, F.H., and Broadbent, F.E., "Influence of Trace Metals on Carbon Dioxide Evolution from a Yolo Soil," Soil Science, Vol. 132, 1981, pp 416-421.

Chang, F.H., and Alexander, M., "Effect of Simulated Acid Precipitation on Algal Fixation of Nitrogen and Carbon Dioxide in Forest Soils," Environmental Science and Technology, Vol. 17, 1983, pp 11-13.

Chang, F.H., and Ehrlich, G., "Microbial Reconnaissance at the Site of a Crude-Oil Spill at the Bemidji, Minnesota Research Site," Report 84-4188, 1984, U.S. Geological Survey Water Resources Investigation, Toxic Waste Ground Water Contamination Study, pp 97-107.

Chang, F.H. et al., "Microbial Degradation of Crude Oil and Some Model Hydrocarbons," U.S. Geological Survey, Toxic Waste--Ground Water Contamination Study Technical Meeting, 1985, Water Resources Investigation (in press).

Colwell, R.R., and Walker, J.D., "Ecological Aspect of Microbial Degradation of Petroleum in the Marine Environment," Critical Review of Microbiology, Vol. 5, 1977, pp 423-445.

Cook, F.D., and Westlake, D.W.S., "Microbiological Degradation of Northern Crude Oils," Report No. 74-1, 1974, Environment-Social Committee, Northern Pipelines Task Force on Northern Oil Development Information, Canada.

Davis, J.B., Petroleum Microbiology, Elsevier, 1967, New York, New York.

Dibble, J.T., and Bartha, R., "Effect of Environmental Parameters on the Biodegradation of Oil Sludge," Applied Environmental Microbiology, Vol. 37, 1979, pp 729-739.

Gibbs, C.F., "Quantitative Studies on Marine Biodegradation of Oil, I. Nutrient Limitation at 14°C," Proceedings of Royal Society of London, Vol. 188, 1975, pp 61-82.

Gibbs, C.F., Pugh, K.B., and Andrews, A.R., "Quantitative Studies on Marine Biodegradation of Oil, II. Effect of Temperature," Proceedings of the Royal Society of London, Vol. 188, 1975, pp 83-94.

Jobson, A. et al., "Effect of Amendments on the Microbial Utilization of Oil Applied to Soil," Applied Microbiology, Vol. 27, 1974, pp 166-171.

Kincannon, C.B., "Oily Waste Disposal by Soil Cultivation Process," EPA-R2-72-100, 1972, U.S. Environmental Protection Agency, Washington, D.C.

Parkinson, D., "Oil Spillage on Microorganisms in Northern Canadian Soils," Report No. 73-25, 1973, Environmental Social Committee, Northern Pipelines Task Force on Northern Oil Development, Information, Canada.

Raymond, R.C., Hudson, J.O., and Jamison, V.W., "Oil Degradation in Soil," Applied Environmental Microbiology, Vol. 31, 1976, pp 522-535.

Tauson, V.O. and Aleshina, V.I., "Reduction of Sulfates by Bacteria in the Presence of Hydrocarbons," Mikrobiologiya (U.S.S.R.), Vol. 1, 1932, p 229.

Tauson, V.O. and Veselow, I.Y., "Bacterial Decomposition of Cyclic Compounds During the Reduction of Sulfates," Mikrobiologiya (U.S.S.R.), Vol. 3, 1934, p 360.

Walker, J.D, and Colwell, R.R., "Microbial Degradation of Model Petroleum at Low Temperatures," Microbial Ecology, Vol. 1, 1974, pp 63-95.

Ward, D.M., and Brock, T.D., "Anaerobic Metabolism of Hexadecane in Sediments," Geomicrobiological Journal, Vol. 1, 1978, pp 1-9.

ZoBell, C.E., "The Occurrence, Effects, and Fate of Oil Polluting the Sea," Advanced Water Pollution Research, Vol. 3, 1964, pp 85-119.

ZoBell, C.E., "Microbial Modification of Crude Oil in the Sea," Proceedings of the API/FWPCA Conference on Prevention and Control of Oil Spills," American Petroleum Institute, Publication No. 4040, 1969, Washington, D.C., pp 317-326.

ZoBell, C.E., "Microbial Degradation of Oil: Present Status, Problems and Perspectives," Microbial Degradation of Oil Pollutants, Workshop Proceedings, D.G. Ahearn and S.P. Meyers, (eds), Atlanta, Georgia, December, 1972, pp 3-16.

CHAPTER 21

INTERACTIVE SIMULATION OF CHEMICAL MOVEMENT IN SOIL

by

D.L. Nofziger and A.G. Hornsby

Managing chemicals which are applied to soils is a growing challenge for many people. Specific concerns in the management process include the protection of ground water quality. Farmers are concerned that the chemical remain in the root zone for sufficient time to complete the function for which it was applied. If it does not, another application may be necessary. Knowledge of the movement of the chemical can serve as a decision aid for farmers contemplating additional chemical applications. If the material has moved out of the root zone, another application may be desirable for the crop. If it has not, the cost of the application can be saved and risks of adverse environmental impacts can be minimized. Water management practices also influence the movement of chemicals in soils. The manner in which these practices impact chemical movement is difficult to predict in that it is dependent upon the soil and chemical properties, weather conditions, and their interactions.

The model presented here was developed as a management tool and as an educational aid for people managing soil-applied chemicals. It was designed for use with readily available soil, chemical, and weather data. It enables the user to enter relevant information for their conditions on popular microcomputers. Values entered by the user are checked to determine that they are reasonable and consistent. Results of the computations may be displayed in graphical and tabular form. Repeated simulations are easily made to enable the user to gain insight into the influence of specific soil and chemical properties on the movement of the chemical. The model supports simulations of up to 15 years for those interested in long-term movement.

In the following sections the use of the model will be presented followed by a description of the computational process and the basic equations involved. Finally the major assumptions in the model will be discussed.

MODEL USE

The software for this model enables users to perform two primary tasks. One task is to simulate chemical movement and display graphical or tabular results. The second task is that of entering the necessary soil, chemical, and weather data into disk files for repeated use and editing and displaying the data as needed. The user selects the task of interest from a menu.

Simulation of chemical movement in soils requires the selection of the soil and chemical of interest, the depth of application of the chemical, the depth of the root zone for the crop being grown on the site, daily evapotranspiration and infiltration records for the period of time being simulated, and the selection of English or metric units for computed results. The user is prompted for these parameters.

When the scenario for the simulation of interest has been defined, the software displays the depth of the chemical as a function of time after application. The graphical presentation also includes a graph of the effective rainfall as a function of time after application of the chemical. Effective rainfall is synonymous with the amount of water entering the soil surface. Graphical results for the movement of dicamba in Alpin sand are shown in Figure 21.1.

Following the graphical display, the user may select other output forms. These outputs include tabular listings of the depth of the chemical on each day in which infiltration occurred. In addition, the relative amount of the chemical remaining in the soil profile on that day is displayed. Calculation of the relative amount remaining is based on the degradation rate for the selected chemical. An abbreviated form of the table is shown in Table 21.1.

In addition to the tabular output, the user may obtain graphical displays of two chemicals applied on the same date. This is useful in comparing the movement patterns for different chemicals under the same conditions. The second chemical may be a nonadsorbed chemical such as nitrate or a chemical from the chemical data file. Figures 21.2 and 21.3 illustrate this type of graphical output for dicamba and atrazine in Alpin sand and Apalachee clay, respectively. Tables 21.2 and 21.3 show the required chemical and soil properties for these chemicals and soils.

Entry and editing of soil, chemical, and weather data files are facilitated by means of a built-in full-screen editor. The editor enables the user to move through the file to enter or edit data at the place of interest. Its operation is similar to that of a spreadsheet program. Each entry is checked for consistent and valid values.

The complete system enables the user to easily enter soil and chemical parameters and weather information for the site of interest. The user may then simulate the movement of the chemical at that site under natural rainfall or under different irrigation schemes. These results will aid in water management decisions. The user may also use the model to monitor movement of the chemical and as an aid in determining the need for additional application of the chemical. If several chemicals are available for the same purpose, the manager may compare their movement and degradation and include this information in the selection process.

COMPUTATIONAL METHODS

The model used to estimate the depth of the chemical in the soil is an extension of the model by Rao, Davidson, and Hammond (1976). In this model chemicals move only in the liquid phase in response to soil-water movement. Specifically, the change in depth of a chemical, Δ, for a particular infiltration event is given by

$$\Delta d = q/(R\ \Theta_{FC})\ \text{if } q > 0,$$

$$\text{or}\quad \Delta d = 0 \qquad\qquad \text{if } q = 0,$$

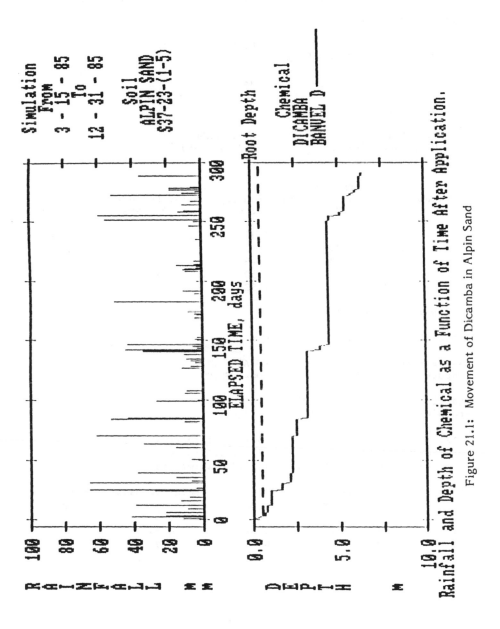

Figure 21.1: Movement of Dicamba in Alpin Sand

Table 21.1:   Tabular Output for Simulation of Movement of Dicamba in Alpin Sand

___

Chemical Selected:

    Common Name: DICAMBA
    Partition Coef. (ml/g OC): 2
    Half-Life (days: 14
    Trade Name: BANVEL D

SOIL SELECTED:

    Soil Name: ALPIN SAND
    Soil Identifier: S37-23-(1-5)

Horizon	Depth (m)	Organic Carbon %	Bulk Density Mg/cubic meter	Water Content (% by Volume) at -0.1 MPa	at -1.5 MPa
1	0.10	0.73	1.33	13.7	2.9
2	0.43	0.31	1.43	11.0	2.1
3	1.02	0.12	1.49	8.0	1.2
4	1.40	0.06	1.50	8.1	0.9
5	2.29	0.04	1.52	8.0	0.5

Root Depth: 0.50 meters

Depth of Application: 0.00 meters

Rainfall File: LOCAL 85.R
Evapotranspiration File: LOCAL 85.ET

Starting Date: 3-15-1985
Stopping Date: 12-31-1985

Total Rainfall: 1304.0 millimeters
Total Evapotranspiration: 775.4 millimeters
Potential Evapotranspiration: 1242.3 millimeters

Month	Day	Year	Rainfall mm	Solute Depth m	Relative Mass	Elapsed Time Days
3	16	1985	11	0.070	0.95	1
3	17	1985	41	0.404	0.91	2

Table 21.1: (continued)

Month	Day	Year	Rainfall mm	Solute Depth m	Relative Mass	Elapsed Time Days
				Chemical Movement Below Root Zone		
3	18	1985	22	0.621	0.86	3
3	21	1985	22	0.760	0.74	6
3	24	1985	8	0.760	0.64	9
3	27	1985	39	1.034	0.55	12
3	31	1985	13	1.034	0.45	16
4	3	1985	7	1.034	0.39	19
4	8	1985	27	1.034	0.30	24
4	9	1985	65	1.637	0.29	25
4	10	1985	4	1.637	0.28	26
4	15	1985	65	2.025	0.22	31
4	16	1985	6	2.078	0.21	32

Table 21.2: Partition Coefficient and Half-life for Chemicals

Common Name	DICAMBA	ATRAZINE
Partition Coefficient (ml/g OC)	2	163
Half-life (days)	14	48

where q represents the volume of water per unit area passing the depth of the solute, θ_{FC} is the soil-water content on a volume basis at "field capacity", and R is the retardation factor for the chemical in the soil. The retardation factor for the linear and reversible equilibrium adsorption model is given by

$$R = 1 + (\rho K_D)/\theta_{FC},$$

where ρ is the bulk density of the soil and K_D is the partition coefficient or the linear sorption coefficient of the chemical in this soil. In this model, the partition coefficient is given by

$$K_D = K_{OC} \ OC,$$

where K_{OC} is the linear sorption coefficient normalized for the organic carbon content of the soil. The use of K_{OC} and organic carbon content (OC) for determining the K_D value for nonionic organic compounds in any soil or any layer in a soil is supported by the research of Hamaker and Thompson (1972) and Karickhoff (1981, 1984).

For each day in the period simulated, the following steps are carried out:

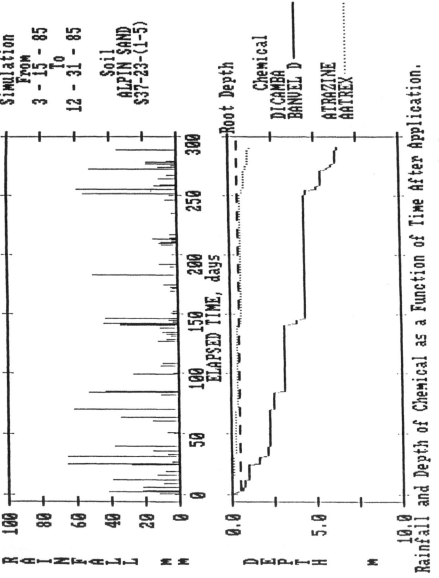

Figure 21.2: Movement of Dicamba and Atrazine in Alpin Sand

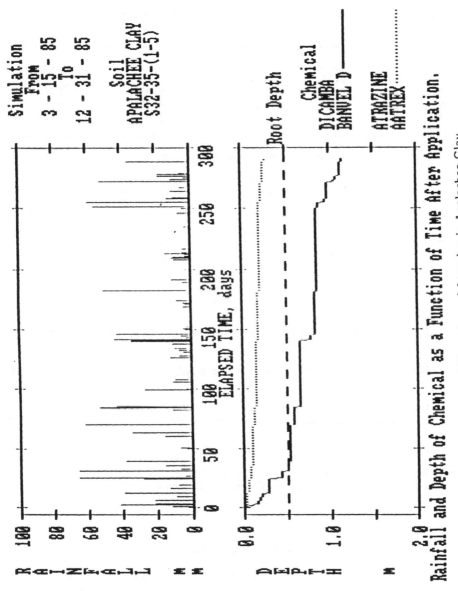

Figure 21.3:   Movement of Dicamba and Atrazine in Apalachee Clay

Table 21.3:   Required Soil Properties for Alpin Sand and Apalachee Clay Soils

Soil Name: Alpin Sand

Horizon	Depth (m)	Carbon %	Bulk Density Mg/cubic meter	Water Content (% by Volume) at -0.1 MPa	at -1.5 MPa
1	0.10	0.73	1.33	13.7	2.9
2	0.43	0.31	1.43	11.0	2.1
3	1.02	0.12	1.49	8.0	1.2
4	1.40	0.06	1.50	8.1	0.9
5	2.29	0.04	1.52	8.0	0.5

Soil Name: Apalachee Clay

Horizon	Depth (m)	Carbon %	Bulk Density Mg/cubic meter	Water Content (% by Volume) at -0.1 MPa	at -1.5 MPa
1	0.25	1.65	1.22	45.5	37.0
2	0.51	1.05	1.18	46.6	37.0
3	0.64	1.08	1.08	52.6	36.1
4	1.14	0.71	1.23	52.2	37.5
5	1.52	0.50	1.36	45.9	27.3

(1)   the water content in the root zone is adjusted for evapotranspiration;

(2)   the water content in the profile is adjusted for infiltration on that day;

(3)   the quantity of water passing the present solute depth is determined; and

(4)   the new solute depth is determined.

Detailed descriptions of the manner in which these computations are carried out for a layered soil are given by Nofziger and Hornsby (1986).

ASSUMPTIONS IN MODEL

The following assumptions are used in this model:

(1)   Chemicals move only in the liquid phase in response to movement of soil water. The model does not estimate movement in the vapor phase. Losses of the compound due to vapor movement to the atmosphere are also ignored.

(2)   All soil water participates in the transport process. Water initially in the soil profile is completely displaced ahead of infiltrating water. If this is not true, the model will tend to underestimate the depth of the chemical front.

(3)   Water entering the soil redistributes instantaneously to "field capacity". This assumption is employed to reduce the need for hydraulic properties of the unsaturated soil layers. It is approached for coarse textured soils. For fine textured soils, the depth predicted in this model will likely be associated with a time which is a few days later than that specified here.

(4)   Water is removed by evapotranspiration from each layer in the root zone in proportion to the relative amount of water available in that layer. A uniform root distribution is assumed. Although this will not be strictly valid in many situations, more precise schemes will require additional information on root densities and soil hydraulic properties.

(5)   Upward movement of water does not occur anywhere in the soil profile. Water lost in the root zone due to evapotranspiration is not replenished by water below the root zone.

(6)   The adsorption process can be represented by a linear, reversible, equilibrium model. If the sorption process is described by a nonlinear isotherm, the partition coefficient decreases with increasing concentration of the chemical. This may be significant for very high application rates but is probably not significant for most agricultural applications. If sorption is irreversible, this model will overestimate the depth of movement. If the adsorption equilibrium is not instantaneous, the model will underestimate the depth.

(7)   The half-life for biological degradation of the chemical is constant with time within a soil layer. The model will accommodate different degradation rates for each soil layer if they are known.

Persons using this model should be aware of these assumptions and assess their validity for the system under consideration.

SELECTED REFERENCES

Hamaker, J.W., and Thompson, J.M., "Adsorption," in Goring, C.A.I., and J.W. Hamaker, (eds), Organic Chemicals in the Environment, 1972, Marcel Dekkar, New York, New York, pp 49-143.

Karickhoff, S.W., "Semi-empirical Estimation of Sorption of Hydrophobic Pollutants on Natural Sediments and Soils," Chemosphere, Vol. 10, 1981, pp 833-846.

Karickhoff, S.W., "Organic Pollutant Sorption in Aquatic Systems, Journal Hydraulic Engineering, Vol. 110, 1984, pp 707-735.

Nofziger, D.L. and Hornsby, A.G., "A Microcomputer-based Management Tool for Chemical Movement in Soil," Applied Agricultural Research, Vol. 1, 1986, pp 50-56.

Rao, P.S.C., and Davidson, J.M., and Hammond, L.C., "Estimation of Nonreactive and Reactive Solute Front Locations in Soils," in Proceedings of Hazardous Wastes Research Symposium, Tucson, Arizona, EPA-600/19-76-015, 1976, pp 235-241.

CHAPTER 22

REGULATION OF THE AGRICULTURAL UTILIZATION
OF SEWAGE SLUDGE IN NEW JERSEY

by

Akos Fekete and Helen Pettit-Chase

New Jersey is a small but densely populated state with over 500 domestic and privately owned sewage treatment plants, which produce over 2.3 million dry pounds of sludge per day. Production is expected to rise to 3.5 million dry pounds per day by July 1, 1988, the U.S. Environmental Protection Agency (EPA) water quality limiting deadline for sewage treatment plants. Presently, 56 percent of the sludge production is dumped in the Atlantic Ocean; 13.6 percent is applied to the land as liquid, dewatered, composted, or processed sludge, and the remainder (30.4 percent) is either incinerated, stored on site, or trucked out of state.

Traditional sludge management disposal practices are slowly disappearing, and being replaced by resource recovery management alternatives. The EPA has changed the sludge ocean dumping limit from 12 miles to 106 miles offshore. This represents the first step leading to the proposed elimination of all ocean dumping activities by 1990. In response to a solid waste crisis caused by diminished landfill capacity and an increased awareness of the utilization potential of sludge, sludge landfilling in New Jersey was effectively prohibited on March 15, 1985. Since that date, land application has increased from .6 percent of the state's total production to 13.6 percent, or 316,043 dry pounds per day. Clearly, the emphasis on land-based sludge management alternatives has increased.

Ground water is used extensively throughout New Jersey for public, industrial, domestic, and agricultural supply. Nearly 3.5 million people (45 percent of the population) depend on ground water. For this reason, most activities which may affect the quality of this natural resource are strictly regulated, including the agricultural utilization of sewage sludge.

STATUTORY AUTHORITY

The New Jersey Solid Waste Management Act (NJSA 13:1E) provides for the maximum practical processing of all sludge into energy, fertilizers, and other useful products. It is the policy of the New Jersey Department of Environmental Protection that land application of "clean" domestic sludge constitutes beneficial reuse and resource recovery, and that such application shall be encouraged.

The agricultural utilization of sludge is regulated under the authority of the New Jersey Water Pollution Control Act (NJSA 58:10A). New Jersey Pollutant Discharge Elimination System (NJPDES) Permits are used to control sludge farming activities under the definition of discharges to ground water.

The quality of sludge being applied, the site where the sludge is applied, and the rate and manner in which it is applied are all controlled through these permits. Considered by some to be excessively restrictive, the regulations and permits were developed in response to a combination of technical concerns and to the public's emotional misconception that sludge application is a "dumping" practice.

PERMITTING OPTIONS

The New Jersey Residuals Management Program provides two opportunities of resource recovery and reuse through agricultural use: the Land Application Program and the Distribution Program. Both require a NJPDES Permit and both assure the protection of ground water quality through site evaluation, sludge quality control, control of application rate and crop control. The distinction between the two programs hinges on the use of Approval Letters and Permits. The Letter of Approval sanctions short-term site use while the Permit allows long-term site use. Where long-term uses are provided, public comment requirements are incorporated into the process.

Distribution Program

Under the Distribution Program a specific sludge or sludge product is identified and evaluated for its suitability for distribution. The sludge must have been treated by a Process to Further Reduce Pathogens (PFRP), as described in 40 CFR Part 257. The facility processing the sludge is then issued a NJPDES Permit which assures that the process requirements and sludge quality are maintained and establishes a fixed application rate for the material generated by the facility. The application rate is based on its nutrient content. The specific agricultural distribution sites are determined by a Residuals Management Site Screening Process as they are submitted for review. The sites are evaluated for suitability using the same criteria that are applied during Land Application permitting and a Letter of Approval is issued for the qualifying site. The Approval allows permitted distribution material to be applied at the site up to the total application rate, which is usually satisfied after two years of applications. Therefore, the Distribution Program provides for long-term sludge permitting and short-term site approval. Distribution Programs may also involve commercial use of processed sludges in horticultural, landscaping and nursery uses. A bagging program is currently being developed which will make these products available to the general public.

Land Application Program

Under the Land Application Program, a specific site is identified and evaluated for its suitability for long-term land application of sludge. The unsuitable portions of the site are eliminated from consideration for sludge applications and a NJPDES Permit is issued for the qualifying portions of the proposed site. Under the terms of a permit, sludge application rates are limited to the nutrient requirements of the crop, and the types of crops that can be grown are restricted. The specific sludges that can be applied on the site are determined by a Residuals Management Sludge Quality Screening Process.

Sludges which have undergone a Process to Significantly Reduce

Pathogens (PSRP), and meet chemical quality criteria, are identified by Letter of Approval; these approvals must be renewed annually. Therefore, the Land Application Program provides for long-term site permitting and short-term sludge approval.

PERMIT CONTROLS

NJPDES Permit conditions are related to the impact that land application of sludge might have on public health and on ground and surface water quality. Concerns arise from the potentially detrimental characteristics and quality of sludges; namely, heavy metals, industrial organics, pathogens and odors. Land application of sludge can be carried on in an environmentally safe manner by adhering to a management scheme which includes controls of these parameters: proper site selection, sludge quality analysis, monitoring of ground water and soils, and proper operations.

Site Evaluation

Prospective sites are investigated thoroughly to determine their suitability for land application. A plot plan is prepared showing boundaries, access roads, private homes, buildings, and other features. Also included is the ground water flow direction beneath and adjacent to the site, as well as an inventory of all public and industrial water supplies, shallow domestic wells, streams, lagoons, and other water courses within 1500 feet of the site.

Soil borings are taken to verify the soil survey map and determine the soil properties for the soil series at that particular location. Soil and site characteristics are evaluated according to permeability, soil drainage class, and flooding hazard, slope, depth to bedrock and depth to seasonal high water table (Table 22.1). Where severe soil or site limitations are present, special precautions must be developed to prevent environmental degradation, or these areas may be excluded from use.

Sludge Quality

The environmental safeguards provided in land application permits are based in large part on sludge quality determinations. Specifically, the degree of pathogen reduction the sludge has undergone and the chemical composition will determine its suitability for agricultural utilization.

Sludge testing and reporting is governed by New Jersey's Sludge Quality Assurance Regulations. All New Jersey treatment plants must test and report at intervals based on the size of the plant and the volume of industrial flow into the plant. Table 22.2 presents the reporting schedule for the five categories of treatment plants. Sludge reports (daily production, volatile solids reduction, pH, percent solids, and sludge management site) must be submitted monthly for all plants, while sludge analyses (heavy metals, selected chemical parameters, and toxic organics) are reported with increasing frequency as plant size increases.

Table 22.1:   Soil and Site Limitations for Land Application of Sludge

	DEGREE OF LIMITATION		
CHARACTERISTIC	Slight	Moderate	Severe
Permeability of the more restricting layer w/in 5 feet of the surface	0.6 to 6.0 inches/hour	6-20 and 0.2 - 0.6 inches/hr	>20, or <0.2 inches/hr
Soil drainage class	Well to moderately well drained	Somewhat excessively drained and somewhat poorly drained	Excessively drained poorly and very poorly drained
Flooding	None	Flooded only during non-growing season	Flooded during growing season
Slope	< 6%	6-12%	> 12%
Depth to seasonal high water table	> 4 ft	2-4 ft	< 2 ft
Depth to bedrock	> 4 ft	2-4 ft	< 2 ft

Table 22.2:   Reporting Frequency for New Jersey Treatment Plants

STP Size	Sludge Report	Heavy Metals	Selected Chemicals	Toxic Organics
<0.1 MGD	monthly	annually	annually	annually
0.1- <1 MGD	monthly	semi-annually	semi-annually	annually
1- <5 MGD	monthly	quarterly	quarterly	semi-annually
>5 MGD	monthly	monthly	monthly	monthly
>10% flow from industry	monthly	monthly	monthly	monthly

Pathogen Reduction

According to NJPDES regulations, only stabilized sludges may be distributed or land applied. Sludges must have been treated by a PFRP or a PSRP, as described in 40 CFR Part 257. Many New Jersey treatment plants which land apply their sludge are not meeting stabilization requirements during the digestion process. As a consequence, they must lime stabilize to meet the pathogen reduction criteria. For lime stabilization to meet PSRP requirements, adequate lime must be added and thoroughly mixed with the sludge in order to produce a pH of 12 for a two-hour period, or a pH of 12.5 for a period of one-half hour. A 2-log reduction of fecal streptococci must be demonstrated as proof of the efficacy of the treatment.

Chemical Characteristics of Sludge

Table 22.3 lists the chemical parameters for which testing is required by the Sludge Quality Assurance Regulations. In addition, a list of suggested parameters is presented. They were developed following review of the July, 1985, EPA document "Summary of Environmental Profiles and Hazard Indices for Constituents in Municipal Sludge", and are under consideration for revision of the Sludge Quality Assurance Regulations. All parameters are reported on a dry weight basis.

Testing Methods

Since the promulgation of the Sludge Quality Assurance Regulations in 1980, it has become clear that existing water testing methods are inappropriate for analysis of the complex sludge matrix. No standard methods existed for use by the Certified New Jersey Laboratories. In response to this need, the New Jersey Department of Environmental Protection (NJDEP) created the Sludge Methods Task Force, composed of chemists from academia, industry, sewage treatment plants, EPA and NJDEP. This Task Force has been developing laboratory methods for testing the existing parameters identified in the Sludge Quality Assurance Regulations. The first series of sludge methods will be identified as interim methods in the amendments to the New Jersey Laboratory Certification Regulations, which were publically announced in May, 1986. Methods validation will be conducted during the effective period of the interim methods, and final adoption will be delayed until satisfactory validation results are received. Interim methods have been developed for testing of pH, total residue, volatile ash content of total residue, phenols, oil and grease, metals (except mercury), and kjeldahl nitrogen. The Task Force will address nitrate/nitrite, organics, and cyanide methods during their next phase of work and will address additional parameters, including biological analyses for pathogens, in the future.

In order to assure the validity of the sludge data, each of the methods developed by the Task Force includes sample preservation and quality assurance and quality control requirements, which the laboratories must perform in order to secure and maintain Certified Laboratory standing. These stringent controls have evolved as a direct result of the litigious history surrounding sludge management in New Jersey.

Table 22.3:   Sludge Quality Assurance Regulations Required Chemical
Parameters

PARAMETER	EXISTING	SUGGESTED
Metals	Arsenic Cadmium Chromium Copper Lead Mercury Nickel Zinc	Arsenic Cadmium Chromium Copper Lead Mercury Nickel Zinc Beryllium Iron Molybdenum Selenium
Conventionals	Oil & Grease Phenols Cyanide	Oil & Grease Phenols Cyanide Flouride
Nutrients	Total Nitrogen Ammonia N Nitrate N Phosphorus Calcium Magnesium	Total Nitrogen Ammonia N Nitrate N Phosphorus
Pesticides	Aldrin Chlordane Dieldrin DDT Heptachlor Lindane PCBs Toxaphene Endrin Heptachlor epoxide Methoxychlor Mirex pp'-DDE pp'-TDE	Aldrin Chlordane Dieldrin DDT Heptachlor Lindane PCBs Toxaphene
Base Neutrals and Acids		Benzidine Benzo(a)pyrene Bis (2-ethylhexyl) pthalate Hexachlorobenzene Hexachlorobutadiene N-nitrosodimethylamine

Table 22.3: (continued)

PARAMETER	EXISTING	SUGGESTED
Purgeables		Benzene
		Carbon Tetrachloride
		Chloroform
		Methylene Chloride
		Tetrachloroethylene
		Trichloroethylene
		Vinyl Chloride

Sludge Quality Approvals

Through a procedure known as the Generic Quality Determination, sewage treatment plants can have their sludge quality evaluated for its suitability for agricultural utilization. This procedure is a prerequisite for approval to go to a specific land application site. All requests for a generic determination must include at least 3 full analyses performed pursuant to the Sludge Quality Assurance Regulations, a list of all industrial dischargers into the sewer system, and an operator's statement describing the sludge handling and digestion process. Frequently, analysis for additional parameters will be required based upon industrial input.

The quality limitations imposed by New Jersey on land application are presented in Table 22.4. The basis for these limitations is indicated in the footnotes to Table 22.4. Generally, metals limits have been set by back calculation of the maximum cumulative metals loadings to determine annual loading over 40 years (Class A sludge) and 20 years (Class B sludge).

Upon completion of the review, the treatment plant is issued a letter which indicates that their sludge is suitable or unsuitable for land application and specifies if additional stabilization will be required in order to meet PSRP standards. The letter restricts applicability according to the physical state of the sludge (liquid or dewatered) upon which the analyses were performed. Once in possession of a generic approval, the treatment plant can shop around for a suitable land application facility.

Ground Water Monitoring

NJPDES Permits for land application of sludge require the preparation of a ground water monitoring program capable of determining the impact of residuals applications on the quality of ground water in the site vicinity. This monitoring program requires, at a minimum, the installation of at least 4 monitoring wells--1 hydraulically upgradient, and 3 downgradient from the application site.

Upgradient wells are installed outside the limit of approved application areas. Their number, location, and depth must ensure that samples are

Table 22.4:   Land Application Sludge Quality Criteria (1)

Metals (ppm, dry weight basis)

	Class A	Class B
Cadmium (2)	20	40
Copper (2)	600	1200
Lead (2)	2400	4800
Nickel (2)	625	1250
Zinc (2)	1200	2400
Chromium (3)	1000	1000
Mercury (3)	10	10
Arsenic (3)	10	10

Pesticides and PCBs (ppm, dry weight basis) (4)

Aldrin	0.10	0.10
Chlordane	0.10	0.10
Dieldrin	0.10	0.10
Endrin	0.10	0.10
Heptachlor	0.10	0.10
Heptachlor epoxide	0.10	0.10
Lindane	0.10	0.10
Methoxychlor	0.25	0.25
Mirex	0.25	0.25
p,p'-TDE (DDD)	0.25	0.25
p,p'-DDT	0.25	0.25
p,p'-DDE	0.25	0.25
Toxaphene	1.0	1.0
PCBs (total)	0.5	0.5

Miscellaneous

Phenols (ppm, total) (5)	5.0	22.0
Oil and Grease (%)(5)	5.0	5.0

(1)   Sludge must be PSRP or PFRP
(2)   Class A can be land applied for 40 years and Class B for 20 years before cumulative metals loading limits are reached.
(3)   From EPA 430/9-77-004, 1977.   Municipal Sludge Management: Environmental Factors.   Controlled Municipal Sludge.
(4)   Current recommended levels in New Jersey Sludge Quality Assurance Regulations.
(5)   "New Jersey Median Sludge Quality," A. Singh, H. Pettit-Chase, and M.L. Morris, 1983.

representative of background water quality near the facility and that the samples are not affected by the facility.

The number of downgradient wells, their location and depth must ensure that they immediately detect any pollutants that migrate from the land application operation to the ground water.

All monitoring wells must be installed by a licensed New Jersey well driller under a valid New Jersey well permit. Wells must be constructed according to NJDEP Monitor Well Specifications which prescribe the materials, construction, security, packing, sealing, and development of wells in consolidated or unconsolidated formations. In addition, a Ground Water Monitoring Well Certification Form must be submitted, signed by a New Jersey Professional Engineer, a licensed New Jersey Well Driller, or a licensed Geologist. All wells must be surveyed by a professional land surveyor. These specifications and submissions are required in an effort to obtain "litigation quality" monitoring data.

Ground Water Monitoring Plans must include the procedures and techniques for sample collection, sample preservation, analytical procedures and chain of custody control. Each well must be sampled for the parameters listed in Table 22.5. Sample collection, preservation, handling, and analysis must conform to procedures approved under 40 CFR Part 136. Wells must be sampled in accordance with NJPDES Regulations which address the flushing or pumping of wells to insure a representative sample, and appropriate sampling techniques for volatile versus nonvalatile parameters. A chain of custody record must be maintained for each sample, from point of sampling to final analysis at the New Jersey Certified Laboratory.

Table 22.5: Ground Water Monitoring Requirements—Parameters and Limitations

PARAMETER	LIMITATION		AFFECTED SLUDGE FARMS
Cadmium	0.01	ppm	All
Zinc	5.0	ppm	All
Copper	1.0	ppm	All
Lead and compounds	0.002	ppm	All
Zinc and compounds	5.0	ppm	All
Ammonia-N	0.5	ppm	All
Chloride	250.0	ppm	Some
Nitrate-N	10.0	ppm	All
Sodium	50.0	ppm	Some
Sulfate	250.0	ppm	Some
pH	–		All
Total Dissolved Solids	500.0	ppm	All
Total Organics	–		Rare
Fecal Coliform	–		All

In an effort to ensure reproducible, verifiable data, a Quality Assurance and Quality Control Program is presently being developed for ground water monitoring, which will involve analysis of field and trip blanks whenever wells are sampled. In addition, a laboratory QA/QC Program is being developed which will expand on the QA/QC requirements of 40 CFR Part 136.

Soil Monitoring

Soils are monitored for the parameters listed in Table 22.6. Three soil cores are taken per acre, with a minimum of eight cores per application area. Samples for each area are composited and submitted for analysis at a Certified New Jersey Testing Laboratory.

Table 22.6:   Soil Quality Monitoring Requirements

PARAMETER	UNITS	AFFECTED SLUDGE FARMS
pH	–	All
Plant Available Phosphorus	lb/ac	All
Plant Available Potassium	lb/ac	All
Total Cadmium	mg/Kg	All
Total Copper	mg/Kg	All
Total Lead	mg/Kg	All
Total Nickel	mg/Kg	All
Total Zinc	mg/Kg	
Cation Exchange Capacity (1)	meq/100 gms	All

(1)   Tested only initially.
       All values reported on a dry-weight basis.

Operational Constraints

Residual application sites must be managed to minimize risks associated with (1) nitrogen, phosphorus, and pathogen contamination of ground waters, (2) soil degradation by heavy metals overloading, (3) pathogen transmission via insects and animals, and (4) production of nuisance conditions such as offensive odors. The degree of site management will be dependent on such factors as site size, planned site lifetime, site properties, sludge transportation modes, application rates and methods, type of crop and the ability to restrict public access (U.S. Environmental Protection Agency, 1983).

Application Method

In general, odors can be minimized through proper sludge application. Liquid sludges may be subsurface injected or surface applied. If they are surface applied to a field with a vegetative cover less than 75 percent, or if dewatered sludge is surface applied, then the sludge must be incorporated to a depth of 3 to 8 inches within 24 hours, field conditions permitting. Sludge may not be applied to frozen or saturated ground.

Soil pH

Maintenance of a soil pH of 6.5 is required for the life of the site and usually maintained for a closure period of two years or more following the last sludge application.

Crops

Crops for direct human consumption (without processing prior to distribution), where the edible portion of the crop comes in direct contact with the residuals, are not considered suitable for growth on residuals amended lands and may not be grown on such lands for a period of at least 18 months after the last residuals application. Tobacco, leafy vegetables and root crops grown for human consumption may not be grown on the residuals amended land for a period of two years after the last residuals applications.

Application Rates

Application rates are controlled to maintain the integrity of the farming operation. Applications in excess of agronomic requirements of the crop are viewed not as normal farming operation, but as a dumping activity.

In order to prevent hydraulic overloading, the maximum daily liquid sludge application rate cannot exceed 25,000 gallons per acre. Subsequent application may not be made on an area of prior application without providing for a minimum rest period of at least three days from the time of last application.

The following information is needed to calculate the rate of sludge application:

(1)   sludge composition;

(2)   soil pH and cation exchange capacity;

(3)   productivity index for the soil from Soil Conservation Service Soil Surveys;

(4)   history of sludge and manure applications to the site for the preceding four years; and

(5)   crops to be grown.

Crop   nitrogen   requirements   are   generally   determined   based   on productivity indices for soils and expected yields from New Jersey Agricultural Extension Service and SCS publications, or from the Cooperative Extension Service. The amount of sludge needed to meet the crop's nitrogen requirement is calculated using available nitrogen in the sludge, and by accounting for any inorganic fertilizer additions, and residual nitrogen from previous sludge or manure applications.

The effect of residual nitrogen in decreasing future application rates is estimated using mineralization rates. Presently, New Jersey utilizes mineralization rates of 20 percent, 8 percent, 3 percent, and 3 percent for the first four years following application. These average mineralization rates do not account for differences in nitrogen mineralization for sludges which undergo differing treatments such as anaerobic or aerobic digestion. To further complicate the situation, we have found that many treatment plants which utilize land application are not meeting stabilization requirements for PSRP, and consequently must lime stabilize prior to land application. There is very little information on the effect of this stabilization process on nitrogen availability the first or subsequent years after application. New mineralization rate calculations are being investigated.

If food chain crops are grown, an annual cadmium loading limit must not be exceeded. The present annual limit of 1.25 kg/ha/yr will decrease to 0.5 kg/ha/yr on January 1, 1987. An annual limit of 100 lb/ac must be observed for sludge applications on pasture lands where animals whose products enter the food chain graze.

Storage

A fundamental problem in land application has been one of timing. Farmers need large amounts of fertilizers at specific times during the growing season, but sludge is produced daily at the treatment plants, in relatively small quantities. The solution to this dilemma has generally been the construction of storage facilities at the land application sites, either aboveground slurry type storage tanks, or concrete or clay-lined lagoons or bunkers. Storage facilities give the flexibility of making sludge available as it is required, which is essential for farming operations. Since these storage facilities are generally aerated and mixed, the nutrient content of the sludge is more uniform over the year, also simplifying farm operations. Additionally, sampling of each incoming load, and storage of the material prior to application lessens the likelihood of a "hot load" being delivered to the site, easing a major public concern. Storage is limited to the amount of sludge which the site can accommodate in one year.

Farm Plans

The Residuals Management Section believes the support and involvement of the agricultural community is essential for the success of the Sludge Management Program. Farm Plans, produced through a cooperative effort between the farmer and the local Soil Conservation District Office is a mandatory feature of NJPDES land application permits. These plans address problems associated with site limitations, particularly erosion potential, and implement control measures such as diversions, terraces, contour stripping, and buffer zones to prevent the movement of sludge and sludge amended soil offsite. Equally important, plans are used to insure the use of prescribed Best

Management Practices which may protect farms from frivolous nuisance lawsuits under the New Jersey Right to Farm Act. Such suits are common in densely populated New Jersey, where urban, suburban, and rural lifestyles exist in close proximity to one another.

Reporting Requirements

Permittees are required to submit quarterly monitoring reports to the Residuals Management Section. Data from these reports are entered into a computerized system which tracks any violations which may have occurred during the reporting period, such as excess nutrient applications and metals overloading.

Monitoring reports include:

(1)   the amount of sludge applied from each approved source;

(2)   the quantity of plant nutrients supplied by the sludge;

(3)   metals loading over the reporting period, and the total applied to date;

(4)   status of the crop plan, such as type of crop grown, dates of seeding and harvesting, yields, and use of harvested crop;

(5)   soils monitoring data;

(6)   sludge quality data; and

(7)   ground water monitoring data.

This monitoring program insures compliance with permit conditions. Through a vigorous enforcement program, fines may be levied for violation of established restrictions.

GROUND WATER DATA

The ground water quality parameter of major concern has been shown to be nitrate-nitrogen. The metals have not been detected at New Jersey's sites and will not be discussed here. Higgins (1984) found that applications at agronomic rates caused nitrate levels of less than 50 ppm in ground water beneath test plots in New Jersey. Monitoring data from permitted New Jersey sites have not shown any ground water nitrate-N levels in the 50 ppm range.

Figures 22.1 and 22.2 illustrate the ground water monitoring data from two of New Jersey's well established sludge farms. The mean nitrate-N levels for upgradient and downgradient wells are graphed by quarterly reporting date - January, April, July, and October of each year. Farm 1 (Figure 22.1) is a sod farm which has been operating under permit since 1981. The upgradient wells show that the background quality of ground water at this site may be affected by ongoing conventional agricultural practices in the vicinity. The mean values for the downgradient wells illustrate that a properly managed farm can operate for extended periods without significantly affecting ground water quality.

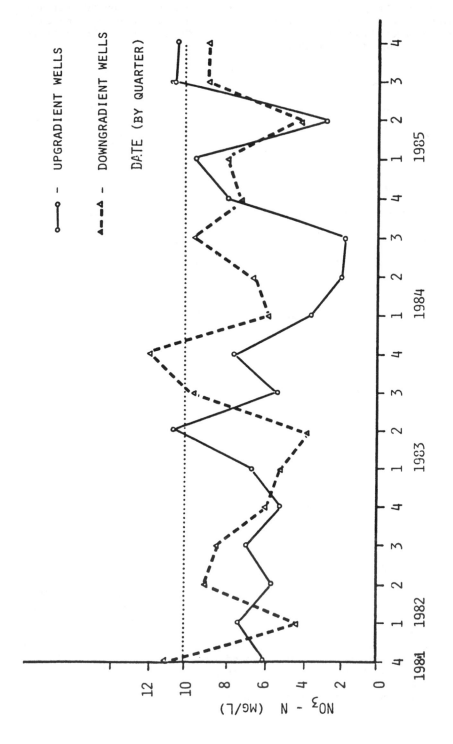

Figure 22.1:   Ground Water Monitoring Data, Farm 1

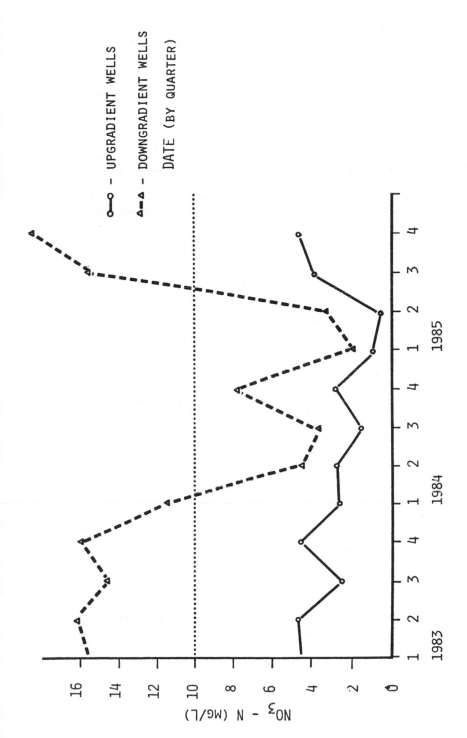

Figure 22.2:   Ground Water Monitoring Data, Farm 2

Ground water beneath Farm 2 (Figure 22.2) periodically exhibits excessive nitrate-N. The high nitrate-N in 1983 and early 1984 may be due to previous unpermitted sludge and septage applications. The third and fourth quarter, 1985, results exceed New Jersey's 10 ppm nitrate-N ground water quality standard, due to several factors. Application of inorganic fertilizers in addition to sludge may have resulted in nitrogen loading in excess of calculated agronomic rates. This farmer has also been cited for improper crop management.

Insufficient data presently exists for the majority of the 21 permitted sludge farms in New Jersey since most have only been operating since the March 15, 1985, landfill deadline. At several older permitted facilities lysimeters were installed. These permits are being revised to require the drilling of monitoring wells. With the ongoing operation of these sites, a computerized data base is being compiled for the continued monitoring of ground water.

OUTLOOK FOR THE FUTURE

Looking ahead, the problems surrounding sludge use on New Jersey's farms are not getting easier. Sludge production, especially in more urban areas, is increasing, while New Jersey's farmland is being developed at an alarming rate. The public, sensitized by numerous Superfund sites, and cases of surface and ground water pollution, increasingly exhibits the "Not In My Back Yard" syndrome. In response to this public pressure, New Jersey's elected officials pass local ordinances restricting land application or requiring excessive testing of sludge. Environmental groups and municipalities have brought suit against farmers and the NJDEP to halt land application operations. A recent New Jersey policy decision may require the involvement of a public entity in all land application permits, in order to increase accountability of these operations. As a result, the Residuals Management staff believes that the number of permits issued will drop, since sewage treatment plants have traditionally balked at such involvement.

However, the Residuals Management staff sees the Distribution Program as a possible solution to many of these problems. Use of sludge products, such as compost, by homeowners may reduce the fear surrounding sludge and pave the way for greater public acceptance of agricultural usage. Ultimately, greater regulation at the source, that is, regulation of sludge quality and decreased regulation at the farm (similar to conventional fertilizer usage) appears to hold the greatest promise.

SELECTED REFERENCES

Higgins, A.J., "Land Application of Sewage Sludge with Regard to Cropping Systems and Pollution Potential," <u>Journal of Environmental Quality</u>, Vol. 13, 1984, pp 441-448.

U.S. Environmental Protection Agency, "Land Application of Municipal Sludge," EPA 625/1-83-016, October, 1983, Washington, D.C.

CHAPTER 23

INCENTIVES AND INSTITUTIONS TO REDUCE
PESTICIDE CONTAMINATION OF GROUND WATER

by

Brian P. Baker

Farmers have polluted ground water for centuries. Wells were polluted by manure being spread on nearby fields or by livestock buried in nearby pits. Incidents of adverse health effects are not extensively documented but were known to occur. Because of the poor understanding of ground water in the past, one wonders how many contamination incidents occurred without being discovered.

Modern agricultural technology has done much to change farming, rural America, and the nature of ground water contamination. Increased use of chemical fertilizers and pesticides have caused ground water pollution problems in many agricultural areas. The new technology has also increased the scale of operations. As farm size increases, so does the environmental impact of farming. Agricultural pollution of ground water has become a source of conflict between farmers and the growing nonfarm rural population. As a result, ground water contamination, or at least awareness of it, has become more common.

The survey described in Chapter 2 will do much to illuminate the extent and seriousness of the problem of pesticide contamination of ground water. One can only infer the widespread nature of the problem from reports from most states (Francis, Brower, and Graham, 1982; Pye, Patrick, and Quarles, 1983; and U.S. Congress, 1984).

One thing we do not seem to know is how to prevent further contamination incidents from occuring, without putting undue economic stress on agriculture. In places that have had such incidents, existing institutions have failed to balance these two objectives. As more incidents occur, new institutions will need to be developed.

This chapter intends to accomplish several objectives. First, the institutions supposed to protect ground water from agricultural pollution are briefly presented. These institutions are critiqued; their weaknesses are pointed out. Alternative institutions for protecting ground water from pollution are proposed. Finally, the tradeoffs between these institutions are evaluated.

INADEQUACY OF THE COMMON LAW

In this historical context, common law provided the principle institutional framework for protecting ground water. Two theories of common law were employed in cases of ground water contamination: nuisance and negligence (Davis, 1976). Unlike some riparian surface water rights, ground water users

345

have no right to use ground water to discharge waste, particularly when that ground water is used as a water supply source.[1] Nuisance law requires that one's enjoyment of one's property was interfered with by the pollution.[2] Negligence theory regarding ground water pollution requires that a ground water user knows or should have known that his or her activity would be likely to cause the injury which has occurred.[3] Both theories place the burden of proof on the plaintiff to show that the injury was caused by the defendant's actions. In a negligence case the plaintiff must also show that the defendant failed to act in a reasonable and prudent manner.

In an agrarian society, both the plaintiffs and defendants of ground water contamination cases were farmers. This is not surprising. A farmer's neighbor was likely to be a farmer. The fact that one farmer would sue another for polluting his well suggests that some farming activities were beyond the scope of ordinary management practices.

Common law dealing with ground water pollution has changed little since the 19th century; the nature of ground water contamination has not been so static. One does not expect farmers to practice agriculture the way they did one hundred years ago. Why should one expect they be governed by the same rules? The courts have not been capable of handling the increase in contamination problems.

INEFFICACY OF FEDERAL REGULATIONS

The failure of the courts and the market to deal with water pollution lead to the passage of the Federal Water Pollution Control Act Amendments of 1972 (Public Law 92-500). These amendments failed to mention ground water explicitly[4] and were vague in their authority to regulate agricultural pollution (Montgomery, 1976). The Culver Amendments of section 208 of the Federal Water Pollution Control Act directs the Secretary of Agriculture to establish and administer Best Management Practices (BMP) for agricultural and silvicultural lands. BMPs are largely concerned with protecting surface water; some have been developed to protect ground water. BMPs as they are currently implemented, however, are not a consistent and reliable means of protecting ground water by themselves (Hines, 1974).   Much depends on the local implementation of the plans developed under 208 (Holmes, 1979).

Federal regulation of pesticides also was adopted in response to a different set of circumstances than ground water contamination. The intention of original pesticide regulation was to protect farmers and, later, consumers, but not the environment. This focus changed in 1972 with the passage of the Federal Environmental Pesticide Control Act (FEPCA), which superseded much

---

[1]Prosser, *Restatement (2d) Torts*, 849(e) (1970); U.S. v. Alexander, 148 U.S. 186, 191-192 (1892).

[2]Prosser, *Restatement (2d) Torts*, section 821D (1970).

[3]Phillips v. Sun Oil Co., 307 NYH 328, 331, 121 N.E. 2d 259, 251 (1954); Collins v. Chartiers Valley Gas Co., 131 Pa. 143, 159, 18 A. 1012, 1013 (1890).

[4]Exxon v. Train, 554 F.2d 1310 (5th Cir. 1975).

of the older Federal Insecticide, Fungicide, and Rodenticide Act. FEPCA changed the procedure for registering pesticides; that is, making them available for use. The registration procedure is designed to prevent the risk of pesticides to human health and the environment from exceeding the benefits (National Research Council, 1980). While registration procedures place the burden of proof on the pesticide manufacturer to show that benefits outweigh risks, pesticides have been registered for uses which have caused contamination problems beyond those anticipated.

The removal of a pesticide's registration is an important tool to arrest pesticide pollution. The environment is no doubt cleaner as a result of FEPCA. I do not, however, share the opinion that it has solved all environmental problems of pesticides. There are holes in the law, and ground water contamination is evidence of the need for further fine-tuning. As a proactive policy, FEPCA has proven less than perfect; as a reactive policy, it has been slow and cumbersome. The recent amendments to FEPCA sound promising, but it will take some time to see if they are implemented in an effective way.

Also pertinent to protecting ground water from chemical contamination is the Safe Drinking Water Act (SDWA). The SDWA is directed at protecting the nation's drinking water. With the exception of injection wells and oil and gas wells, SDWA does not have the authority to prevent contamination of drinking water. The SDWA gives the U.S. Environmental Protection Agency (EPA) the authority to set standards for acceptable levels of contaminants in drinking water. The burden of meeting these standards is placed on the individual supplier.

The burden of the SDWA is unevenly spread between urban and rural water supply systems, as is the coverage of protection. Agricultural pollution of ground water is largely a rural problem. Small, rural public systems serving 25 or more people face higher per unit costs to comply with the SDWA than large scale systems in densely populated areas. A rural system has more difficulty in raising the capital required to purchase the hardware for treating water.

Congress intended to provide an incentive for small systems to combine into larger systems that would spread the cost of treatment over more consumers. The irony may be that when rural systems are too widely distributed, geographically, the diseconomies of investing in and maintaining a larger system may outweigh the economies of combining system treatment costs for many rural systems. Many rural water supplies do not fall under the SDWA's coverage. Extending SDWA regulations to private wells has met with resistance, both from well owners who worry about government intrusion into their homes, and those who recognize the high administrative costs of broadening the coverage.

A word should be put in for the Comprehensive Environmental Response, Compensation and Liability Act, also known as CERCLA or Superfund. Originally intended to clean up massive environmental problems caused by hazardous waste disposal, CERCLA may be increasingly used to correct pesticide contamination problems. The addition of six wells drawing from ground water contaminated by normal use of pesticides to EPAs priority list for cleanup action in October, 1984, has given CERCLA a role in pesticide contamination problems (Conner, 1985). This action has been controversial, prompting some to challenge EPA's authority to do so under the act. Even if

EPA has the authority to use CERCLA for pesticide contamination cleanup, one cannot help but wonder if EPA would be willing to do so. The expense of cleaning up hazardous waste already exceeds the amount earmarked by Superfund. The additional expense of responding to pesticide contamination incidents may be viewed as a drain on resources intended for other purposes.

The patchwork of federal statutes and regulations which tangentially deal with agricultural pollution of ground water provides an impressive amount of material to develop a ground water protection strategy. This array of institutions fails to coordinate different ground water protection needs. The Office of Technology Assessment has proposed a comprehensive Federal Groundwater Act to remedy these shortcomings. To be politically viable, the act would have to reconcile the needs of the arid parts of the country with the humid, populous ones. Until such an act is passed, the responsibility for developing and implementing ground water protection programs will reside at the state and local level.

THE INCONSISTENCY OF STATE AND LOCAL PROGRAMS

Given the local nature of many ground water problems, it makes some sense to have a flexible, decentralized program for managing ground water problems (Allee and Powell, 1985). All state governments have statutes which correspond to the FWPCA, FEPCA, and SDWA. Some of the institutional structures which perform the functions mandated by federal law predate the passage of the statutes. An example of these are soil and water conservation districts which are responsible for implementing the planning requirements of section 208 of the FWPCA. Most soil and water conservation districts concentrate on surface water and soil erosion problems and have had little to do with ground water. This is changing in some places, but the point is that programs do not always emerge where problems exist. Few soil and water conservation districts have the resources necessary to implement programs capable of dealing with serious and extensive contamination problems (Tripp and Jaffe, 1979).

States vary widely in the power delegated to the districts. Some have strong regulatory powers over land use; others are restricted to rely on voluntary compliance and financial incentives.

BMP strategies for protecting water can reduce the quantity of all pollutants. Some BMP practices may, on the other hand, involve a tradeoff between different types of pollutants and the sinks where they end up. With the proper rotation, fertilizer, pesticide, and erosion runoff can all be reduced.

The authority of soil conservation districts vary from state to state. Most midwestern states grant the districts authority to prohibit certain uses of land next to water courses (Massey, 1983). Such delegation is the exception in the Northeast. Pesticide use and crop rotations are usually not regulated. Land disturbing uses are.

Local ordinances have a long history in protecting drinking water supply sources (Hennigan, 1981). These had their origins when the primary health threat via drinking water was bacteriological, rather than chemical. chemical technology changed the nature of water contamination in two ways. First, through chlorination, it virtually eliminated pathogens from public water supplies. Second, evolving and widespread use of chemicals in industry and agriculture have caused a wide range of chemical pollutants to enter public water supplies.

With the resurgence of drinking water contamination as a public health threat, water supply source protection is again being considered as a strategy for the management of drinking water quality. To be effective, however, regulations must take into account modern technology and an increased understanding of hydrogeology.

State aquifer classification systems have also been used to protect and manage aquifers. Most classification systems separate aquifers which need to be protected from those which do not, usually on the basis of their current condition and use. New York State has one of the more extensive classification systems. The agency charged with the protection of ground water, the state Department of Environmental Conservation, is severely restricted in its regulation of agricultural practices.

PROPOSED POLICIES

Alternatives to the current policy described above are analyzed herein. These policies are no regulation (laissez faire), taxes on pesticides, subsidies for low-input crops, control of agricultural practices, and pest control districts. Two alternatives considered, but not analyzed, were a change in the structure of liability and exclusion of agriculture from critical recharge areas. To analyze the effect of different policy on farm income and environmental quality, it is necessary to understand the effect of policy on behavior which controls agriculture and the activities which pollute ground water.

Taxes and Subsidies

Economists have long favored taxes and subsidies as an efficient means for dealing with externalities. A tax on externalities would make the private cost of those inputs more closely reflect the social cost which they inflict (Baumol and Oates, 1975). The government may choose to recognize property rights of individuals and corporations, either polluters or the victims of pollution, over those public goods which are polluted (Dales, 1968). The government could conceivably assert rights over those same public goods by charging those same people for the use of those resources.

The implementation of a tax on agricultural inputs has advantages over other measures to remedy their pollution. Because many producers are involved, the cost of enforcing practices would be high compared with the cost of setting and collecting a tax. A tax provides an incentive for the farmer to reduce the amount of the input. If the farmer reduces chemical inputs, the amount of pesticide or fertilizer which reaches the saturated zone of the soil is reduced. Another way to use a tax to reduce agricultural pollution is to tax crops requiring polluting inputs.

On the other hand, crops which do not use polluting inputs can be subsidized. The subsidization of low input crops can be likened to a purchase of chemical rights. The specification of the purchase of chemical rights program could also be called a purchase of crop rights program with as much accuracy.

Rather than pay farmers to grow crops without chemicals, or 'organically' if you will, farmers would receive payments for planting low-input conservation crops. Like taxation, this provides a benefit to ground water users in that it decreases pollution. Unlike taxation, it does not place an extra burden on farm income; it can, in fact, enhance farm income.

The drawbacks of taxes and subsidies are that they require a great deal of information to be optimally set. One could argue that the information needed to set optimal standards can never be achieved, because this implies an optimal rate of contamination. The effectiveness of the tax or subsidy depends on the level they are set at. Once they are set, taxes and subsidies are difficult to adjust.

In the analysis below, subsidies were set to compensate farmers for the loss of income from not being able to grow certain high input crops. Taxes were set at the level required to make the use of banned pesticides unprofitable.

Modification of Agricultural Practice

Cultural practices can be modified to limit impact on ground water quality. Integrated Pest Management (IPM) combines biological, mechanical and agronomic methods to reduce the amount of chemicals needed to control pest populations. IPM is not to be confused with organic farming, or the complete abandonment of chemical pest control. Rather than being used in a programmed, periodic manner, chemical pesticides are used only when other methods fail or are inappropriate for the pest situation. Crucial to the success of an IPM strategy is a scouting system which gives growers information about the status of pests attacking a given crop. Once a pest problem is identified, the farmer may spray or may adopt another integrated pest control strategy.

The need for a coordinated regional IPM program is created by the external effects caused by individual management decisions. Biological and cultural control frequently are public goods and are, therefore, not provided at efficient levels by the market. For example, crop rotation as a pest control strategy also requires regional cooperation. One farm's attempt to control a pest can be undermined by a neighboring farm's cultivation of host plants for that pest. At a much broader level, the market fails to provide IPM with the research necessary to develop alternatives to chemical methods. While a private market exists for the development of chemical control products and for certain biological agents, the development of pest resistant varieties and many biological techniques cannot be completely captured by the market. Investment in basic research would very likely be suboptimal. Intervention in the market is justified where the public benefits of such research go beyond the private benefits. This has long been one justification for the land grant college system.

This program would be susceptible to free riders if the program was voluntary or the sanctions for breaking rotations were not strong enough. If the return on investment to planting potatoes is greater than the return on the other crops, the farmer will have a strong market incentive to not follow the rotation. It is for this reason that crops which require large amounts of inputs are grown in preference over less intensive crops. The more intensive crops have a higher net return. Anytime a farmer is growing alfalfa on a field where he could be growing potatoes, he is suffering an immediate loss of income. If the crops permitted fail to produce enough income to provide a 'reasonable rate of return', then the farmer could challenge the district on the grounds that it constituted a taking without compensation.

A change in nonpoint pollution regulation to a nondegradation policy, or even to a performance standard would almost certainly require a change in the way that agricultural chemicals are used in sensitive areas. Management practices for controlling chemical application can be implemented in such a way as to minimize water pollution or meet certain performance standards. These practices must be tailored to local conditions. Because of this and the administrative difficulty of monitoring farm practices, a high degree of voluntary compliance would be necessary.

Chemical Control Districts

Voluntary compliance threatens a free-rider problem. This problem could be alleviated by a service which would apply fertilizer and pesticides in a way which reduces the threat to ground water. Even if the price of the service was low, farmers could increase yields and probably incomes by cheating.

A case could be made for the provision of subsidies for such a service. Because the service is benefitting the consumers of water, without them paying for that benefit, the provision of a tax on water would offset the opportunity loss which farmers face by limiting production. The optimal tax would equate the marginal value of the pollution abated with the marginal cost of abating the pollution.

If local control over pesticide application is to succeed, it must have approval from the state. A town ordinance restricted the use and transportation of pesticides, required the registration of applicators and banned three types of insecticides. This ordinance was struck down because state and federal laws have preempted the field.[5]

States do have the power to delegate monitoring and enforcement of pesticide laws to the counties. California has created the authority for agricultural areas to form pest abatement districts. Oregon takes unprecedented account of the use of pesticides and their use on the ecosystem. The objective of the legislation in Oregon is to protect crops, insects, wildlife, and forests.

Support for pest control districts from farmers could be increased by subsidies. One form of subsidy is a purchase of chemical rights. Another way to provide an incentive for cooperation is for government to absorb all or most of the costs of administering and operating the districts.

POLICY ANALYSIS

The proposed policies were analyzed for the effect on farm income and ground water contamination. The analysis methodology was described in Baker

---

[5]Long Island Pest Control Association v Huntington, 341 N.Y.S. 2d 93, (1973).

(1985). The area used in the economic and biological model was the Long Island region of New York State, an area subjected to pesticide contamination of ground water.   The model assumes that farmers are competitive, profit-maximizing profit takers and that their production functions are well defined and linear. The results of the model are generalized to reflect their relative rank over a broad range of conditions. The author would like to emphasize that measures and ranks are qualitative, not quantitative.

Table 23.1 presents the relative gross farm income which would result from the different policies, ceteris paribus.

Table 23.1: Alternative Policies Ranked by Gross Farm Income

1.   Subsidies for Low Input Crops

2.   Laissez Faire

3.   Pesticide Tax

4.   Current Policy

5.   Pest Control District

6.   Cultural Control

The most profitable policy for farmers is to subsidize the growing of crops without pesticides. If farmers are given the choice of either growing with pesticides and not receiving payments, or receiving payments for growing low-input crops, the income of profit maximizing farmers cannot be any lower than under the alternatives. The laissez faire policy is next, followed by the tax on pesticides. These two results are also consistent with the profit maximizing activity assumed in the model. The current policy outlined above came in at fourth place. Policies further constraining farm activity and choice came out the worst in terms of gross farm income.

Two measures of ground water contamination were used to rank the policies for environmental impact.   One is the leach index developed by Laskowski et al., (1982). The other takes the results of the leach index and divides through by the toxicity of the array of pesticides included in the model, measured by their LD_{50}. The results of these two indices are shown in Tables 23.2 and 23.3.

One reason for presenting both of these tables is to highlight the inconsistency of attempts to quantify environmental quality. Tables 23.2 and 23.3 are not identical, but agree on both the best and worst policies for environmental quality. These two policies were distinct in environmental quality to an extreme degree. The laissez faire policy was, by far, the worst of the alternatives for protecting the environment. If farmers are not prohibited from using pesticides which have deleterious effects on the environment, they will use them. Environmental quality will suffer accordingly.

Table 23.2: Alternative Policies Ranked by Leaching Potential

1. Cultural Control

2. Subsidize Low Input Crops

3. Pest Control Districts

4. Current Policy

5. Pesticide Tax

6. Laissez Faire

Table 23.3:   Alternative Policies Ranked by Hazard from Acute Toxicity

1. Cultural Control

2. Pest Control Districts

3. Subsidize Low Input Crops

4. Pesticide Tax

5. Current Policy

6. Laissez Faire

At the other extreme, imposition of strict regulations on agricultural practices with environmental quality as an objective will do the most to protect ground water from pesticide contamination. Subsidization of low input crops again comes out ahead of the current policy, as does pest control districts. Beyond this, the current policy and a tax on pesticides are virtually identical in both environmental indices.

CONCLUSIONS

Now for the final task laid out at the beginning of the chapter; that is, evaluating the tradeoffs between the different policies. There is no single magic formula that policymakers can punch into a computer to evaluate the 'ideal' policy. With that in mind, there is no perfect policy which will resolve the protection of ground water with loss of farm income. This is as true of current policy as it is for any of the other policies presented.

The price of food does not reflect the social costs of agriculture. The price of clean, safe ground water does not reflect the cost of keeping it that way. If ground water supplies are to be maintained, someone will have to pay. consumers can pay by higher water rates, higher taxes, or higher food prices. Farmers could pay by having restrictions placed on their practices or taxes levied on inputs. The economic solutions of taxes and subsidies offer more efficient ways to assure clean ground water, but at varying levels of 'clean'. Because consumers and the public at large benefit from clean ground water, the author is favorably disposed toward subsidies for reduced-input set asides for critical recharge areas. This is viewed as a short run solution.

For the long run, technology will have to adapt to an agriculture which does not seriously threaten or harm the environment. Agricultural research needs to focus on reducing inputs, not on increasing output. This will not only reduce the out-of-pocket costs of the farmer, it will also help to correct the overproduction problems facing agriculture today. Reduced input farming will enhance profitability as well as improve environmental quality. This solution to the farm problem is not new, but lacks the institutional support required to make it a viable policy option.

Farmers can change their practices to reduce the impact to ground water, but will not do so on their own accord. The police power can be used by government to impose environmentally sound agricultural practices, but this is undesirable for social and economic reasons. If pesticide contamination of ground water is to be eliminated, farmers will need assistance in the form of subsidy payments for low-impact crops in the short-run, subsidized technical assistance over the medium-run, and, over the long-run, research for reduced input production techniques. This should be paid for by the beneficiaries of clean ground water, which is everyone, to be equitable and efficient.

SELECTED REFERENCES

Allee, D.J., and Powell, J., "Local Government and Groundwater Quality Management," Staff Paper 85-19, 1985, Department of Agricultural Economics, Cornell University, Ithaca, New York.

Baker, P., "The Protection of Groundwater from Agricultural Pollution: Institutions and Incentives," Unpublished Ph.D. Dissertation, Cornell University, 1985, Ithaca, New York.

Baumol, W.J., and Oates, W.E., The Theory of Environmental Policy, Prentice-Hall, Englewood Cliffs, New Jersey, 1975, pp 152-190.

Conner, J.L., II, "Using CERCLA to Clean Up Groundwater Contamianted Through Normal Use of Pesticides," Environmental Law Reporter News & Analysis, Vol. 15, 1985, pp 10100-10108.

Dales, J.H., Pollution, Property and Prices, University of Toronto Press, Toronto, Ontario, Canada, 1968, pp 77-100.

Davis, P.N., "Groundwater Pollution: Case Law Theories for Relief," Monthly Law Review, Vol. 39, 1976, pp 17-159.

Francis, J.D., Brower, B.L., and Graham, W.F., "National Statistical Assessment of Rural Water Conditions," 1982, U.S. Environmental Protection Agency, Office of Drinking Water, Washington, D.C.

Hennigan, R.D., "Water Supply Source Protection Rules and Regulations Project: Final Report," 1981, New York State College of Environmental Sciences and Forestry, Syracuse, New York.

Hines, N.W., "Farmers, Feedlots and Federalism: The Impact of the 1972 Federal Water Pollution and Control Act Amendments on Agriculture," South Dakota Law Review, Vol. 19, 1974, pp 540-566.

Holmes, B.H., "Institutional Bases for Control of Nonpoint Source Pollution," 1979, U.S. Environmental Protection Agency, Office of Water and Waste Management, Washington, D.C.

Laskowski, D.A. et al., "Terrestial Environment," in R.E. Conway (ed.), Environmental Risk Analysis for Chemicals, Van Nostrand Reinhold, New York, 1982, pp 198-240.

Massey, D.T., "Land Use Regulatory Power of Conservation Districts in the Midwestern States for Controlling Nonpoint Source Pollutants," Drake Law Review, Vol. 33, 1983, pp 35-111.

Montgomery, J.E., "Control of Agricultural Water Pollution: A Continuing Regulatory Dilemma," University of Illinois Law Forum, 1976, pp 533-559.

National Research Council, Regulating Pesticides, National Academy of Science, Washington, D.C., 1980, pp 20-28.

Pye, I., Patrick, R., and Quarles, J., Groundwater Contamination in the United States, University of Pennsylvania Press, 1983, Philadelphia, Pennsylvania.

Tripp, J.T.B. and Jaffe, A., "Preventing Groundwater Pollution: Towards a Coordinated Strategy to Protect Critical Recharge Zones," Harvard Environmental Law Review, Vol. 3, 1979, pp 364-410.

U.S. Congress, Office of Technology Assessment, "Protecting the Nation's Groundwater From Contamination," 1984, OTA Reports, Washington, D.C.

CHAPTER 24

POULTRY MANURE MANAGEMENT AND GROUND WATER QUALITY: THE DELAWARE SOLUTION

by

J. Ross Harris, Jr.

Ground water quality studies have been conducted in selected areas of Delaware during the last decade. As early as 1972, areas with elevated levels of nitrate nitrogen above the interim drinking water standard of 10 mg/l were identified (Miller, 1972). Ten years later, the findings of a more extensive survey showed that 24 percent of all the wells tested had nitrate nitrogen levels above 10 mg/l nitrogen; and in the sandy soils of the coastal areas, the percentage increased to 32 percent of the wells tested (Ritter and Chirnside, 1982). The most significant finding was that elevated nitrate nitrogen levels in the ground water above the federal interim drinking water standard had resulted from overuse of commercial fertilizers and improper storage and use of poultry manure and septic tank wastewater. The highest nitrate concentrations occurred on excessively drained soils under intensive agricultural production.

Most of Delaware lies within the coastal plain. The soils are alluvial deposits of sands and silts, typically moderately to excessively drained with a minor amount of clay. The unconfined water table aquifer is the primary source of water for all domestic and agricultural uses. The depth to ground water below the soil surface is from 0.6-3.6 m (2-12 feet) in most areas. Many of the domestic water supply wells are 12-18 m (40-60 feet) deep. Recharge to the aquifer is approximately 30.5 cm/yr. The potential for water soluble pollutants to leach into the ground water is high.

PROGRAM DEVELOPMENT

Recommendations from the state, federal, and university officials encouraged the development of a strong voluntary program to control ground water pollution.

The Delaware Department of Natural Resources, a designated 208 agency by the U.S. Environmental Protection Agency (EPA) received funding through Section 208 of the Clean Water Act Amendments to implement water quality management plans. The creation of a position to address the problem of agricultural nonpoint source pollution came about through an interagency agreement between the U.S. Department of Agriculture (USDA) and EPA. The Delaware Cooperative Extension acting as the agent of the USDA and the Delaware Department of Natural Resources and Environmental Control (DNREC) by agreement with EPA, entered into a Memorandum of Understanding regarding participation in the state water quality management plan. Much work had been done on cause/effect relationships of agriculturally-related activities and water quality. However, for the program to be successful, these results had to be understandable and acceptable to the professional and farm communities, and workable solutions needed implementing. A void between research and implementation existed, and the Delaware Cooperative Extension was asked to fill the void.

The development of the Environmental Quality Specialist position in the Delaware Extension System occurred in late 1980. The goals were to plan, develop, and conduct educational programs to assist farmers and landowners with planning and implementing practices related to the control of rural nonpoint source pollution.

The program was not well received the first year. It was extremely difficult to demonstrate that this position was educational. Many farmers felt that the position was a regulatory position. The Extension and Soil Conservation Service staff were even apprehensive. In their viewpoint there were insufficient data to prove conclusively that agriculture polluted ground water or surface water (even in light of two published reports). The reaction by the agricultural community to the research findings was an attempt to discredit the research. The Environmental Quality Specialist conducted programs with key agricultural officials to educate farmers on the cause and solutions to the problem. Simultaneously, programs were also conducted with DNREC officials on methods of agricultural production and limitations placed upon the agricultural community. A list was developed of viable Best Management Practices to address the problem. The Environmental Quality Specialist acted as a liaison between the DNREC and the agricultural community. This liaison development has had significant positive impact in the long-term success of the programs. This retooling process cost the program approximately nine months of time. Gradually the energies of the concerned parties focused on solving the problem. From that point forward, broad-based support from state agencies, the agricultural community, and environmental groups laid the groundwork for program acceptance.

During 1982, a plan of action to address the major source of nitrate contamination to ground water was finalized. The program became known as the MANURE Program.

> **M**-anaging
> **A**-gricultural
> **N**-utrients
> **U**-tilizing
> **R**-esources
> **E**-ffectively

The program was targeted to the broiler industry of Delaware. The broiler industry is vital to the economy of Delaware. Annual production in 1985 exceeded 204 million broilers which, in turn, produce over 228,000 mt of manure. The broiler industry is located entirely within the coastal plain of Delaware. Through overuse of manure and fertilizer, as well as improper manure storage techniques, poultry farmers were contributing significantly to nitrate contamination of ground water. The MANURE Project emphasized that with proper management, use, and storage, broiler manure could be eliminated as a source of pollution.

The first step was a survey of existing poultry operations. Results of the survey are presented in Table 24.1. The conclusion from the 1981 survey showed that over 48 percent of the growers applied manure at a rate equal to or greater than 13,440 kg/ha (6 tons/ac) (Chaloupka, 1981). When manure could not be directly applied to the land, it was stockpiled for an average of 3 to 4 months. Only 5 percent of the growers had the manure analyzed for nutrient content, but 97 percent of the growers felt that the manure was a valuable source of plant nutrients. This seemed to be a contradiction because farmers felt manure was a valuable source of plant nutrients; yet few reduced commercial fertilizer application on fields receiving manure. The rationale was a remnant from the time of cheap fertilizer and high crop prices. The Extension Service was partially to blame for overfertilization. The need for further information on the nutrient content of manure, application rates, and the extent of the nutrient losses were frequently cited from the survey (Chaloupka, 1981). Concurrently, a review of broiler management techniques revealed that management of broiler houses had changed dramatically over the past twenty years. During the 1960s and 1970s all the litter/manure was removed from the broiler house every eight weeks. During the late 1970s to early 1980s, the litter/manure was not completely removed after each flock. The procedure called caking or crusting out became the new management tool. This procedure only removes the upper layer of litter/manure, the "caked" material. The cake was different in nutrient levels and total volume produced from that which had been removed every eight weeks. The material remaining in the house after each crusting was finally removed after a year or two. This material had a nutrient value several times higher than the litter that was cleaned out after each flock. Literature published during the 1960s and 1970s placed nitrogen values of broiler/litter manure at approximately one to two percent. Through extensive sampling in 1981-82, it was determined that broiler litter/manure contained nitrogen values of greater than 4 percent (Harris, 1982a).

Before the beginning of the MANURE Program, broiler manure was looked upon as a material that had to be disposed. It was thought to have little or no fertilizer value. The data proved this wrong. Broiler manure along with additional commercial fertilizer was applied to meet crop needs. The average grower was applying 540 kg/ha (480 lbs/ac) of nitrogen for dryland corn resulting in over fertilization by 393 kg/ha (350 lbs/ac). Therefore, it was no surprise that nitrate levels in ground water were above the interim drinking water standard. Increasing numbers of farmers were reporting crop failures accentuated by droughty conditions. The corn crop was burning up from extremely high nitrogen applications. Some farmers were experiencing high levels of nitrogen in drought stressed corn silage resulting in the death of some dairy cows.

PROGRAM IMPLEMENTATION

During the following year, the program began to address the needs of the poultry growers and grain farmers. A manure testing program was developed along with fact sheets on manure spreader calibration techniques. Farmers were encouraged to enter the program through communications made at local meetings and through mass media. The manure analysis program ran out of money within three months and was refunded twice during the first year.

The purpose of the program was two-fold. The first purpose was to develop an extension database on the nutrient value of manure for University of Delaware soil test recommendations and fact sheets on manure management.

Table 24.1: Results of 1981 Poultry Manure Management Questionnaire
(Chaloupka, 1981)

		Yes	No
1.	Do you calibrate manure spreader?	0%	100%
2.	Do you analyze the manure for plant nutrient?	5%	95%
3.	Application rate to meet crop needs 4500-9000 kg/ha (2-4 tons/ac)	52%	48%
4.	What information would be most helpful in planning a manure management program?		
	Nutrient content	24%	
	Application rates	17%	
	Nutrient losses	17%	
	Economic value of manure	24%	

The second purpose was to demonstrate to the farmer the nutrient value of the manure. The data not only showed the value of the manure as a fertilizer but also showed the dramatic change that had occurred in the nutrient content of the manure due to management changes over the last decade.

Calibrating manure spreaders was the second step in the MANURE Program. As of 1981, no one was calibrating manure spreaders in Delaware. There were not any fact sheets available on calibrating manure spreaders. Two fact sheets were developed to meet the need (Harris, 1982b and c). Calibration workshops were conducted around the state with limited success. Distribution of the fact sheets to all poultry growers was accomplished through the help of the Delmarva Poultry Industry, Inc. The fact sheets have received wide circulation throughout the area. Although they do not cover every detail of manure spreader calibration, they are simplistic and were a first approximation that have proven highly effective. Results of the 1986 poultry growers survey (Table 24.2) showed that 39 percent of those responding now calibrate manure spreaders (Harris et al., 1986). A 39 percent increase in farmers calibrating spreaders is encouraging. Manure testing did not have as high an acceptance rate. Five percent of those surveyed in 1981 tested the nutrient content of manure as compared to 12 percent in 1986. The poor adaption rate may be a result of data being generated by the University of Delaware on the nutrient value of broiler manure.

The cost of manure analysis (approximately $30) was considered high by most farmers. They were reluctant to spend that amount of money when a handout with the approximate nutrient value was easily accessible. Soil test recommendations from the University of Delaware recommended fertilizer application rates based upon type and volume of manure to be applied.

Demonstrations on numerous farms compared the use of broiler manure and commercial fertilizer in both no-till and conventional-till corn production. The results demonstrated that broiler manure could be used as a substitute for commercial fertilizer without a net loss of profit (Harris, 1982d, 1983a). The demonstrations have been duplicated on numerous farms throughout Delaware and Maryland over the last four years.

Table 24.2: Results of 1986 Poultry Manure Management Questionnaire (Harris et al., 1986)

		Yes	No
1.	Do you calibrate manure spreader?	39%	61%
2.	Do you analyze the manure for plant nutrient?	12%	88%
3.	Application rate to meet crop needs 4500-9000 kg/ha (2-4 ton/ac)	90%	10%
4.	Do you cover stockpiled manure with plastic?	15%	85%

During 1982-83, an animal waste inventory of the entire state was conducted to locate and quantify all animal waste producing farms. The farms were located on U.S. Geological Survey quadrangle maps and color coded for animal type. Determination of the size of the operation and the amount of land available for manure disposal was qualified. The most dramatic finding was that over 30 percent of the poultry operations did not have land available for proper utilization of the manure they produced, therefore, requiring off-site disposal or "dumping" of the manure on a small amount of acreage (Harris, 1982d, 1983a and b). Somewhat successful attempts were made to assist farmers in locating larger farms that could utilize the manure. Currently localized brokerage and storage sites for manure are being considered.

The next step addressed the problem of manure storage. Because broiler houses are crusted out every seven weeks with a total cleanout every one to two years, when crops are in the field, manure must be stockpiled. The manure is windrowed and remains exposed to the elements for up to 10 months. The practice wastes valuable nutrients, contributes to ground water contamination, and renders the site useless for cropping for at least two years. Through development of inexpensive and simple manure storage practices, farmers were able to eliminate the problem with manure piles.

The poultry manure storage method involved covering the manure with 6 mil polyethylene tarp, anchored with old tires, rope, or dirt. The plastic keeps the manure dry, preventing runoff and leaching of nitrogen and other nutrients (Sims, 1983; and Harris, no date).

A cost-share program was developed to bring these findings to the agricultural community. The cost-share program offered through the Agricultural Stabilization and Conservation Service (ASCS) was a poultry manure management program consisting of three components. Table 24.3 provides tasks and agency roles. The program cost shared at 75 percent the cost of the plastic and nutrient analysis. The farmers were obligated to calibrate their manure spreaders and develop and implement nutrient management programs. The farmer is obligated to the program for 5 years but only receives cost sharing in 3 of the 5 years. Technical and educational assistance was handled by the Delaware Cooperative Extension Service with the financial assistance from ASCS. The Soil Conservation Service was not allowed to work on the program because of the lack of an engineered technical standard. All technical signoff for USDA forms was handled by the Cooperative Extension Service and ASCS. The program has been in place for two years. Based upon the 1986 survey of broiler growers, 15 percent of the respondents now cover their manure with plastic. It is important to note that the cost-share program was developed as an educational tool to expand the awareness and acceptance of the storage practice and nutrient management program.

PROGRAM EVALUATION

The evaluation of the program is documented by the changes in farm management practices over the 4-year period between surveys. Table 24.4 compares the survey responses from 1981 and 1986. There has been an acceptable level of adopting manure spreader calibration and manure storage. The acceptance of manure testing is low. This may be due to the availability of published literature from the University of Delaware.

Overall, the Department of Natural Resources and Environmental Control, as well as other agencies involved, are pleased with the initial results. Reduced inputs of nitrogen to the ground water have been documented by significant reductions of nitrogen application to agricultural fields. A 38 percent change in application rates of broiler manure to levels that do not exceed crop needs is a dramatic improvement in four years.

Based upon the initial success of the program, an Environmental Quality Assistant Specialist's position has been created in 1986. The role of this person will be to expand the existing program which currently requires one third of the time of the Environmental Quality Specialist. The funding source for the Assistant Specialist is Section 205j of the Clean Water Act through the Delaware Department of Natural Resources and Environmental Control. Beginning in 1985, the Environmental Quality Specialist was funded by the University of Delaware.

Best Management Practice evaluation projects on ground water monitoring under fields receiving broiler manure and on stockpiled manure sites are being conducted by the Extension Service through funding support from Section 205j. Through the support of various state agencies, agricultural community and environmental groups, the program on poultry manure management has been successful. Without the financial support for the program by the Department of Natural Resources and Environmental Control, the program would not exist. The program is proof that funding for educational programs in reducing nonpoint pollution problems is well spent.

WHAT IS AHEAD?

Facilitating change in management techniques is a slow and laborious process. The rate of adoption of various practices is a function of many factors. The initial group of participants are the innovators and will adopt new practices quickly. The great percentage of the impacted group will be a very

Table 24.3:   Agency Roles for ACP Poultry Manure Management Practice

TASK	AGENCY & ROLES
1. Awareness Program	ASCS – monthly newsletter CES – announcement at public meetings and news packet
2. Sign up	ASCS County Office
3. Needs and Site Selection	CES – determination of need and site selection. Verification that proper placement and installation has been completed. Send verification to ASCS. Fact sheet on proper site selection and installation of plastic cover.
4. Manure Spreader Calibration	CES – fact sheets and program on manure spreader calibration. Verification of farmer participation in manure spreader calibration sent to CES.
5. Manure Analysis	CES – fact sheet on manure sampling. Copy of manure analysis sent to CES.
6. Soil Test	Farmer – copy sent to CES
7. Nutrient Management Program	CES – fertilizer program developed based upon soil test, manure test, and cropping practice.
8. Verification of Annual Completion of Practice	CES – sends verification to ASCS
9. Payment to Farmer	ASCS – upon receipt of CES notification of completion and farmers's receipts for plastic and manure analysis.

Note: Long-term monitoring and evaluation coordinated by the CES (Cooperative Extension Service)

364 PROTECTION AND MANAGEMENT

slow steady trickle of adoptors. As always, there will be those who will not change their management practices. The question that must be addressed is what length of time is considered reasonable for people to adopt new techniques--5, 10, 20 years? What percent of the impacted group must change their practices to consider the program successful? How do you deal with the reluctant farmer? These are the questions we are now facing in Delaware. A task force on regulations governing agricultural wastes was developed to try to address the difficult question of what next? The task force is comprised of various state agencies, soil and water conservation district members, Delaware Farm Bureau, State Grange, and private citizens. The greatest challenge is to come up with a solution to the question of a voluntary vs regulatory program. The solution will probably be a blend of the two.

Table 24.4:   Comparison of Farmers' Responses from 1981-1986

	1981	1986	% Change
Calibration of manure spreaders	0%[1]	39%	+39%
Manure analysis	5%	12%	+ 7%
Poultry manure storage	0%	15%	+15%
Manure application rate (2-4 tons/ac)	52%	90%	+38%

[1]Percent of farmers responding to the questions

CONCLUSIONS

1.   A voluntary educational program on manure management has met with initial success in the Delaware broiler industry.

2.   A volunteer program for manure management requires a one-on-one effort. The use of public media, although helpful, does not impact the targeted audience dramatically enough to effect change. Therefore, continued effort will be needed to accomplish the long-term goals of protecting ground water.

3.   The need for broad-based support from regulatory, educational, and agricultural groups is extremely valuable in guaranteeing success.

4.   The question of how long a program can be voluntary and yet meet environmental goals is difficult to answer; nevertheless, it must be resolved.

5.   The greatest challenge cited in 1986 was not manure management but dead poultry disposal. This problem may be the greatest challenge and threat to ground water from the broiler industry in the years ahead.

SELECTED REFERENCES

Chaloupka, G.W., "Poultry Manure Questionnaire," 1981, University of Delaware, Cooperative Extension Service, Dover, Delaware, Unpublished data.

Harris, J.R., "Nutrient Value of Poultry Manure," 1982a, University of Delaware, Cooperative Extension Service, Dover, Delaware.

Harris, J.R., "Calibrating Manure Spreaders: Procedure I for Solid and Semisolid Manure," Fact Sheet AW1, 1982b, University of Delaware, Cooperative Extension Service, Dover, Delaware.

Harris, J.R., "Calibrating Manure Spreaders: Procedure II for Liquid, Solid, or Semisolid Manure," Fact Sheet AW2, 1982c, University of Delaware, Cooperative Extension Service, Dover, Delaware.

Harris, J.R., "Poultry Manure as Fertilizer," Demonstration Reports, 1982d, 1983a, University of Delaware, Cooperative Extension Service, Dover, Delaware.

Harris, J.R., "Delaware Animal Waste Inventory," 1983b, University of Delaware, Cooperative Extension Service, Dover, Delaware, Unpublished Data.

Harris, J.R. et al., "Poultry Manure and Dead Bird Disposal Management," 1986, University of Delaware, Cooperative Extension Service, Dover, Delaware, in press.

Harris, J.R., "Poultry Manure Storage," Fact Sheet AW3, University of Delaware, Cooperative Extension Service, Dover, Delaware.

Miller, J.C., "Nitrate Contamination of the Water Table Aquifer in Delaware," Report of Investigation No. 20, 1972, Delaware Geological Survey.

Ritter, W.F., and Chirnside, A.E.M., "Groundwater Quality in Selected Areas of Kent and Sussex Counties, Delaware," 1982, Delaware Agricultural Experiment Station, University of Delaware, Dover, Delaware.

Sims, T., "The Effect of Stockpiling Practices on the Fertilizer Value of Poultry Manure," 1983, University of Delaware, Cooperative Extension Service, Dover, Delaware, Project Completion Report to Delmarva Poultry Industry.

CHAPTER 25

NITROGEN AND GROUND WATER PROTECTION

by

S.J. Smith,
J.W. Naney, and
W.A. Berg

The benefits of nitrogen fertilizers are well recognized. Because available soil nitrogen (N) supplies are generally inadequate, application of fertilizer N provides a way to achieve optimum crop production. At present, N represents the mineral fertilizer most applied to agricultural lands (Hargett and Berry, 1985). Fertilizer N applications, however, particularly when excessive for crop needs, may involve certain undersirable side effects. This can occur if the fertilizer N forms, either ammonium or nitrate, find their way into ground water supplies and exceed designated limits. Once this occurs, the N forms are considered to pose an environmental hazard (U.S. Environmental Protection Agency, 1973). The tendency for movement to ground water is much greater for the highly soluble nitrate anion than for the ammonium cation, which attaches to the soil particles. Concurrently, the ammonium is subject to nitrate conversion by the action of certain soil microbes.

In perspective, it is important to note that large amounts of soluble N from various natural sources can exist in the soil environment. In California, for instance, certain soil extracts have been observed to contain as much as 2000 mg/l nitrate-N, primarily due to geologic contributions (Strathouse et al., 1980). The greatest general release of nitrate to the soil environment probably occurred many years ago, when vast areas of the virgin prairies were first placed in cultivation.

For the past several years, part of our research has been to assess the impact of agricultural practices on ground water quality of the Southern Plains (Naney, Smith, and Berg, 1984; and Smith, Naney, and Berg, 1985). Presented here are nitrate-and ammonium-N results and recommendations for wells that are monitored periodically on watersheds in the Cross Timbers (CT), Rolling Red Plains (RRP), and Rolling Red Prairies (RP) major land resource areas (U.S. Soil Conservation Service, 1981).

STUDY SITES AND METHODS

The locations of the approximately 30 watersheds on which the study was conducted are indicated in Figure 25.1. The three major land resource areas involved, CT, RRP, and RP, represent approximately 27,000, 130,000, and 53,000 km^2, respectively, in the three-state area of Kansas, Oklahoma, and Texas (U.S. Soil Conservation Service, 1981). Principal land management, soil, and geologic features of the watersheds, as they relate to ground water quality studies, are given in Table 25.1. For each major land resource area, characteristic settings are represented where ground water quality may be

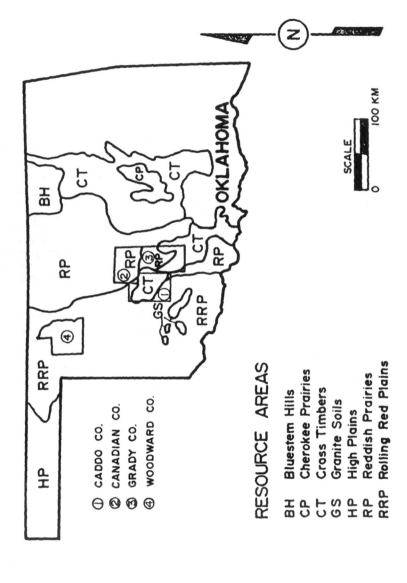

Figure 25.1:   Locations of Major Land Resource Areas in West Central Oklahoma   with Study Sites in Caddo, Canadian, Grady, and Woodward Counties, Oklahoma

Table 25.1: Principal Watershed Features Related to Ground Water Quality Studies for Soluble N Assessment

Site	County	LRA	Wells	Drilled Depth (m)	Major Land Use	Major Soils	Geologic Age	Stratigraphy	Lithology
Chickasha	Grady	RP & CT	11	10-40	Native grass, wheat	Ustolls Ustalfs	Quaternary	Terrace deposits and alluvium	Sand and gravel
El Reno	Canadian	RP	10	10-25	Native grass, wheat, grain sorghum	Ustolls	Quaternary	Terrace deposits	Sand and gravel
							Permian	El Reno group	Shale and sandstone
							Permian	Cloud Chief formation Whitehorse El Reno group	Gypsum, shale and sandstone
Ft. Cobb	Caddo	RRP & CT	2	15-30	Peanuts and grain sorghum	Ustalfs Ustolls	Quaternary	Alluvium	Sand and silt
							Permian	Whitehorse Group	Sandstone
Woodward	Woodward	RRP	22	3-9	Improved grasses, wheat and alfalfa	Ustalfs Ustolls	Quaternary	Terrace Deposits	Sand and gravel
							Permian	Whitehorse Group	Shale and sandstone

LRA - Land Resource Area
RP - Rolling Red Prairies
CT - Cross timbers
RRP - Rolling Red Plains

affected by changes in land use and management. The watersheds ranged in size from 1.6 to 6 ha, and included different crop, grass, tillage, fertilizer, and grazing practices. Fertilizer N applications were made in conjunction with recommended N soil test results. Water table depths generally ranged from 3 to 20 m, with the wells being sampled on a seasonal basis. Additional details about the watersheds and wells may be found in earlier publications (Naney, Smith, and Berg, 1984; and Smith et al., 1983).

After sampling, all well samples for chemical analysis were refrigerated at approximately $4°$ C. Chemical analyses for nitrate- and ammonium-N were made using standard methods described in the Federal Water Pollution Control manual (U.S. Department of Interior, 1971). Publications by the U.S. Environmental Protection Agency (1973, 1976) and the National Technical Advisory Committee (1968) were used as guides for water quality standards.

In the case of the soil samples, multiple borings (generally three of 2.5 cm diameter) were taken in 30 cm increments down to 180 cm. The samples were refrigerated at approximately $4°$ C until extracted with water (10 g soil/50 ml water) and analyzed, using techniques noted above.

RESULTS AND DISCUSSION

Soluble N, nitrate, and ammonium contents, considered to reflect any possible fertilizer contaminants in the wells, are given in Table 25.2. For reference purposes, nitrate-N contents within 10 and 100 mg/l are considered acceptable for human and livestock consumption, respectively. Ammonium-N contents below 0.5 mg/l are considered acceptable for humans, while contents above 2.5 mg/l may be harmful to fish. The results presented here indicate, for the most part, that N contents of the well were within acceptable limits. However, with both N forms there are some noteworthy exceptions.

In the case of nitrate, these exceptions occurred mainly on minimum-till wheat at El Reno and certain improved grass sites at Woodward. The highest concentration observed at El Reno was 17.7 mg/l nitrate-N and 18.8 mg/l nitrate-N at Woodward, so even here the levels are not extremely high. Nevertheless, they clearly point out the potential for contamination under minimum tillage and intensive grass production.

Due to the increased adoption of minimum tillage, particular attention should be given to its environmental implications regarding ground water. Generally, minimum tillage provides a wetter, cooler environment that may enhance nitrate leaching potentials. This is because more undisturbed large soil pores and burrows may exist for nitrate movement. In addition, there is less evaporation of water from the soil profiles and, consequently, less movement of nitrate upward toward the soil surface. Practices that may reduce the nitrate leaching potential under minimum tillage, and also conventional tillage, include appropriate banding, splitting, positioning, and timing of fertilizer N applications.

In the case of the improved grasses at Woodward, the monitored sites were on shallow water-table sandy soils and had received N fertilizer at annual rates of 67-134 kg N/ha for several consecutive years. Under the appropriate conditions, this type situation can be conducive to nitrate leaching (Kissel, Bidwell, and Kientz, 1982).

Table 25.2: Nitrate and Ammonium Concentrations in Ground Waters at Various Locations in Oklahoma as Affected by Land Use

Location	Period	Use	Wells	Number of Observations	NO₃-N(mg/l)		NH₄-N(mg/l)	
					Range	$\bar{x} \pm sd$	Range	$\bar{x} \pm sd$
Chickasha	1979-82	Native Grass	2	9	0.1 - 0.9	0.5 ±0.4	0.0 -1.99	0.34 ±0.38
El Reno	1983-86	Native Grass	3	21	0.0 -11.9	1.8 ±1.1	0.01-0.08	0.04 ±0.01
Woodward	1983-85	Native Grass	3	15	1.4 - 8.6	3.6 ±2.8	0.0 -0.65	0.09 ±0.16
Woodward	1981-85	Alfalfa	1	12	0.8 - 8.3	6.6 ±2.7	0.05-1.60	0.22 ±0.46
Woodward	1981-85	Bermuda Grass	4	46	0.1 -10.4	6.5 ±7.2	0.0 -0.44	0.12 ±0.15
Woodward	1980-85	Eastern Gama Grass	6	70	0.0 -12.9	2.4 ±3.2	0.0 -1.04	0.16 ±0.16
Woodward	1980-85	Love Grass	2	20	2.6 -13.5	8.3 ±2.8	0.0 -0.14	0.07 ±0.05
Woodward	1982-85	Old World Bluestem	1	10	11.3 -18.8	16.1 ±2.2	0.01-0.28	0.09 ±0.08
El Reno	1983-86	Grain Sorghum/Wheat	3	14	0.2 - 8.8	2.6 ±1.3	0.01-0.27	0.05 ±0.04
Ft. Cobb	1983-85	Grain Sorghum	1	5	0.7 - 2.5	1.7 ±0.4	0.0 -0.17	0.06 ±0.08
Ft. Cobb	1983-85	Peanut	1	5	0.0 - 3.5	1.8 ±1.5	0.0 -0.12	0.04 ±0.05
Chickasha	1979-82	Wheat	5	9	0.1 - 4.1	1.2 ±1.4	0.0 -2.83	0.50 ±0.78
El Reno	1979-86	Wheat	1	16	0.0 - 2.5	0.7 ±0.8	0.05-1.61	0.46 ±0.44
Woodward	1981-85	Wheat	2	25	2.2 - 7.0	3.5 ±2.3	0.01-0.58	0.10 ±0.11
El Reno	1983-86	Min-Till Wheat	3	21	1.5 -17.7	8.6 ±4.3	0.0 -0.09	0.03 ±0.03
Woodward	1983-85	Min-Till Wheat	3	15	0.0 - 6.9	1.0 ±1.4	0.0 -0.11	0.07 ±0.04
Chickasha	1979-82	Farmsteads	3	9	0.17-18.4	4.1 ±2.0	0.0 -3.30	0.5 ±0.78
Chickasha	1979-82	Oil Field	1	9	0.0 - 0.36	0.25±0.11	0.0 -0.11	0.05 ±0.05

As a precautionary aid to "pin-pointing" sites more susceptible to nitrate leaching, some states are now classifying soils according to their leaching potential. Such classification can be made through the use of available soil survey and analytical data. In Kansas, for instance, soils have been placed recently into different leaching classes on the basis of soil profile characteristics (Kissel, Bidwell, and Kientz, 1982). Placement is made by consideration of the finest textured and least permeable horizon in the soil profile. Using such an approach, the major agricultural soils of the state have been designated into leaching classes 1 through 4, with 1 being the most susceptible to leaching. Coarse, sandy soils tend to be in Class 1, because they have little capacity to hold water and are very permeable. Also, swelling clay soils (e.g., vertisols) may pose special cases during extreme dry periods, when they develop large cracks from the surface down through the profile.

Another factor that merits special attention in minimizing nitrate leaching potential is the N status of the soil. While this status has generally declined since soils were first brought under cultivation, all soils have a certain capacity to supply N to crops. The goal is to add only enough N fertilizer to satisfy crop needs above that which the soil does not provide. To this end, recommended N soil tests and guidelines should be utilized in planning fertilizer N programs. Moreover, available N contributions from preceding cover crops, legumes, and manure applications should be given appropriate credits. Recently, precision application fertilizer spreaders have been introduced to gear N rates to specific soil areas within a field.

In the case of ammonium N, the high contents (Table 25.2) were traced typically to improperly installed and maintained well casings that allowed surface runoff to flow directly into the wells. A proper well seal is essential to prevent direct surface flow contamination into the annulus between the borehole and well casing. Improved well protection techniques eliminated this problem and reduced the ammonium contents (as well as a high nitrate content in the case of one farmstead) to acceptable levels.

Because ground water seepage tends to follow the same general direction as that of surface water, to the extent possible, agricultural field, feedlot, farmyard, and septic tank drainage should be directed away from wells. In the case of feedlots, nitrate problems may occur when they are not in use. Then, the surface seal breaks down, conditions become more aerobic, and nitrate production from the accumulated ammonium is initiated. Also to be avoided around wells are the storage, mixing, and loading of N fertilizer.

PERSPECTIVE

The results obtained here are generally encouraging in that they indicate no major problems presently exist regarding nitrate and ammonium contents in ground waters analyzed from a range of agricultural watersheds and settings. However, these findings are not a justification for complacency, because ground water problems may take several years to become established. Therefore, as a precautionary measure, we plan to continue periodic surveillance of the watershed wells, in addition to occasional soil profile sampling. To this end, several of the watershed soils were sampled recently in 30-cm increments down to the 180-cm depth, to ascertain whether nitrate was accumulating below the root zone. As shown in Figure 25.2, there is no evidence to date of a nitrate buildup. Nevertheless, in order to obtain a more complete handle on the

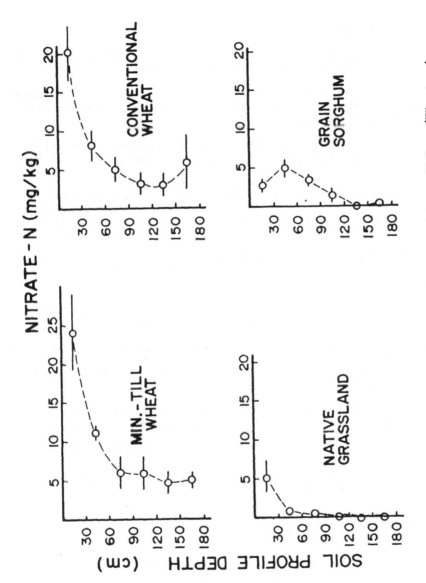

Figure 25.2: Distributions of Nitrate-N in Soil Profiles at El Reno (Wheat and Grassland) and Ft. Cobb (Grain Sorghum) in March, 1986. The Horizontal Bars Represent ±1 Standard Deviation of the Means

subject, deeper profile sampling is envisioned in the future. Should a buildup of deep-positioned nitrate be found, plans are to use a deep-rooted crop, such as alfalfa (Mathers, Stewart, and Blair, 1975), to "denitrate" the profile.

SELECTED REFERENCES

Hargett, N.L., and Berry, J.T., "Fertilizer Summary Data," Bulletin Y-189, 1985, Tennessee Valley Authority, Muscle Shoals, Alabama.

Kissel, D.E., Bidwell, O.W., and Kientz, J.F., "Leaching Classes of Kansas Soils," Bulletin 641, 1982, Kansas State University Agricultural Experiment Station, Manhattan, Kansas.

Mathers, A.C., Stewart, B.A., and Blair, B., "Nitrate-nitrogen Removal from Soil Profiles by Alfalfa," Journal of Environmental Quality, Vol. 4, 1975, pp 403-405.

Naney, J.W., Smith, S.J., and Berg, W.A., "Management and Geologic Effects on Ground Water Quality of Agricultural Lands in the Southern Plains," Water for the 21st Century: Will It Be There?, M.A. Collins, ed., Center for Urban Studies, Southern Methodist University, Dallas, Texas, 1984, pp 638-648.

National Technical Advisory Committee, "Water Quality Criteria," Report to the Secretary of the Interior, April 1, 1968, U.S. Government Printing Office, Washington, D.C.

Smith, S.J. et al., "Nutrient and Sediment Discharge from Southern Plains Grasslands," Journal of Range Management, Vol. 36, 1983, pp 435-439.

Smith, S.J., Naney, J.W., and Berg, W.A., "Nonpoint Source Impacts on Ground Water Quality in Major Land Resource Areas of the Southwest," Perspectives on Nonpoint Source Pollution, EPA 440/5-85-001, 1985, U.S. Government Printing Office, Washington, D.C., pp 121-124.

Strathouse, S.M. et al., "Geologic Nitrogen: A Potential Geochemical Hazard in the San Joaquin Valley, California," Journal of Environmental Quality, Vol. 9, 1980, pp 54-60.

U.S. Department of Interior, Federal Water Pollution Control Act Methods for Chemical Analysis of Water Wastes, FCN 16020-7171, 1971, National Environmental Research Center, Analytical Control Laboratory, Cincinnati, Ohio.

U.S. Environmental Protection Agency, "Water Quality Criteria," 1973, U. S. Government Printing Office, Washington, D.C.

U.S. Environmental Protection Agency, "Quality Criteria for Water," 1976, U.S. Government Printing Office, Washington, D.C.

U.S. Soil Conservation Service, "Land Resource Regions and Major Land Resource Areas of the United States," Agricultural Handbook 296, 1981, U.S. Department of Agriculture, Washington, D.C.

CHAPTER 26

GROUND WATER AND AGRICULTURE: ADDRESSING THE INFORMATION NEEDS OF PENNSYLVANIA'S CHESAPEAKE BAY PROGRAM

by

Joseph R. Makuch
and
Mitchell D. Woodward

During the environmental movement of the late 1960s and early 1970s, concern was expressed about pollution of the Chesapeake Bay. Maryland Senator Charles Mathias was the political champion for efforts to protect the ecological viability of the bay. In 1975 Congress agreed that something should be done and directed the U.S. Environmental Protection Agency (EPA) to conduct a five-year, $25 million study of the bay's resources. Along with the research, EPA was also instructed to determine trends and provide recommendations as to how any adverse ecological impacts could be eliminated (U.S. Environmental Protection Agency, 1983a).

A key finding of the study was that levels of nutrients (nitrogen and phosphorus) were increasing, especially in the middle and upper bay, and the consequent eutrophication problems were contributing to a decline in the population of several of the bay's finfish and shellfish species. The EPA report stated that nonpoint sources of pollution contribute 67 percent of the nitrogen and 39 percent of the phosphorus to the water. Point sources account for 33 percent of the nitrogen and 61 percent of the phosphorus (U.S. Environmental Protection Agency, 1983b).

Political action was taken on the EPA study in December, 1983, at the Chesapeake Bay Conference. During the conference the governors of Maryland, Pennsylvania, and Virginia, the administrator of the EPA and others acted on the findings and recommendations of the EPA study and pledged resources to a concerted Chesapeake Bay cleanup effort. This commitment was formalized in the Chesapeake Bay Agreement of 1983 (Citizens Program, 1984). A month later, in his State of the Union address, President Reagan said:

"Though this is a time of budget restraints, I have requested for EPA one of the largest percentage budget increases of any agency. We will begin the long, necessary effort to cleanup a productive, recreational area and a special national resource--the Chesapeake Bay." (Reagan, 1984)

PENNSYLVANIA AND THE CHESAPEAKE BAY

The Chesapeake Bay basin covers 63,390 square miles. Pennsylvania makes up 35 percent of the basin; it has more land in the basin than any other state. Nearly half of Pennsylvania lies in the basin, mostly within the Susquehanna River subbasin (U.S. Environmental Protection Agency, 1983c).

376 PROTECTION AND MANAGEMENT

Fifty percent of the fresh water entering the Chesapeake Bay does so via the Susquehanna River. Considering only the upper bay, the Susquehanna provides 90 percent of the fresh water (U.S. Environmental Protection Agency, 1983a).

The EPA study indicated that 40 percent of the nitrogen and 21 percent of the phosphorus enter the Chesapeake by way of the Susquehanna River. Furthermore, 85 percent of the nitrogen and 60 percent of the phosphorus delivered by the Susquehanna come from cropland. Modeling studies have shown that 41 percent of this nonpoint source pollution comes from the lower Susquehanna River basin in southeast Pennsylvania (U.S. Environmental Protection Agency, 1983a). This area has high livestock densities and intensive cropping practices.

Most of the land area is used for agricultural production. Due to adequate rainfall and highly productive soils, crop yields are considerably higher when compared to other parts of the state. The amount of land in production requires that farmers apply large quantities of commercial fertilizers. There is also a large and diverse animal production industry that includes poultry, beef, hogs, dairy, and sheep. This industry produces tremendous quantities of manure which greatly increase the nutrient load applied to fields in southeastern Pennsylvania.

PENNSYLVANIA'S CHESAPEAKE BAY PROGRAM

On April 12, 1985, Pennsylvania's State Conservation Commission adopted the policy statement and guidelines for the Chesapeake Bay Nonpoint Source Pollution Abatement Program. The intention of the program is to improve water quality in the Chesapeake Bay and the Susquehanna River basin by reducing agricultural nonpoint source pollution. The guidelines officially established the manner in which funds for the program would be used. For fiscal year 1984-85, $2 million in state and federal appropriations were allocated (Pennsylvania Bulletin, 1985). The appropriation for fiscal year 1985-86 is $4.35 million (Commonwealth of Pennsylvania).

Pennsylvania's program is administered by the State Conservation Commission through the Pennsylvania Department of Environmental Resources (DER). Initially, six southeastern Pennsylvania counties (Figure 26.1) will be targeted by the program. The program involves many agencies and organizations operating in four arenas: planning and research, technical assistance, financial assistance, and education.

The planning and research component encompasses work by DER, the U.S. Geological Survey (USGS), the Susquehanna River Basin Commission, and Pennsylvania State University. Research projects are studying such things as the effect of Best Management Practices on surface water and ground water, the complexities of nitrogen in the soil, and the flow of nutrients into and out of farms.

Technical assistance provided by the Soil Conservation Service, conservation districts, and the Cooperative Extension Service has been supplemented by additional DER personnel. Four individuals have been hired to aid farmers in the six target counties with soil conservation and nutrient management decisions.

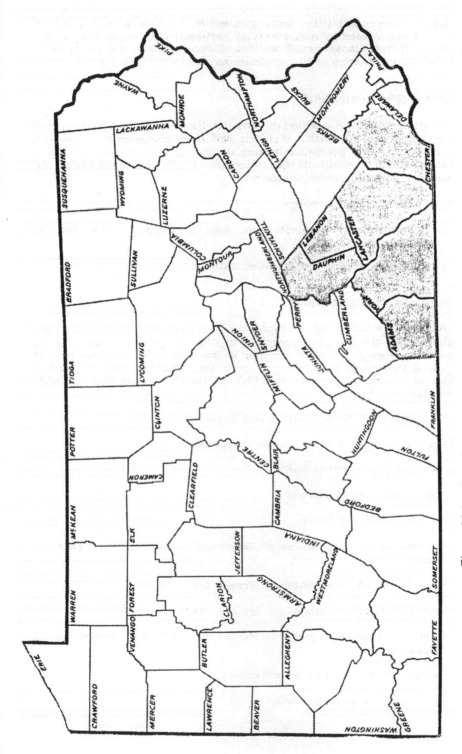

Figure 26.1: Counties Included in Pennsylvania's Chesapeake Bay Program

Approximately half of the funds allocated for the program to date will be paid to farmers in selected watersheds as cost-share allowances towards the installation of Best Management Practices. Cost-share rates will pay up to 80 percent of a given practice with a maximum payment of $30,000 per farmer.

THE EDUCATION COMPONENT

Twenty percent of Bay Program funds for fiscal year 1985-86 will be used to support educational efforts. Planning and research received 11 percent; technical assistance 14 percent; and financial assistance, for cost-share monies, 55 percent (Commonwealth of Pennsylvania). Under the umbrella of the education component are three categories:

1. a promotional campaign;

2. demonstration projects on soil conservation and nutrient management; and

3. a water quality information/education program.

THE PROMOTIONAL CAMPAIGN

A committee made up of representatives from the various state and federal agencies that are participating in the Bay Program is conducting the promotional campaign. Also represented on the committee are organizations that have a strong interest in program activities. The following is a list of agencies and organizations with members on the Chesapeake Bay Education Subcommittee:

- Agricultural Stabilization and Conservation Service

- DER Bureau of Soil and Water Conservation (2)

- DER Bureau of State Parks

- DER Community Relations

- League of Women Voters

- Pennsylvania Association of Conservation District Directors, Inc.

- Pennsylvania Association of Farmer Cooperatives

- Pennsylvania Department of Agriculture

- Pennsylvania Department of Education, Office of Environmental Education

- Pennsylvania Farmers Association

- Pennsylvania Farmers Union

- Pennsylvania State Grange

- Soil Conservation Service

- Susquehanna River Tri-State Association

- Pennsylvania State University

- U.S. Geological Society

The committee is chaired by the executive director of the Pennsylvania Association of Conservation District Directors, Inc. This organization has received grant money for educational activities.

The purpose of the promotional campaign is to inform the agricultural community (and the general public) about the program and to encourage the adoption of Best Management Practices by farmers in southeastern Pennsylvania. A first step in the campaign was developing a logo and slogan for the program (Figure 26.2). The logo provides a unifying element. This is essential given the fact that the Bay Program has several components and is being conducted by numerous agencies and organizations.

Figure 26.2: Logo for Pennsylvania's Chesapeake Bay Program

The promotional campaign is using a wide variety of delivery methods. Perhaps the most important method is a series of fact sheets and technical notes on Chesapeake Bay Program topics. The following have been published:

- Chesapeake Bay Nonpoint Source Pollution Abatement Program

- Pennsylvania's Mobile Nutrient Lab

- Pennsylvania's Chesapeake Bay Program

- Keystone in the Cleanup

- Animal Manure - What's It Worth?

- Managing Barnyard Runoff

- Testing Drinking Water in Agricultural Areas

- Manure Spreader Calibration

- Types of Manure and Methods of Storage

In addition to the above, two brochures, "Animal Waste Management: Issues and Answers" and "Save Your Soil and Nutrients" have also been produced. All of these materials were designed to either provide general information about the program or to address specific concerns regarding agricultural nonpoint source pollution. The fact sheets are produced with the logo prominently displayed and are distributed via agencies participating in the Bay Program.

An exhibit, in the form of an electronic quiz board, has also been developed. The quiz board contains six questions concerning Pennsylvania and the Chesapeake Bay. The unit is designed so that the questions can be changed depending on the audience; one set of questions is appropriate for farm audiences whereas another is used for the general public. The interactive nature of the exhibit has made it extremely popular at fairs, conservation tours, and other events where it has been used. The aforementioned printed material is used in tandem with the display. Also, novelty items displaying the logo are distributed at appropriate events. These items include lapel pins, stickers, and buttons. In addition, a combination postcard/bumper sticker has been sent to farmers in the target counties. The postcard side contains a short message about the Bay Program; the reverse side is a peel-off bumper sticker with the program's slogan and logo.

Media coverage, both print and electronic, has been good. Major events, such as the signup of the first farmer to participate in the program, have been well covered. Announcements of funding levels or expressions of political support for the program have also received media attention. Press and radio releases have been used and a television public service announcement encouraging participation in the program is in development. Articles on agricultural nonpoint source pollution and related preventative measures have appeared frequently in the farm press.

One information delivery method that is currently being developed is the use of the Pennsylvania Extension Network (PEN). PEN is a computer network that links each department within the Penn State College of Agriculture and all sixty-seven county Cooperative Extension Service offices. A recently added feature to the network is PENpages, an electronic reference source containing information on the wide variety of topics addressed by the College of Agriculture and the Cooperative Extension Service. As of January, 1986, anyone in Pennsylvania with access to a microcomputer (or compatible terminal), modem and computer software can enter the system for the cost of a local call. (People outside of Pennsylvania can also access the system through a long-distance call.) There is no subscription charge. Specific information stored on the system is located through the use of keywords.

Information on a number of different topics related to the Chesapeake Bay Program will be made available on PENpages. Plans include entering (1) selected fact sheets, (2) listings of publications and audiovisual materials, (3) a directory of Bay Program personnel, and (4) a calendar of events.

Presentations to groups and organizations (agricultural or otherwise) have also been an important method of providing information about the Bay Program. To support this effort, films, videotapes, and slide presentations on relevant subjects have been developed or obtained from outside sources. Interagency communication will be facilitated through the use of a monthly news sheet.

DEMONSTRATION PROJECTS

The second broad category of activity under the education component consists of demonstration projects. These projects are designed to show farmers actual examples of ways to manage soil resources and nutrients in a manner which is both environmentally and economically sound.

The demonstration projects conducted by the Penn State Cooperative Extension Service have focused on nutrient management. Producers, for the most part, know that the manures they apply to soil contain nutrients valuable to crop production. Many, however, do not have the knowledge to, or feel uncomfortable with, calculating the nutrient value of manure applications to crops. Few producers use manure analysis or know how many pounds of nitrogen, phosphorus, and potassium are contained in the manure on their farms. many view manure as a waste to be disposed of as quickly and as inexpensively as possible.

This situation leads to over application of nutrients in the form of commercial fertilizers and manures to cropland. This is reflected in the soil samples received for testing at Penn State. Of the samples from the six target counties, nearly 40 percent were considered very high or excessive in phosphorus and more than 50 percent were at these levels for potassium (Merkle Laboratory, 1986). In these cases additional applications of phosphorus and potassium are not cost-effective and will not increase yields.

To help make this point, several demonstration projects were initiated in the spring of 1985 in three counties in southeastern Pennsylvania. The purpose of these research/demonstration plots was to show producers that: (1) manure is a valuable source of plant nutrients; (2) manure should be spread accurately and be evenly distributed on fields throughout the farm; (3) manure and commercial fertilizer should not be applied in excess of crop nutrient needs; (4) the application of manure can be used to reduce fertilizer inputs to cropland; and (5) application rates of crop nutrients in the form of manure or commercial fertilizer, in excess of the quantity needed for crop production, can cause surface and ground water contamination.

Cooperative Extension Service agronomy agents were contacted in early spring and asked to locate fields for demonstration plots. After meeting with the agronomy specialist at Penn State, it was decided that agents should locate two types of fields. The first should be a field which has a history of heavy manure applications. Our objective in this situation was to demonstrate to producers that additional heavy applications of manure or commercial nitrogen fertilizers are not cost-effective and will not boost yields. Since most of the farms in southeastern Pennsylvania are in this situation because of high animal densities, the emphasis would be placed on these fields.

In the second type of situation, agents were to locate fields which had received little or no manure in the last ten years. Our objective here was to demonstrate that manure has significant nutrient value. By determining the nutrient value through manure analysis and applying a rate based on crop nutrient needs, the amount of commercial fertilizer needed to be purchased would be reduced. It was also hoped that producers with excess quantities of manure on their farms would consider "marketing" it to other crop producers needing an inexpensive source of nutrients.

Field plots were located in Lancaster, Lebanon, and York Counties. The Lancaster and Lebanon plots were small plot, randomized block designs with four replications. The York plots were larger plot designs, some as large as one-half acre but were not replicated. Data from the replicated plots will be used for future study. These plots also performed well as producer demonstration plots. The nonreplicated plots in York County were used for demonstation purposes only.

We felt that the producer hosting the plot work on their farm should be a community leader and innovator and should be involved in constructing the plot and applying the treatments. The advantages of doing this were several fold. Other producers in the community respected the cooperators and this added credibility to the projects. Also, with the producer closely involved in our work he could easily explain the results and benefits to other producers.

The treatments were applied that spring and observed throughout the year. Alleyways were cut between individual treatments and signs were located at the front of each plot to aid in identification. Photographs and slides were taken throughout the year for use during the upcoming winter agronomy meetings.

When the corn crop reached maturity that fall, yield data were taken and field days were held to show results to producers. In general, results indicated that on those plots that had received heavy manure applications in previous years, applying additional amounts of manure or commercial fertilizer was not always profitable. However, on plots without a history of heavy manure application, manure was an excellent source of nutrients and increased yields as well as commercial fertilizer. The field days usually took the form of a tour in which the manure management demonstration plot was a component. Other points of interest that encouraged attendance at the tours were a presentation on manure storage, a manure spreader calibration demonstration, the presence of local officials from the DER and Soil Conservation Service who discussed Best Management Practices and special funding available for manure management, and equimpent demonstrations by local dealers.

Farmer reaction to the plots and field days was quite favorable. The discussions centered on cost effectiveness of nutrient applications to grow the highest economic yield. Local drinking water quality was also discussed. We felt that the approach of talking about local water quality concerns and how to save money on fertilizer bills met with a better response than talking about how this effort to reduce nutrient inputs on cropland will help the Chesapeake Bay, although the two approaches achieve the same end.

In order to reach a wider audience than only those who participated in the field days, follow-up news articles were written for local papers and more are planned for the future. The information was also used at winter meetings throughout the region.

We were pleased with producer participation and acceptance of our demonstation plots, especially since this was the first year of the program. Building on the year's successes, we plan to enlarge the demonstration plot program this year by expanding the program into three additional counties. This will provide us with eight to ten plots in total. Improvements in the program include having plots well marked and located along heavily traveled roads. This could result in additional exposure of our research. A manure application rate worksheet has also been developed for producers and will be used in the future at tour and spreader calibration clinics. By using some commonly found items on the farm, producers can accurately determine their rate of application. Few producers actually know at what rate they spread manure. Future work plans also include cooperation with local crop management associations that would help establish plots and then use the results to assist their producers.

In addition to the plots, further emphasis is planned in the areas of soil testing, manure analysis, and the use of an Extension Service Farm Nutrient Management Program on computers at county offices. Training sessions for county agents on the use of this program is slated for fall, 1986. With this comprehensive computer program, agents (using soil tests, manure analyses, and crop production records) will be able to advise producers on rates of nutrients to apply to cropland in order to achieve maximum economic yield, while at the same time reducing water pollution caused by excess nutrients.

Other demonstration projects include several educational/technical assistance programs provided by the county conservation districts. Plans are underway to hire full-time crop management association technicians to assist producers in crop nutrient management. It is hoped that this project will increase nutrient use efficiency. A common problem is poor distribution of manure; manure may be either over- or underapplied to croplands. To address this concern the Lebanon County District will provide equipment and manpower to provide a manure spreading service to apply manure at a known rate to meet crop nutrient needs. It is hoped this project will demonstrate to farmers the economic benefit of using manure to fill their crop nutrient needs.

The production of liquid manure has many problems associated with it. It is expensive to haul due to its weight and has a low nutrient value/weight ratio. Another demonstration project sponsored by the conservation district is a program to encourage producers to use woodlot areas to dispose of wastes. This could benefit the farmer interested in woodlot production, provide an area to dispose of manure other than the often overapplied crop acreage, and reduce nutrient runoff.

One other demonstration project is a joint effort of Penn State University and DER. The project is the "Nutrient Management Mobile Laboratory." The mobile lab will be made available to producers in the target counties and will bring soil testing and manure analysis technology to the farm to help producers to better understand the concept of nutrient management.

THE WATER QUALITY PROGRAM

The third element of the education component is a water quality information/education program. As stated earlier, the ultimate goal of Pennsylvania's Chesapeake Bay Program is to improve the water quality of the Chesapeake Bay and the Susquehanna River Basin. Therefore, it is essential that information about water quality be made available.

The water quality information/education program is being conducted as a Cooperative Extension program in Penn State's Department of Agricultural Engineering. Because of this arrangement, the new Chesapeake Bay Water Quality Program has the advantage of using the infrastructure that serves the Cooperative Extension Service's water quality program that is currently in place. Existing personnel and materials can be called upon to contribute to the new program. Conversely, new materials developed to address the needs of the Chesapeake Bay program in southeastern Pennsylvania may also have applicability in other areas of the state.

The initial focus of the water quality program is on ground water. The four major reasons for this are outlined below.

The current Cooperative Extension Water Quality Program has as its top priority private, individual water systems, virtually all of which are supplied by ground water. This policy was set by the Extension water quality task group and reflects in large measure the fact that individuals who have their own water supplies are solely responsible for the quality of their drinking water. Consequently, these individuals need to be aware of the management actions required to insure a safe water supply.

Second, there is evidence of ground water contamination by nitrate in southeastern Pennsylvania. A survey of forty-two wells and one spring in the Conestoga headwaters area of Lancaster County showed high nitrate levels (Table 26.1) (U.S. Department of Agriculture, 1984).

Table 26.1:   Percentage of Wells Exceeding the U.S. Environmental Protection Agency's Criterion for Nitrate (10 mg/L as N)

	Agricultural Land	Nonagricultural Land
Fall 1982	41	9
Spring 1983	42	18
Summer 1983	66	27
Fall 1983	41	18

Data from Penn State's Merkle Lab, which formerly provided a water testing service, shows similar results (Mooney, 1984). Southeastern (SE) Pennsylvania had the highest percentage of water samples that were found to be contaminated with nitrate (Figure 26.3).

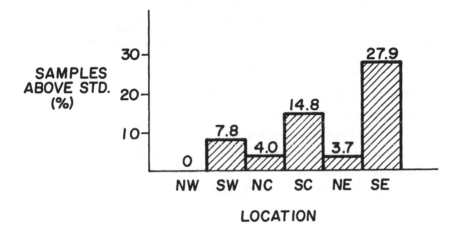

Figure 26.3: Contamination of Drinking Water by Nitrate in Pennsylvania by Region (Mooney, 1984)

Third, ground water is a local concern and may, in fact, be a personal concern. Ground water is in southeastern Pennsylvania, not in a bay between Virginia and Maryland. Ground water is used by the residents of southeastern Pennsylvania. In the six target counties more than one-half million Pennsylvanians get their water from their own wells or springs. Public water supplies that use ground water serve an additional 273,265 people. In total, 825,307 people--50 percent of the area's population--rely on ground water for domestic use (Ward, 1985).

If this water is contaminated, the contamination becomes a personal concern for those, including farmers, who use it. The thrust of Pennsylvania's Chesapeake Bay Program is to encourage farmers to adopt Best Management Practices. This usually means that an individual farmer will have to alter, at least to some extent, his current methods of operation. This change is much more likely to occur if a personal benefit, e.g., an uncontaminated water supply for home and farm, is perceived.

The fourth reason for developing an information program on ground water is that ground water is poorly understood. Many people do not realize that ground water is an integral part of the hydrologic cycle and that ground water and surface water are intimately connected. In addition, linkages between land-use activities and ground water quality are sometimes not fully understood. The feeling is that if farmers are aware of how ground water behaves and how their agricultural practices can affect ground water quality--their own water--they will be more inclined to adopt Best Management Practices.

INFORMATION DELIVERY METHODS

A variety of media are being used to disseminate information on ground water and drinking water quality. All materials are geared towards a popular, i.e., nontechnically oriented audience. Although some materials are aimed specifically at the agricultural community, many nonfarmers will find the information relevant. Current publications being relied upon include:

Agricultural Engineering Fact Sheets:

- How to Interpret a Water Analysis Report
- Disinfecting a Water System
- Water Problems and Treatment
- Bacteriological Treatment of Farm Water Supplies
- Removal of Nitrates
- Where to Have Your Water Tested
- Water Testing

Special Circulars:

- Your Water Test Report--What Do the Numbers Mean?
- Nitrate: Its Effect on Families and Livestock

New print materials, designed specifically for Pennsylvania's Chesapeake Bay Program, have (or will be) developed. As previously mentioned, two fact sheets, "Keystone in the Cleanup," which offers ecological justification for Pennsylvania's participation in the bay cleanup effort (based on the EPA reports), and "Testing Drinking Water in Agricultural Areas," which gives basic information about water testing, have been published in the Chesapeake Bay fact sheet series.

A small foldout brochure, "A Quick Look at Pennsylvania Groundwater," is being cooperatively produced with the USGS. This brochure will provide the reader with a brief description of the nature of ground water as well as providing some facts on the type and extent of ground water use in the state.

A second publication, "Groundwater and Agriculture in Pennsylvania," (also in cooperation with the USGS) will be larger in scope. It will be a Cooperative Extension Service circular that discusses the ground water resources of Pennsylvania; the current state of agriculture in Pennsylvania; and the occurrence, prevention, and treatment of three common agricultural contaminants of ground water (bacteria, nitrate, and pesticides). A slide/tape presentation covering the publication material is also being produced as is a slide show on "Nitrate and Your Water Supply" which is designed to answer the questions most frequently asked about this contaminant. A television public service announcement on nitrate contamination has been produced and distributed to television stations in the target area.

As an additional means of conveying information about ground water, a floor-standing exhibit is being used. Through the use of photographs, a chart, a minimum of text, and two lighted diagrams, the display gives the viewer a quick overview of ground water. When the ground water brochure (mentioned above) is completed, it will be used as a handout with the display.

"Make Your Water Supply Safe" is an Extension Service Publication that is being revised through the Bay Program. This publication focuses on protecting wells (with some discussion of springs, ponds, and cisterns) from contamination. In conjunction with this revision, an exhibit on proper well construction is being developed.

The remaining information delivery method being employed is the use of Safe Drinking Water Clinics. These clinics began in 1984 as part of the Extension Service's Rural Drinking Water Program. The clinics are county meetings designed to provide information to owners of private, individual water systems on methods to ensure an adequate and safe water supply. The clinic format involves:

- a program of speakers (extension water specialists and others cover different aspects of water supply management);

- arrangements with one or more local water testing laboratories to provide a customized package (for each county) of water tests at a reduced rate;

- the availability of the Water Supply Records book (a three-ring binder containing information on water supply management and sections to file information specific to an individual system);

- displays of water treatment equipment by commercial vendors; and

- the opportunity to consult with Extension water specialists about particular water supply problems.

To provide maximum accessibility, clinics are held in the afternoon with the event repeated in the evening. During the past year clinics were held in four of the six target counties. The total attendance was 420 people.

The clinics in Lancaster and York Counties will be evaluated by means of a mailed survey (with multiple mailings).

The purpose of the survey is to (a) assess the merits of the various clinic features, (b) measure the amount of new information learned about water supply management, (c) determine the extent and kinds of water supply management actions taken by clinic participants, and (d) solicit information concerning the Water Supply Records book.

CONCLUSIONS

Nutrient runoff from agricultural land in southeastern Pennsylvania has been implicated as one of the factors contributing to ecological perturbations in the Chesapeake Bay. Pennsylvania has made a commitment to help improve the water quality of the Chesapeake Bay through a program aimed at reducing agricultural nonpoint source pollution. Since participation in the program is voluntary, a strong information/education effort is necessary. In order to decide if (or to what extent) they will participate in the program, farmers need to be well informed. Similarly, members of the general public, who as taxpayers are supporting the program, need to be kept abreast of program development.

The three major thrusts of the education component, a promotional campaign, demonstration projects and a water quality program, all are designed to contribute to the success of the overall program. With strong inter- and intraagency cooperation the education component can be an effective tool in helping to improve the water quality in Pennsylvania and in the Chesapeake Bay.

SELECTED REFERENCES

Citizens Program for the Chesapeake Bay, Inc., "Choices for the Chesapeake: An Action Agenda," 1983 Chesapeake Bay Conference Report, 1984, Baltimore, Maryland.

Commonwealth of Pennsylvania, Department of Environmental Resources, Pennsylvania Chesapeake Bay Program Annual Grant FY 1985-86, Harrisburg, Pennsylvania.

Merkle Soil and Forage Testing Laboratory (Unpublished data), 1986, University Park, Pennsylvania.

Mooney, D., "Analysis of Pennsylvania's Rural Drinking Water," Master's Thesis, 1984, Environmental Pollution Control Program, The Pennsylvania State University, University Park, Pennsylvania.

Pennsylvania Bulletin, Vol. 15, No. 10, March 9, 1985, Harrisburg, Pennsylvania.

Reagan, R., "State of the Union Message," House Document No. 98-162, 1984, Washington, D.C.

U.S. Department of Agriculture, "Conestoga Headwaters Rural Clean Water Program 1984 Progress Report," 1984, Berks County, Lancaster County, Pennsylvania.

U.S. Environmental Protection Agency, "Chesapeake Bay: A Framework for Action," 1983a, Philadelphia, Pennsylvania.

U.S. Environmental Protection Agency, "Chesapeake Bay Program: Findings and Recommendations," 1983b, Philadelphia, Pennsylvania.

U.S. Environmental Protection Agency, "Chesapeake Bay: A Framework for Action," Appendices, 1983c, Philadelphia, Pennsylvania.

Ward, J., Personal Communication, 1985, U.S. Geological Survey, Harrisburg, Pennsylvania.

CHAPTER 27

DEVELOPING A STATE GROUND WATER POLICY IN THE
CORN BELT: THE IOWA CASE

by

Brian P. Borofka,
C.S. Cousins-Leatherman, and
Richard D. Kelley

Recent surveys indicate that 68 percent of all municipal drinking water (by volume) in Iowa is derived from ground water as well as approximately 100 percent of the private rural drinking water. As much as 82 percent of Iowa's total population depends on ground water for their drinking water. That makes Iowa one of the most dependent ground water users in the Midwest.

Iowa is also ranked first in the country in corn production and second in soybean production. Many believe that these record crop-producing activities have had a significant impact on Iowa's ground water.

It is important that Iowa's citizens know the extent of the ground water problems associated with current agricultural practices. Consequently, the state is preparing an overall ground water protection strategy which assesses all potential sources of ground water contamination including those associated with agriculture.

THE PROBLEM

Many of Iowa's concerns have been discussed in other chapters in this book. As a quick review, the following agriculturally related ground water problems have been identified.

1. There are several studies which indicate that nitrate contamination is more than a surface water or runoff problem in Iowa--that the problem is getting more serious as rates of fertilizer application increase and that the problem may not be limited to shallow ground water exclusively. Figure 27.1 shows the location of public drinking water supplies with elevated nitrate concentrations. The number of public water supplies exceeding the Safe Drinking Water Standard for nitrates has doubled since 1975 from twenty-seven (27) to fifty (50) violations per month.

2. Due to the presence of fractured bedrock in parts of Iowa, agricultural drainage wells were commonly used for draining wetlands to create cultivated farmland. These injection wells may also serve as conduits for contaminated surface water and tile drainage water to the bedrock aquifer. Figure 27.2 shows the areas where drainage wells are likely to be located in Iowa.

3.   A preliminary survey of public water supplies has found detectable residues of commonly used pesticides in 57 percent of the wells sampled. Several types of pesticides have been routinely detected, including five herbicides and three insecticides: Atrazine, Bladex, Sencor, Dual, Lasso, Dyfonate, Counter, and Bolstar. In addition, other pesticides have been detected periodically.

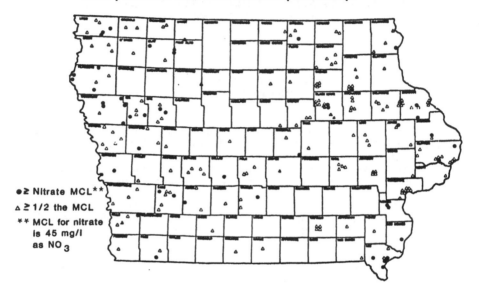

Figure 27.1: Public Water Supplies with High Nitrate Concentrations

BACKGROUND

Following receipt of the 1985 State Water Plan, which primarily addressed water quantity issues, the Iowa Legislature requested the Department of Water, Air, and Waste Management to prepare a Ground Water Protection Strategy. This mandate included four specific directives:

1.   "Consider the effects of potential contamination sources on ground water quality."

2.   "Evaluate the ability of existing laws and programs to protect ground water quality."

3.   "Recommend any necessary additional or alternative laws and programs."

4.   "Develop the plan with the assistance of and in consultation with representatives of agriculture, industry, the public and other interests."

In order to fulfill this mandate the following goal has been set for Iowa's Ground Water Protection Strategy:

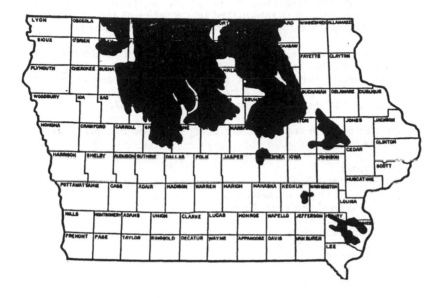

Figure 27.2: Likely Location of Drainage Wells

To develop a unified approach among all federal, state, and local programs in Iowa in order to PREVENT ground water contamination, to RESTORE ground water quality in areas where pollution has occurred, and to MAINTAIN an adequate supply of acceptable quality ground water for Iowa's projected uses and demands.

The audience for this strategy includes the legislature, the public, interest groups, other agencies and other institutions.

STRATEGY DEVELOPMENT STEPS

The first step in developing the strategy was problem identification. Towards this end, staff prepared a series of technical reports describing the extent, magnitude, and priority of the problems relating to ground water quality. They include an assessment of the problems with comparisons to assist decision makers in drawing their own conclusions about the estimated impacts and the urgency of the recommendations.   Public input has been and will continue to be solicited throughout the process, in order to sensitize the public to the issues, to establish the department's creditability, and to solicit information which might otherwise be unavailable.   The department is being assisted in their technical evaluation by the Technical Advisory Committee, an outside peer review group of professionals familiar with ground water technical information.

The next step is the identification of policy and program alternatives. Staff are preparing a second set of papers covering these policy items. Completion of this step will signal an end to the information gathering process.

Most of the important issues will be identified. At this point, the forum will be opened for discussion. It is anticipated that a series of public meetings, public hearings, workshops, and advisory committee meetings will be held.

The end result will be a highly condensed technical and policy presentation of the problems with a select and prioritized list of recommendations. This document and the list of recommendations will be reviewed by senior staff and the Water, Air and Waste Management Commission before submittal to the Iowa Legislature in January, 1987.

Specific problem areas that are being examined in separate technical and policy papers include:

1. Nitrates
2. Agricultural Chemicals
3. Synthetic Organic Chemicals
4. Agricultural Drainage Wells
5. Leaking Lagoons
6. Sanitary Landfills
7. Hazardous Waste Sites
8. Abandoned Dumps
9. Land Application
10. Chemical and Fuel Handling/Storage
11. Well Construction, Maintenance, and Abandonment

In addition to these topical reports, other reports will be prepared including six regional ground water evaluations (Figure 27.3), an overview of all federal, state, and local regulations, an overview of all public information and education programs, a review of current ground water data management systems and monitoring efforts, as well as projected research needs.

Figure 27.3: Iowa's Six Ground Water Quality Regions

Assistance is being provided to the staff by a Technical Advisory Committee consisting of representatives from the following organizations:

Iowa Department of Agriculture

Iowa Geological Survey

U.S. Geological Survey

University Hygienic Laboratory

U.S. Soil Conservation Service

Iowa Department of Soil Conservation

University of Iowa

Iowa State Water Resources Research Institute

Iowa Groundwater Association

The role of the Technical Advisory Committee is to provide additional information, technical expertise, and peer review of papers.

POLICY AND PROGRAM ALTERNATIVE IDENTIFICATION

Policy options and program recommendations will be developed separately, but based upon the technical findings. Essentially, this will involve a three-step process. The first step will be to select an overall ground water policy for the state. This policy will be either a nondegradation policy, a limited degradation policy, a preferential degradation policy, or a hybrid of all three.

The next step will be to select management options, such as ground water standards or classification systems to implement the overall policy. The Water, Air and Waste Management Commission will be addressing these overall policy and management options at a special two-day management retreat in early June. At that time direction will be set for overall protection philosophy and selected management options.

The last step is to apply the policymaker's choices to the problems identified in the technical assessments. Because all of these steps will require interpretation and application of ideas, a Program Advisory Committee has been selected to assist staff. The Committee will consist of representatives from the following organizations:

Iowa Department of Agriculture

Iowa Department of Soil Conservation

Iowa Farm Bureau Federation

American Water Works Association

League of Women Voters of Iowa

Iowa Association of Municipal Utilities

Iowa State Association of Counties

Iowa Fertilizer and Chemical Association

Iowa Irrigation Association

Sierra Club

Iowa League of Municipalities

Iowa Association of Business and Industry

Iowa Water Well Association

Iowa Natural Heritage Foundation

The role of the Program Advisory Committee is to identify and prioritize issues, to identify and discuss alternatives, and to assist in information dissemination and consensus building. More importantly, the Program Advisory Committee will suggest policy and recommendations within the context of socio-economic conditions, and particularly those associated with agriculture and its state of transition in Iowa.

As mentioned before, the final forum within the Department for public input is the Water, Air and Waste Management Commission. The commission consists of nine appointed representatives. Their representation, as specified in the Iowa Code, consists of the following:

three livestock and grain farming representatives;

one person engaged in the management of a manufacturing operation;

one person engaged in the areas of finance and commerce; and

four electors of the state.

The Commission has been updated on pertinent ground water facts and issues during their monthly meetings.

STRATEGY OPTIONS

Strategy options fall under four different categories. They are listed below in the order of preference in terms of feasibility, political acceptability, and cost effectiveness.

I. Information/Education

- Research
- Demonstration
- Technology Transfer

II. Program Changes

- Monitoring
- Assessment
- Public Participation
- Enforcement

III. Statutory Changes

- State Laws
- Administrative Regulations

IV. Bans or Restrictions

- Administrative Review and/or Statutory Changes
- Public Notices
- Enforcement

Specific options are also being developed for each of the problem areas listed earlier. In the case of pesticides, bans, or restrictions may be the first choice option if warranted by compound toxicity or carcinogenicity.

1986 LEGISLATIVE ACTIONS

Several pieces of environmental legislation were passed by the Iowa House and Senate this year. At least twice as many were proposed, including a retail tax on pesticides and fertilizers. There is a strong indication that the people of Iowa are becoming more concerned about ground water quality. Consequently, they may become more receptive to the statutory and program changes necessary to protect human health.

The following bills are currently waiting the governor's signature:

(1)   landfill ban by 1997;

(2)   a one-time test of municipal water supplies for pesticides and SOCs;

(3)   testing of bottled water for contaminants;

(4)   testing/certification of home water treatment devices; and

(5)   proposed use of Exxon Rebate for education and demonstration projects.

The fifth item listed above is a substitute for the proposed tax on fertilizers and pesticides which did not leave committee. The Exxon Rebate consists of $27 million which must be used for demonstration or education in the area of energy conservation. The department proposed the use of $6 million over six years for the study of tillage practices and nitrogen and pesticide management in Iowa. The end results would lead to conservation of natural gas and gasoline, as well as protection of the ground water. This is one innovative technique developed by the department for funding needed research, demonstration, and educational activities.

INDEX

abandoned wells 100
acephate 280
adsorption 166-168, 181-182, 327
Agricultural Research Service (ARS) 1, 3-5, 20, 22
agricultural chemicals   1-3, 7, 182, 185, 204, 273, 282, 285-286,
  290, 392
air injection 105
Alabama 275, 277, 279, 283
alachlor 165, 203, 210, 212-213, 279, 282, 288
aldicarb 163-168, 279, 281, 286-287, 289-290
aldrin 291, 334, 336
allanine 279
ammonium (NH_4^+) 116, 127, 128, 139, 141-142, 155, 370-372
anhydrous ammonia 116, 148, 150, 155
Arizona 67, 153, 155-156, 275, 277, 279, 283
Arkansas 137, 275, 277, 283, 285
arsenic 279, 290, 334, 336
artificial recharge 57, 60-61, 67, 91, 104
atrazine   163, 165, 167, 197, 201, 203-206, 210, 212-213, 219, 229,
  237, 243-244, 279, 281, 286-288, 320, 323-325, 390

BAY SMY 1500   176-177, 180-182
benomyl 281
benzene 335
beryllium 334
Best Management Practices (BMPs)   247, 346, 348, 358, 362, 376, 378-
  379, 382, 385
biochemical oxygen demand (BOD) 113, 314-315
biodegradation 295-296, 299, 301, 303, 314
bladex 219
bromacil 166-167, 282, 287

cadmium 334, 336-338
California   7, 10, 67, 94, 114, 153, 155-156, 275, 277, 279, 283,
  285-286, 351, 367
captan 281
carbaryl 279, 281
carbofuran 7, 281, 287
chemical oxygen demand (COD) 310-311, 314-315
Chesapeake Bay 4, 375-377, 379-382, 384, 387-388
chlordane 279, 289, 291, 334, 336
chloride 100, 102, 158, 291, 337
chlorobenzilate acaricide 165
chloropicrin 280, 287
chlorothalonil 282
chlorpyrifos 280
chromium 334, 336
Clean Water Act Amendments 111, 357
Colorado 22, 36, 47-52, 55, 107, 161, 275, 277, 283